Methods in Molecular Biology

Series Editor
John M. Walker
School of Life and Medical Sciences
University of Hertfordshire
Hatfield, Hertfordshire, UK

For further volumes:
http://www.springer.com/series/7651

For over 35 years, biological scientists have come to rely on the research protocols and methodologies in the critically acclaimed *Methods in Molecular Biology* series. The series was the first to introduce the step-by-step protocols approach that has become the standard in all biomedical protocol publishing. Each protocol is provided in readily-reproducible step-by-step fashion, opening with an introductory overview, a list of the materials and reagents needed to complete the experiment, and followed by a detailed procedure that is supported with a helpful notes section offering tips and tricks of the trade as well as troubleshooting advice. These hallmark features were introduced by series editor Dr. John Walker and constitute the key ingredient in each and every volume of the *Methods in Molecular Biology* series. Tested and trusted, comprehensive and reliable, all protocols from the series are indexed in PubMed.

Gapmers

Methods and Protocols

Edited by

Toshifumi Yokota

Department of Medical Genetics, University of Alberta, Edmonton, AB, Canada

Rika Maruyama

Department of Medical Genetics, University of Alberta, Edmonton, AB, Canada

Humana Press

Editors
Toshifumi Yokota
Department of Medical Genetics
University of Alberta
Edmonton, AB, Canada

Rika Maruyama
Department of Medical Genetics
University of Alberta
Edmonton, AB, Canada

ISSN 1064-3745 ISSN 1940-6029 (electronic)
Methods in Molecular Biology
ISBN 978-1-0716-0770-1 ISBN 978-1-0716-0771-8 (eBook)
https://doi.org/10.1007/978-1-0716-0771-8

This Humana imprint is published by the registered company Springer Science+Business Media, LLC, part of Springer Nature.
The registered company address is: 1 New York Plaza, New York, NY 10004, U.S.A.

Preface: Gapmers

> "*When Researchers Work Together, Miracles Can Happen.*"
> ——Jesse Davidson (1980–2009): lived with Duchenne muscular dystrophy, a champion for people with disabilities, who fought by raising millions of dollars for research and to help others.

Gapmers, chimeric antisense oligonucleotides that cleavage RNA through the recruitment of RNase H, are an emerging tool for gene silencing. The concept of gapmer antisense oligonucleotides was pioneered by Hideo Inoue and colleagues, which was published in May 1987 [1]. The first patent describing the antisense oligonucleotide gapmer concept was filed by Joseph Walder and colleagues in November 1987 [2, 3]. To date, gapmers have found widespread use as research reagents for inducing RNA degradation. In addition to basic research, gapmer-based technologies are often used in therapeutics [4]. The FDA-approved mipomersen (brand name Kynamro, marketed in the USA by Genzyme, a Sanofi company), a gapmer antisense oligonucleotide inhibitor of apo B, is an orphan drug for the treatment of homozygous familial hypercholesterolemia in 2013 [5]. More recently, a couple of gapmers have been approved for the treatment of devastating genetic diseases. First, the FDA has approved inotersen (brand name Tegsedi, marketed by Akcea Therapeutics and Ionis Pharmaceuticals) for the treatment of polyneuropathy in adults with hereditary transthyretin amyloidosis [6]. Second, in May 2019, volanesorsen (brand name Waylivra, marketed in the EU by Akcea and Ionis) was conditionally approved in the EU for the treatment of adult patients with familial chylomicronemia syndrome (FCS) [7]. Several more gapmer therapeutics have reached clinical trials, including Ionis-HTTRx for Huntington disease [8, 9]. Importantly, in dominant genetic diseases, gapmer-mediated degradation can selectively target the mutated mRNA [10]. In our laboratory, we are developing gapmer-mediated therapies for facioscapulohumeral muscular dystrophy (FSHD) and other forms of musculoskeletal disorders [11]. The cover image shows FSHD muscle cells differentiating into myotubes with gapmer treatment (see [11]).

Despite the widespread use, there has been no review or protocol book specifically written about the use of gapmer antisense oligonucleotides to the best of our knowledge. Our goal is to provide the latest and comprehensive information needed to design robust studies with gapmer oligonucleotide technology. This book begins with a historical and contemporary perspective of gapmers, and then presents a comprehensive collection of detailed state-of-the-art protocols from leaders in the field, from both academia and industry. Thanks to all of the authors who have contributed to this volume, we demonstrate the diverse applications of gapmers along with protocols that will assist readers in moving the frontier.

We would like to thank all contributors for openly sharing their detailed methods. These detailed methods and tips are hard to find elsewhere and will play an important role in new projects. We also deeply thank Dr. John Walker, the series editor, for his excellent guidance, careful paper reading, and encouragement throughout the development of this book. We are very grateful to the Friends of Garrett Cumming Research, Muscular Dystrophy Canada, Henri M. Toupin Chair in Neurological Sciences, and the University of Alberta for the funding to start gapmer studies in our laboratory and thankful to the Canadian Institutes of Health Research (CIHR), Canada Foundation for Innovation (CFI), Japan Society for the

Promotion of Science (JSPS), Alberta Enterprise and Advanced Education, BC Children's Hospital Foundation, FSH Society, Canadian FOP Network, FSHD Canada Foundation, Gilbert K. Winter Fund, International FOP Association, Rare Disease Foundation, and Jesse's Journey for their support in expanding our research in developing antisense therapies for devastating genetic diseases. Lastly but most importantly, we are deeply grateful to the patients and family members; their support, inspiration, and participation in research have been critical to the development of novel therapies.

We hope that gapmers will continue to be key tools in new scientific discoveries, advancing our understanding, and the development of drugs. Jesse Davidson (1980–2009), who lived with DMD and was a champion for people with disabilities, raising millions of dollars for research to help others, firmly believed that "When researchers work together, miracles can happen." That is our ultimate goal: to make miracles happen together.

Edmonton, AB, Canada

Toshifumi Yokota
Rika Maruyama

References

1. Inoue H, Hayase Y, Iwai S et al. (1987) Sequence-dependent hydrolysis of RNA using modified oligonucleotide splints and RNase H. FEBS Lett 215(2):327–330. https://doi.org/10.1016/0014-5793(87)80171-0
2. Lundin KE, Gissberg O, Smith CE (2015) Oligonucleotide therapies: the past and the present. Human gene therapy 26(8):475–485
3. Walder JA, Walder RY (1995) Nucleic acid hybridization and amplification method for detection of specific sequences in which a complementary labeled nucleic acid probe is cleaved. Google Patents
4. Lee JJ, Yokota T (2013) Antisense therapy in neurology. J Pers Med 3(3):144–176. https://doi.org/10.3390/jpm3030144
5. Rader DJ, Kastelein JJ (2014) Lomitapide and mipomersen: two first-in-class drugs for reducing low-density lipoprotein cholesterol in patients with homozygous familial hypercholesterolemia. Circulation 129(9):1022–1032. https://doi.org/10.1161/CIRCULATIONAHA.113.001292
6. Keam SJ (2018) Inotersen: first global approval. Drugs 78(13):1371–1376
7. Paik J, Duggan S (2019) Volanesorsen: First Global Approval. Drugs 79(12):1349–1354
8. Yang X, Lee SR, Choi YS et al. (2016) Reduction in lipoprotein-associated apoC-III levels following volanesorsen therapy: phase 2 randomized trial results. J Lipid Res 57(4):706–713. https://doi.org/10.1194/jlr.M066399
9. Tabrizi SJ, Leavitt BR, Landwehrmeyer GB et al. (2019) Targeting Huntingtin Expression in Patients with Huntington's Disease. N Engl J Med 380(24):2307–2316. https://doi.org/10.1056/NEJMoa1900907
10. Skotte NH, Southwell AL, Ostergaard ME et al. (2014) Allele-specific suppression of mutant huntingtin using antisense oligonucleotides: providing a therapeutic option for all Huntington disease patients. PLoS One 9(9):e107434. https://doi.org/10.1371/journal.pone.0107434
11. Lim KRQ, Maruyama R, Echigoya Y et al (2020) Inhibition of DUX4 expression with antisense LNA gapmers as a therapy for facioscapulohumeral muscular dystrophy. Proc Natl Acad Sci U S A. https://doi.org/10.1073/pnas.1909649117

Contents

Contributors

SARA AGUTI • *The Dubowitz Neuromuscular Centre, Molecular Neurosciences Section, Developmental Neurosciences Research and Teaching Department, Great Ormond Street Institute of Child Health, University College London, London, UK*
YOSHITSUGU AOKI • *Department of Molecular Therapy, National Institute of Neuroscience, National Center of Neurology and Psychiatry (NCNP), Tokyo, Japan*
TEJAL ASLESH • *Neuroscience and Mental Health Institute, Faculty of Medicine and Dentistry, University of Alberta, Edmonton, AB, Canada; Department of Medical Genetics, Faculty of Medicine and Dentistry, University of Alberta, Edmonton, AB, Canada*
MARK A. BEHLKE • *Integrated DNA Technologies, Inc., Coralville, IA, USA*
YUNPENG CAI • *The Interdisciplinary Nanoscience Center (iNANO), Department of Molecular Biology and Genetics, Aarhus University, Aarhus C, Denmark*
LEANNA CHAN • *Department of Medical Genetics, Faculty of Medicine and Dentistry, University of Alberta, Edmonton, AB, Canada; Faculty of Arts and Science, University of Toronto, Toronto, ON, Canada*
MOBASHAR HUSSAIN URF TURABE FAZIL • *Lee Kong Chain School of Medicine, Nanyang Technological University, Singapore, Singapore*
DAVID S. GREENBERG • *The Life Sciences Institute and The Edmond and Lili Safra Center of Brain Science, The Hebrew University of Jerusalem, Jerusalem, Israel*
RINTARO IWATA HARA • *Department of Neurology and Neurological Science, Graduate School of Medical and Dental Sciences, Tokyo Medical and Dental University, Tokyo, Japan*
YASUMASA HASHIMOTO • *Department of Molecular Therapy, National Institute of Neuroscience, National Center of Neurology and Psychiatry (NCNP), Tokyo, Japan*
SHIN-ICHIRO HORI • *Graduate School of Pharmaceutical Sciences, Osaka University, Osaka, Japan; Medicinal Chemistry Research Laboratory for Medium Molecular Drug Discovery, Shionogi & Co., Ltd., Osaka, Japan*
KENNETH A. HOWARD • *The Interdisciplinary Nanoscience Center (iNANO), Department of Molecular Biology and Genetics, Aarhus University, Aarhus C, Denmark*
SHINZO IWASHITA • *Modality Research Laboratories, Daiichi Sankyo Co., Ltd., Tokyo, Japan*
ATISH KIZHAKEYIL • *Lee Kong Chain School of Medicine, Nanyang Technological University, Singapore, Singapore*
MAKOTO KOIZUMI • *Modality Research Laboratories, Daiichi Sankyo Co., Ltd., Tokyo, Japan*
KIM A. LENNOX • *Integrated DNA Technologies, Inc., Coralville, IA, USA*
KENJI ROWEL Q. LIM • *Department of Medical Genetics, Faculty of Medicine and Dentistry, University of Alberta, Edmonton, AB, Canada*
CHENGUANG LOU • *Department of Physics, Chemistry and Pharmacy, Biomolecular Nanoscale Engineering Center, University of Southern Denmark, Odense M, Denmark*
MARIA MAHFOUZ • *Department of Medical Genetics, Faculty of Medicine and Dentistry, University of Alberta, Edmonton, AB, Canada; School of Medicine, Royal College of Surgeons in Ireland, Dublin, Ireland*
RAQUEL MANZANO • *Department of Paediatrics, Anatomy and Genetics, University of Oxford, Oxford, UK*

ELENA MARROSU • *The Dubowitz Neuromuscular Centre, Molecular Neurosciences Section, Developmental Neurosciences Research and Teaching Department, Great Ormond Street Institute of Child Health, University College London, London, UK*
RIKA MARUYAMA • *Department of Medical Genetics, Faculty of Medicine and Dentistry, University of Alberta, Edmonton, AB, Canada*
FRANCESCO MUNTONI • *The Dubowitz Neuromuscular Centre, Molecular Neurosciences Section, Developmental Neurosciences Research and Teaching Department, Great Ormond Street Institute of Child Health, University College London, London, UK; NIHR Great Ormond Street Hospital Biomedical Research Centre, London, UK*
QUYNH NGUYEN • *Department of Medical Genetics, Faculty of Medicine and Dentistry, University of Alberta, Edmonton, AB, Canada*
SATOSHI OBIKA • *Graduate School of Pharmaceutical Sciences, Osaka University, Osaka, Japan*
CHAITRA SATHYAPRAKASH • *Nuffield Department of Clinical Neurosciences, University of Oxford, Oxford, UK*
TAKAO SHOJI • *Modality Research Laboratories, Daiichi Sankyo Co., Ltd., Tokyo, Japan*
HERMONA SOREQ • *The Life Sciences Institute and The Edmond and Lili Safra Center of Brain Science, The Hebrew University of Jerusalem, Jerusalem, Israel*
KEVIN TALBOT • *Nuffield Department of Clinical Neurosciences, University of Oxford, Oxford, UK*
YONAT TZUR • *The Life Sciences Institute and The Edmond and Lili Safra Center of Brain Science, The Hebrew University of Jerusalem, Jerusalem, Israel*
MIGUEL A. VARELA • *Department of Paediatrics, Anatomy and Genetics, University of Oxford, Oxford, UK*
NAVIN KUMAR VERMA • *Lee Kong Chain School of Medicine, Nanyang Technological University, Singapore, Singapore; Skin Research Institute of Singapore, Singapore, Singapore*
FUMITO WADA • *Graduate School of Pharmaceutical Sciences, Osaka University, Osaka, Japan; Department of Molecular Innovation in Lipidology, National Cerebral and Cardiovascular Center Research Institute, Osaka, Japan*
JESPER WENGEL • *Department of Physics, Chemistry and Pharmacy, Biomolecular Nanoscale Engineering Center, University of Southern Denmark, Odense M, Denmark*
MATTHEW J. A. WOOD • *Department of Paediatrics, Anatomy and Genetics, University of Oxford, Oxford, UK*
TSUYOSHI YAMAMOTO • *Graduate School of Pharmaceutical Sciences, Osaka University, Osaka, Japan; Graduate School of Biomedical Sciences, Nagasaki University, Nagasaki, Japan*
TAKANORI YOKOTA • *Department of Neurology and Neurological Science, Graduate School of Medical and Dental Sciences, Tokyo Medical and Dental University, Tokyo, Japan*
TOSHIFUMI YOKOTA • *Department of Medical Genetics, Faculty of Medicine and Dentistry, University of Alberta, Edmonton, AB, Canada; Faculty of Arts and Science, University of Toronto, Toronto, ON, Canada; The Friends of Garrett Cumming Research and Muscular Dystrophy Canada HM Toupin Neurological Science Research Chair, Edmonton, AB, Canada*

KOTARO YOSHIOKA • *Department of Neurology and Neurological Science, Graduate School of Medical and Dental Sciences, Tokyo Medical and Dental University, Tokyo, Japan*
HAIYAN ZHOU • *NIHR Great Ormond Street Hospital Biomedical Research Centre, London, UK; Genetics and Genomic Medicine Research and Teaching Department, Great Ormond Street Institute of Child Health, University College London, London, UK*

Part I

Basics and Introduction

Chapter 1

Invention and Early History of Gapmers

Kenji Rowel Q. Lim and Toshifumi Yokota

Abstract

Gapmers are antisense oligonucleotides composed of a central DNA segment flanked by nucleotides of modified chemistry. Hybridizing with transcripts by sequence complementarity, gapmers recruit ribonuclease H and induce target RNA degradation. Since its concept first emerged in the 1980s, much work has gone into developing gapmers for use in basic research and therapy. These include improvements in gapmer chemistry, delivery, and therapeutic safety. Gapmers have also successfully entered clinical trials for various genetic disorders, with two already approved by the U.S. Food and Drug Administration for the treatment of familial hypercholesterolemia and transthyretin amyloidosis-associated polyneuropathy. Here, we review the events surrounding the early development of gapmers, from conception to their maturity, and briefly conclude with perspectives on their use in therapy.

Key words Chimeric oligonucleotides, Early development, Antisense therapy, Nucleic acid analogues, Ribonuclease H (RNase H), Gene knockdown, DNA–RNA hybrids, Target RNA, Mipomersen (brand name Kynamro), Inotersen (brand name Tegsedi)

1 Introduction

The ability to control gene expression has a central role in research and medicine. The concerted participation of numerous genes and their products is required for the success of biological processes. Disrupting this tightly regulated system by causing the absence or overabundance of involved players could lead to changes in how a process works, which are exploited to determine the functions of affected genes. By extension, such studies can help not only in understanding the molecular etiology of genetic disorders, which are due to similar disruptions, but also in treating them. Knowing how gene expression is affected in disease guides us in accomplishing the reverse, that is, restoring gene expression to healthy levels.

Here, we focus on manipulating gene expression in one direction: down. Knockdown and knockout studies, in vitro and in vivo, provide a wealth of information in elucidating gene function. Based on these efforts, online databases and consortia were formed that offer a large number of resources for working on different model

Toshifumi Yokota and Rika Maruyama (eds.), *Gapmers: Methods and Protocols*, Methods in Molecular Biology, vol. 2176, https://doi.org/10.1007/978-1-0716-0771-8_1,

organisms, from information on mutant phenotypes to providing the actual models themselves; some major model organism databases are part of the Alliance of Genome Resources (https://www.alliancegenome.org/). Targeting gene downregulation serves to treat diseases caused by gene overexpression [1]. Excessive protein production can cause pathway overactivation and/or dysregulation, disruption of protein complexes, and sequestration of cellular factors [2]. In some cases, this can lead to protein aggregation, which is the basis for a host of neuromuscular disorders [3]. Knockdown therapies are also potentially beneficial for diseases of gene overdosing, as is seen for most chromosomal aneuploidies [4], and diseases where a gene is expressed at the wrong place and/or time, for example *DUX4* in facioscapulohumeral dystrophy [5]. Indeed, the translation of gene knockdown techniques into the clinic has been demonstrated to be useful for treating such genetic disorders.

There is an entire suite of tools available for knocking down gene expression, the main ones being those employing antisense oligonucleotides (AOs), ribozymes, small molecule inhibitors, or RNA interference (RNAi) with small interfering RNAs (siRNAs) [6]. Perhaps the earliest approach developed in this list involves the use of AOs, which were first shown to achieve gene knockdown in two seminal back-to-back studies in 1978 by Zamecnik and Stephenson [7, 8]. AOs can decrease gene expression in one of two ways, both relying on their ability to recognize and bind targets through Watson-Crick base-pairing [9] (Fig. 1). The first is

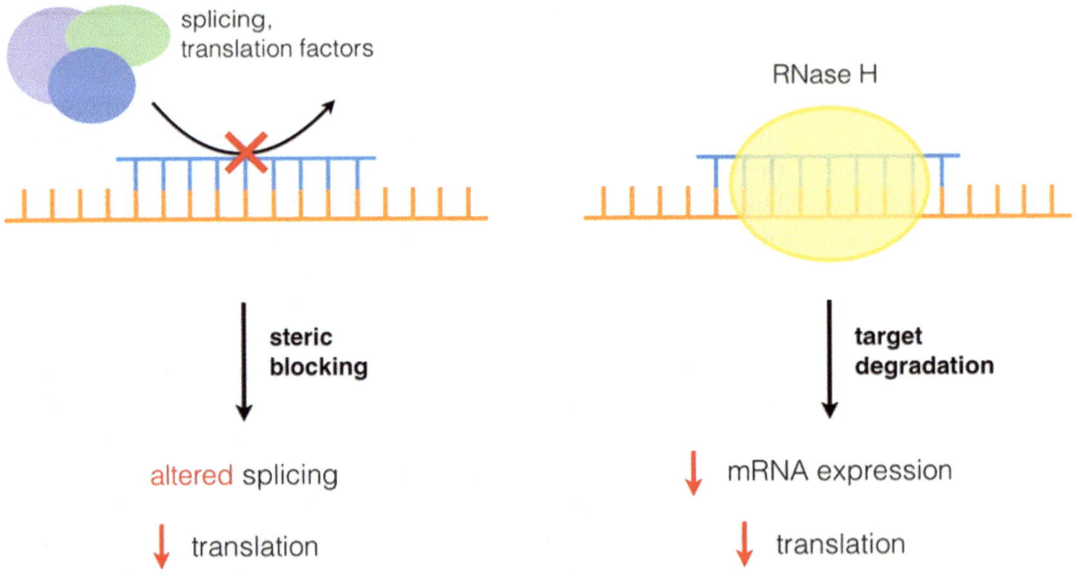

Fig. 1 Antisense oligonucleotide mechanisms of action. Antisense oligonucleotides (blue) act on their target RNAs (orange) by one of two mechanisms: steric blocking of splicing and translation factors (left) or target degradation by recruiting and inducing ribonuclease H/RNase H activity (right). In the context of gene knockdown, use of both will inhibit protein synthesis

primarily physical, by steric blocking of splicing factors or translational machinery from interacting with important RNA sequences such as splice sites or initiation codons, respectively. The second is mainly chemical, through the recruitment of ribonuclease H (RNase H), which recognizes amenable AO–RNA hybrids and cleaves the RNA portion of the complex, resulting in transcript degradation.

We have previously reviewed the early history of AOs working with the first approach [10]. We now direct our attention to the RNase H-mediated knockdown of genes by AOs, particularly those of the gapmer configuration. Gapmers are a class of AOs characterized by having a central DNA segment flanked by short stretches of nucleotides with varying chemistries on either end [9]. In a gapmer, the central segment is the RNase H-recruiting portion and hence determines the cleavage site of the target RNA; the modified ends are responsible for providing stability and target-binding affinity to the AO. At present, many different types of gapmers have been developed, each with their own chemistries, structural properties, pharmacological behaviors, and gene targets. Clinical development has also proceeded quite rapidly with these AOs, with two gapmers having been approved by the U.S. Food and Drug Administration (FDA) for familial hypercholesterolemia in 2013 [11] and for hereditary transthyretin amyloidosis-associated polyneuropathy recently in 2018 [12]. A sizeable number of gapmers are currently in clinical trials, for diseases such as cancer and Huntington's disease, among others [13].

In this chapter, we review the events leading to the invention of gapmers and their early development for use in basic and translational research. We briefly introduce the enzyme responsible for gapmer activity, RNase H, as a prelude and then end with some brief perspectives on the current state and future of antisense gapmers in the field, particularly in therapy. We recognize that such a survey of the literature necessarily restricts us to be selective. In writing this work, we would like to gratefully acknowledge the work of all researchers who have contributed and are contributing to the development of gapmer technology.

2 Ribonucleases H and DNA–RNA Hybrids

2.1 Ribonucleases H

RNases H are a group of enzymes that cleave the RNA parts of DNA–RNA hybrids, through 5′-phosphodiester bond hydrolysis [14–16]. They were first discovered in 1969 by Stein and Hausen, while purifying RNA polymerase from calf thymus extracts [17]. Stein and Hausen looked at the potential of native and denatured DNA to serve as templates for their RNA polymerase preparations. High-purity RNA polymerase gave higher template activity with denatured DNA, whereas a more crude extract resulted in RNA synthesis with native DNA and not denatured DNA. They later discovered an enzyme in the crude preparation that was responsible for specifically degrading the RNA half of the DNA–RNA hybrid formed during RNA synthesis with denatured DNA. Nucleic acids with uniform composition, either DNA or RNA, and in single-stranded or double-stranded forms were not preferred substrates for this enzyme.

It was found 6 years later that the enzyme in this study was not one but actually three, two of which were classified as RNase H [18]. Presently, we classify RNases H into two types, 1 and 2, with mostly slight differences in structure and function [14]. Both types function to regulate the presence of DNA–RNA hybrids in a cell, promoting genome stability [14, 15]. DNA–RNA hybrids typically form as part of lagging strand synthesis during DNA replication, of R loops (3-stranded complex with single-stranded DNA and a DNA–RNA duplex) during transcription, and of reverse transcription during retroviral replication in cases of infection. They are generated by the misincorporation of ribonucleotides by DNA polymerase during DNA replication as well. The importance of RNases H for survival is highlighted by observations that mice null for RNase H1 or H2 are embryonic lethal and suffer defects in genomic structure or integrity [15, 19, 20]. Mutations in RNase H2 also result in Aicardi–Goutières syndrome, a damaging, and mostly fatal neurological disorder with autoimmune-like features [21].

Structurally, RNase H1 is a monomer with both hybrid binding and nuclease activities [14]. On the other hand, RNase H2 exists as a heterotrimer [14, 22]. Only the A subunit possesses catalytic activity, but both the B and C subunits, which provide structural support, are required for RNA hydrolysis [16]. RNases H act as dimers, and have divalent metal ions (Mg^{2+}, Mn^{2+}) as cofactors [15, 16]. For target recognition and activity, RNase H1 requires at least four consecutive 2′-OH groups from the RNA strand in the hybrid, whereas RNase H2 only requires a single ribonucleotide [14, 16, 23]. Because of this, RNase H2 is primarily responsible for helping proofread the actions of DNA polymerase and is considered the more active of the two RNases H in humans [24]. Finally, both RNases H rely heavily on structure rather than sequence to recognize their targets [14, 15]—as long as a DNA–RNA hybrid meets the criteria for being bound and cleaved by RNase H, it serves as a suitable target.

2.2 Considerations with Regard to AOs

The main concerns associated with AO–RNA hybrids, in the context of promoting RNase H-mediated gene knockdown, are if they can (1) be recognized and bound, and (2) be acted upon by RNase H. As these are tightly linked with one another, especially the first as it leads to the second, AOs should preserve as much of either ability as possible. Recognition of AO–RNA hybrids depends on how closely they resemble DNA/RNA helices and if the conformation of the hybrid promotes interaction with RNase H [25]. As previously mentioned, the 2′-OH groups are critical for target binding and hydrolysis by RNase H. Modifications at this position, of which there are a considerable number in the literature, will highly influence the capacity of an AO to recruit RNase H. There is also the length of the DNA–RNA hybrid segment to consider, knowing that this must be at least four nucleotides long to engage both RNase H1 and H2 activities.

Furthermore, the AO–RNA hybrid has to be flexible enough to "fit" into the RNase H active site [25]. Some AO chemistries, especially bridged nucleic acids, sacrifice conformational freedom for enhanced stability and RNA target binding affinity, which negatively affects RNase H binding [25, 26]. Not only is a certain extent of malleability advantageous for a given AO–RNA hybrid, but it must also possess the appropriate assortment of chemical moieties to interact with the enzyme catalytic site [25]. An example would be the capacity to support hydrogen-bonding interactions to ensure proper placement of the substrate and success of the catalytic reaction. Finally, space is also a concern, as the binding of RNase H to the AO–RNA hybrid should leave enough room for the cleavage site to undergo hydrolysis.

At this point, we now explore the early stages of gapmer development. As we will see, the abovementioned considerations played a huge role in explaining much of the events that culminated in the invention of gapmers. Figure 2 is a timeline summarizing the covered events, and serves as a useful reference throughout the chapter.

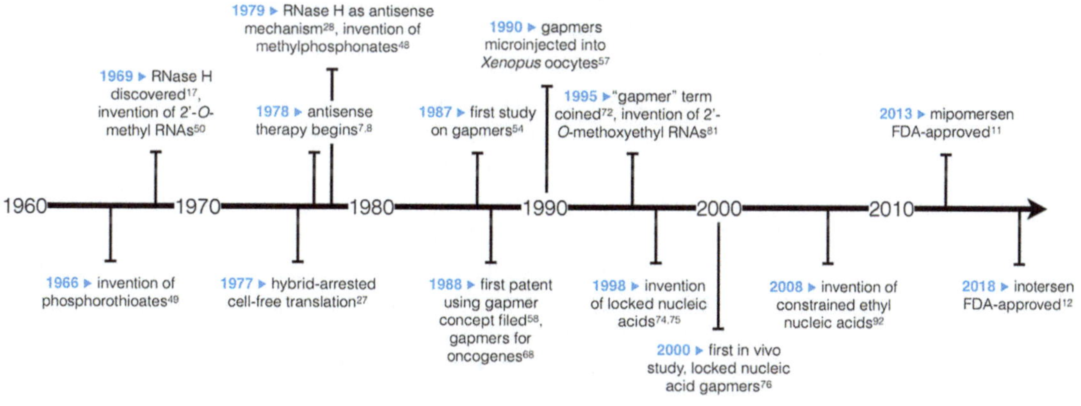

Fig. 2 Timeline of key events in the history of gapmer development. The different events considered critical to the invention and development of gapmers, and which are discussed in the main text, are summarized

3 Before Gapmers: Discovery of Antisense Oligonucleotides

In 1977, Paterson et al. described how DNA from a recombinant plasmid can be hybridized to complementary β globin mRNA and inhibit its translation [27]. They called the process "hybrid-arrested cell-free translation" and was then seen as a means for identifying and mapping gene sequences. This was followed in 1978 by two papers from Zamecnik and Stephenson that are considered as having started the entire field of antisense therapy. Using a synthetic 13-mer oligodeoxynucleotide that recognizes the Rous sarcoma virus (RSV) 35S RNA sequence, they were able to inhibit the translation of RSV proteins in vitro [7] as well as the replication of the virus itself upon infection of chick fibroblasts [8].

At that time, Zamecnik and Stephenson speculated that the oligodeoxynucleotide they introduced was acting as a physical block for processes such as proviral DNA circularization, viral DNA integration, transcription, translation, and ribosomal association [7, 8]. However a year later, in 1979, it was reported that the inhibitory effect is due to the action of RNase H. Donis-Keller found that hybridizing an oligodeoxynucleotide to complementary RNA sequences will allow for specific cleavage of those sites using RNase H from calf thymus or *E. coli* [28]. Two important observations were made in this study: firstly, that the secondary structure of the target RNA determines the efficiency of RNase H activity at the site, and secondly, that at least a DNA tetramer was sufficient to direct RNase H-mediated cleavage. The report also mentions how, unlike restriction endonucleases that had sequence-based targets, oligodeoxynucleotides coupled with RNase H used "recognition structures." The results from this study all make sense in light of the properties of RNase H that we have earlier described. In later years, the RNase H-based mechanism of oligodeoxynucleotide activity would be confirmed through other in vitro systems [29, 30].

These discoveries launched a myriad of studies using AOs to inhibit the translation of various proteins in vitro [31–34]. In vivo work was conducted mostly using microinjection of AOs into *Xenopus* oocytes as a study system [32, 33, 35, 36]. The use of AOs for therapy was also gaining traction, as supported by studies showing their antiviral capacity [37–39] as well as their potential to treat parasitic infections such as those by trypanosomes [40–42]. However, a problem with using plain oligodeoxynucleotides that was becoming increasingly recognized at the time was their poor stability, which made them easy targets for degradation by nucleases [36, 43]. One study showed that oligodeoxynucleotides had a half-life of 30 min in HeLa postmitochondrial cytoplasmic extracts and were completely degraded within 15 min in bovine calf serum [44]. Because of this and also due to their limited ability to

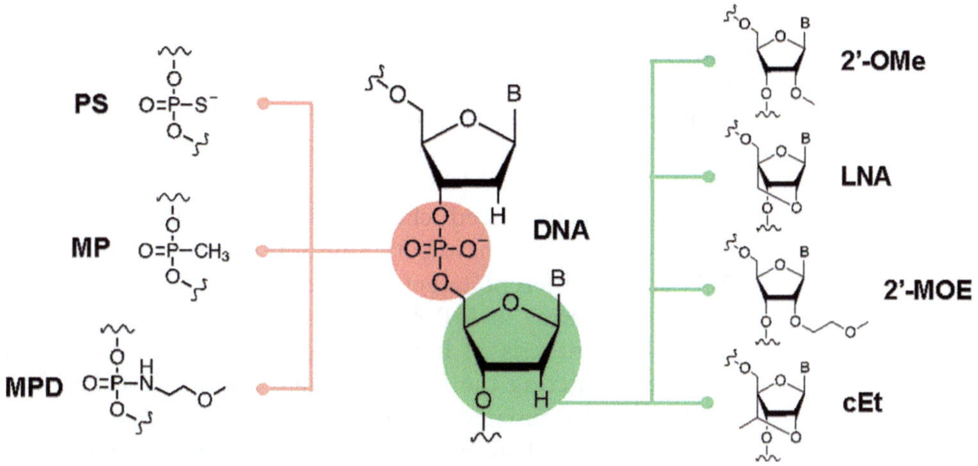

Fig. 3 Gapmer chemical modifications. Backbone (left) and sugar (right) modifications applied to gapmers that are discussed in the main text are shown. *PS* phosphorothioate, *MP* methylphosphonate, *MPD* methoxyethyl-phosphoramidate, *2′-OMe* 2′-*O*-methyl, *LNA* locked nucleic acid, *2′-MOE* 2′-*O*-methoxyethyl, *cEt* constrained ethyl, *B* base

penetrate cell membranes [45], oligodeoxynucleotides showed limited in vivo efficacy.

In response to these challenges, groups began looking into whether or not chemically modifying AOs would help improve their behavior in biological systems. These modifications involved altering the backbone and/or sugar chemistry of the AOs (Fig. 3) as well as the conjugation of moieties to increase AO stability, uptake, or both [46, 47]. While it seemed that this chemical revolution greatly improved the physical characteristics of AOs themselves, it was soon realized that it would negatively impact a different aspect of their performance altogether: the ability to induce RNase H-mediated gene knockdown.

4 Developing Antisense Gapmer Oligonucleotides

4.1 The Problem Between Chemical Modifications and RNase H

Three AO chemical modifications in particular preceded development of the first gapmers: the methylphosphonate (MP) and phosphorothioate (PS) backbone modifications, and the 2′-*O*-methyl RNA (2′-OMe) sugar modification (Fig. 3). In the MP and PS modifications, one of the non–sugar-binding oxygen atoms in the backbone is replaced with a methyl residue or a sulfur atom, respectively [9, 48, 49]. This makes MP AOs nonionic compared to the unmodified phosphodiester backbone, whereas PS AOs retain the negative charge. On the other hand, in the 2′-OMe modification, the 2′-hydroxyl group on the ribonucleotide ring is substituted with a methoxy group [50]. All three modifications were found to increase AO stability and cellular uptake.

MP AOs were demonstrated to inhibit protein synthesis using a wide array of study systems. For instance, some early studies used MP AOs to achieve up to 76% inhibition of rabbit globin mRNA translation in vitro [51], or to decrease herpes simplex virus type 1 viral load in infected cells from 50% to 99% depending on the amount of AO used [38]. It was soon revealed, however, that despite their enhanced characteristics as oligonucleotides, MP AOs did not always have better antisense activity than unmodified AOs across different targets. Maher and Dolnick compared the ability of an array of MP AOs versus that of phosphodiester AOs to inhibit synthesis of human dihydrofolate reductase (DHFR), both sets targeting the 5′ untranslated region (5′ UTR) and part of the initial translated region of the *DHFR* mRNA [52]. They found that an MP array of three or four AOs at a total concentration of 400 μM was only half as effective in inhibiting DHFR production than an array of three unmodified AOs at a total concentration of 25 μM. The unmodified AOs, in fact, completely abolished synthesis. It was then revealed that MP AOs had lower binding affinities to their target RNAs, and that their antisense activity was not due to RNase H.

A similar situation occurred for 2′-OMe AOs, which likewise are unable to knock down gene expression through RNase H [53, 54]. They have enhanced target-binding affinity unlike MP AOs, however. On the other hand, PS AOs still retain the ability to induce RNase H cleavage activity although to a lesser extent than phosphodiester AOs [55, 56]. This makes PS AOs more desirable than MP AOs [57] and may have influenced their choice for use in gapmers later on.

Overall, these findings on the relationship between chemically modified AOs and RNase H cemented the understanding that AOs work with two mechanisms of action—physical inhibition in relation to steric effects and chemical inhibition via RNase H-mediated target degradation. This became clearer when more and more studies started looking at the transcript level as opposed to just the protein level to observe AO efficacy. More importantly, the limitations imposed on AOs by chemical modifications were recognized. One such caveat is that these modifications narrow down the applicability of AOs for other targets where degradation would be more appropriate (e.g., viral transcripts and toxic gain-of-function RNA). This was the situation into which the first gapmers would enter the antisense scene.

4.2 The First Gapmers

In 1987, Inoue et al. published a four-page article in *FEBS Letters* that could perhaps be considered as the earliest appearance of gapmers in the scientific literature [54]. Inoue and colleagues were interested in developing a technique to cleave RNA at specific points, much like how a restriction enzyme cleaves DNA. They knew that RNase H could not cleave 2′-OMe AO–RNA hybrids because the conformation of the sugars in 2′-OMe and RNA nucleotides was similar. With this knowledge, they made chimeric "oligonucleotide splints" that were essentially AOs having a mixture of 2′-OMe and DNA nucleotides. They put the 2′-OMe nucleotides at the ends of the AOs and left DNA nucleotides at the center—the gapmer configuration—reasoning that upon binding to a target RNA molecule the flanking RNase H-incompetent regions would direct cleavage at the central RNase H-competent site. Upon in vitro testing, they did indeed achieve site-specific cleavage of their target RNA, finding that a four-nucleotide DNA gap was sufficient for RNase H activity. A five-nucleotide gap retained single-site specificity, whereas a three-nucleotide one induced cleavage at a second site. The position of the DNA gap was likewise important for the specificity of cleavage, that is, gaps positioned more to the side resulted in cleavage at two sites.

Inoue et al. concluded, "This [method] will contribute to our understanding of the biological function of RNA." Indeed, at the time gapmers were first being developed, they were seen more for their potential as a research tool, akin to restriction enzymes. That same year, Walder and Walder filed a patent describing how the gapmer concept can be applied to design oligonucleotide probes for the detection of DNA or RNA molecules [47, 58]. As researchers were figuring out how best to integrate non-RNase H-cleavable residues with RNase H-cleavable ones in the same oligonucleotide stretch, the mixmer class of AOs was also developed [59, 60], that is, AOs with alternating RNase H-incompetent and -competent residues.

Multiple efforts evaluating the use of gapmers rapidly emerged the following decade. In 1990, a follow-up to the study of Inoue et al. was published using similar 2′-OMe gapmers but this time against *E. coli* formylmethionine tRNA [61]. That same year, Agrawal et al. published a study testing a variety of modified 15-mer MP gapmers and found that those with a 2- or 4-nucleotide central DNA gap elicited less RNase H activity than that with a 6-nucleotide gap [62]. A series of studies in *Xenopus* also helped solidify the place of gapmers in the field. The first such study was in 1990 by Baker et al. who wanted to knock down the expression of the histone H4 gene in *Xenopus* oocytes through AO microinjection [57]. The authors compared the ability of AOs having 1 or 2 of the terminal nucleotides at each end modified with an MP or PS linkage, full-PS AOs, and unmodified AOs, all 20-mer in length, to decrease H4 mRNA levels. All modified AOs were injected at 0.1 mg/ml and worked as well as the unmodified AO that was given at a tenfold higher 1 mg/ml dose in knocking down *H4*, as assessed by Northern blot. They also found that the full-PS AO had slower cleavage activity than other AOs; however, they had a longer duration of effect, as observed with longer incubation times. Other studies in *Xenopus* used gapmers containing methoxyethyl phosphoramidate ends (Fig. 3) with a central PS or phosphodiester DNA gap to successfully knock down targets such as An2 (codes for a mitochondrial ATPase subunit), cyclins [63], and transforming growth factor β [64].

The focus of the *Xenopus* studies was largely on using gapmers to generate gene knockdown models for investigating biological processes. On the other hand, different groups were realizing the potential of gapmers in another direction, that is, the treatment of human diseases. A sizeable amount of work in those days (late 1980s to early 1990s) developed gapmers against gene targets related to cancer, perhaps as inspired by earlier studies using oligodeoxynucleotides for *c-myc* [65, 66]. MP gapmers were designed against the oncogenes *NRAS*, *HRAS*, and, as before, *c-myc* [67–71]. Varying rates of success were obtained, as the studies largely involved gapmer design optimization in terms of stability, antisense activity, and off-target toxicity. Information on how long the central DNA segment should be in gapmers was also generated, with one pointing out that a 7-nucleotide gap was optimal for antisense activity [70]. Overall, the studies indicated that the choice of gap length is a balance between specificity and efficacy, and must be decided upon depending on the aims of gapmer use.

Throughout this period, gapmers have been called splints, chimeric nucleic acid analogues, modified oligonucleotides, etc. Interestingly, based on our search, the name "gapmer" did not appear in the literature until 1995 in a report by Crooke et al., some years after the explosion of studies relating to their development [72]. There, "gapmer" was used for "chimeric oligonucleotide" and its chemically modified ends were called "wings."

4.3 New Gapmer Chemistries

Naturally, with their broadening applicability in research, gapmers became subject to an extensive amount of chemical modification to further improve their efficacy. Such efforts continue to this day, with numerous gapmer chemistries having been developed. Of these, we believe two in particular made a large impact on the future of gapmers: the locked nucleic acid (LNA) and 2′-*O*-methoxyethyl (2′-MOE) modifications.

LNAs are synthetic bicyclic nucleotides that have the 4′-C of the sugar ring bridged to the 2′-O atom by a methylene bridge (Fig. 3). This "locks" the sugar into a 3′-*endo* conformation that dramatically improves binding affinity with target RNAs; LNAs also have enhanced resistance to nuclease attack [9, 73]. LNAs were independently developed and described by the groups of Imanishi and Wengel in 1998 [74, 75]. Wahlestedt et al. were the first ones to test the therapeutic potential of LNA gapmers in vivo in rats, targeting the knockdown of the delta opioid receptor (DOR) gene [76]. LNA gapmers were stable in serum, did not induce detectable toxicity when intracerebrally injected into the caudate-putamen, and significantly decreased DOR-mediated antinociception upon treatment at reduced intrathecal doses, as low as 0.02 μg. These promising results launched LNA gapmers further into the therapeutic scene. Despite their outstanding potency, however, LNA gapmers are currently handicapped by their toxicity (contradicting the results from the 2000 study), which limits their clinical utility [77–80].

The more successful of the two, and also considered as one of the most successful antisense chemistries overall in terms of clinical development, is the 2′-MOE modification developed in 1995, where the 2′-OH group in the sugar is essentially replaced with an alkyl substituent [81] (Fig. 3). Similar to the LNA, the 2′-MOE modification increases AO stability and target binding affinity [9]. Importantly, though, 2′-MOEs are safer than LNAs as observed experimentally. Early studies with 2′-MOE gapmers showed that they are effective antisense effectors in vitro [82, 83]. Numerous 2′-MOE gapmers now serve as drug candidates in clinical trials for various genetic disorders [9, 13]. As a testament to its success, the only two FDA-approved gapmers at present are 2′-MOE gapmers: mipomersen (brand name Kynamro; Ionis, Genzyme) for familial hypercholesterolemia [11] and inotersen (brand name Tegsedi; Ionis, Akcea) for hereditary transthyretin amyloidosis-associated polyneuropathy [12].

5 Conclusion: Current State and Challenges for the Future

The early days of gapmers were clearly filled with innovation and excitement. From starting out as an elegant solution to an RNase H sensitivity problem, gapmers have grown to carve out their own niche in the antisense community, especially in therapy. Gapmers are now being developed as treatments for a variety of disorders (e.g., Huntington's disease, myotonic dystrophy, familial chylomicronemia, cancer, and cardiovascular diseases, among others) and are part of numerous clinical trials [9, 13, 84–87]. Their range of targets has also diversified, with gapmers already targeting long noncoding [88, 89] and enhancer RNAs [90]. The chemical modification of gapmers continues, with the list growing as new chemistries are created and integrated in different combinations.

Gapmers have indeed made large strides in development since their initial appearance as an RNA counterpart of restriction endonucleases, and are now mostly being developed as therapies. However, they are facing two major hurdles that are challenging their utility as therapeutics, i.e., toxicity and uptake. Interestingly enough, these are the same issues that have concerned researchers about gapmers since the early 1990s [59, 71]. Gapmer toxicity is categorized as hybridization-dependent or hybridization-independent [9]. Hybridization-dependent toxicity mainly stems from gapmer activity on nontarget sites, or what we call off-target effects. Meanwhile, hybridization-independent toxicity comes from the binding of gapmers to cellular proteins, which may elicit unwanted consequences—examples include the activation of Toll-like receptors, inflammation, and negative effects on coagulation [9]. Hepatotoxicity is especially associated with gapmer treatment [77, 78]. Uptake is another concern that has to be taken into consideration. Gapmers, at least those of the 2′-MOE chemistry, accumulate primarily in the kidney, liver, and mesenteric lymph, with limited availability in other tissues [91]. This may impact efficacy, as has been observed for other AOs.

Overcoming these challenges calls for more extensive studies on the nature, behavior, and effects of gapmers in biological systems. As chemistry likely has the most influence on both toxicity and uptake, a systematic evaluation of the various chemical modifications in gapmers would be helpful. One such study was published recently, reporting how constrained ethyl (cEt) gapmers [92] (Fig. 3) can be made significantly less toxic through the conversion of a single nucleotide in the central gap from a PS-DNA to a PS-2′OMe residue [93]. In silico tools are also being developed to enhance gapmer sequence design and minimize off-targeting [94–96]. Additionally, such tools may be helpful in predicting how well gapmer–RNA hybrids form [97], and in providing insight as to how they interact with RNase H and other proteins. In terms of gapmer uptake, studies developing better delivery strategies (e.g., through the use of cell-penetrating peptides, fatty acid moieties, or nanoparticles, to name a few) or looking into the mechanisms surrounding gapmer entry into cells are certainly needed; it is encouraging to see that there are currently studies available on the subject [98–102]. Finally, careful interpretation of in vitro results is recommended, as they are in most cases not completely representative of the case in vivo, especially as it concerns gapmer pharmacological behavior. Much progress has been made in the last three decades in understanding, designing, developing, and refining the use of gapmers in basic and translational research. Using these lessons from the past, we can perhaps expect another revolution in gapmer research—this time, in our near future.

References

1. Prelich G (2012) Gene overexpression: uses, mechanisms, and interpretation. Genetics 190:841–854. https://doi.org/10.1534/genetics.111.136911
2. Bolognesi B, Lehner B (2018) Protein overexpression: reaching the limit. Elife. https://doi.org/10.7554/eLife.39804
3. Aguzzi A, O'Connor T (2010) Protein aggregation diseases: pathogenicity and therapeutic perspectives. Nat Rev Drug Discov 9:237–248. https://doi.org/10.1038/nrd3050
4. Jackson M, Marks L, May GHW, Wilson JB (2018) The genetic basis of disease. Essays Biochem 62:643–723. https://doi.org/10.1042/EBC20170053
5. Wang LH, Tawil R (2016) Facioscapulohumeral dystrophy. Curr Neurol Neurosci Rep 16:66. https://doi.org/10.1007/s11910-016-0667-0

6. Burnett JC, Rossi JJ (2012) RNA-based therapeutics: current Progress and future prospects. Chem Biol 19:60–71. https://doi.org/10.1016/j.chembiol.2011.12.008
7. Stephenson ML, Zamecnik PC (1978) Inhibition of Rous sarcoma viral RNA translation by a specific oligodeoxyribonucleotide. Proc Natl Acad Sci U S A 75:285–288. https://doi.org/10.1073/pnas.75.1.285
8. Zamecnik PC, Stephenson ML (1978) Inhibition of Rous sarcoma virus replication and cell transformation by a specific oligodeoxynucleotide. Proc Natl Acad Sci U S A 75:280–284. https://doi.org/10.1073/pnas.75.1.280
9. Bennett CF, Swayze EE (2010) RNA targeting therapeutics: molecular mechanisms of antisense oligonucleotides as a therapeutic platform. Annu Rev Pharmacol Toxicol 50:259–293. https://doi.org/10.1146/annurev.pharmtox.010909.105654
10. Lim KRQ, Yokota T (2018) Invention and early history of exon skipping and splice modulation. In: Yokota T, Maruyama R (eds) Exon Skipp. Incl. Ther. Methods Protoc. Springer, New York, pp 3–30
11. Hair P, Cameron F, McKeage K (2013) Mipomersen sodium: first global approval. Drugs 73:487–493. https://doi.org/10.1007/s40265-013-0042-2
12. Keam SJ (2018) Inotersen: first global approval. Drugs 78:1371–1376. https://doi.org/10.1007/s40265-018-0968-5
13. Shen X, Corey DR (2018) Chemistry, mechanism and clinical status of antisense oligonucleotides and duplex RNAs. Nucleic Acids Res 46:1584–1600. https://doi.org/10.1093/nar/gkx1239
14. Cerritelli SM, Crouch RJ (2009) Ribonuclease H: the enzymes in eukaryotes. FEBS J 276:1494–1505. https://doi.org/10.1111/j.1742-4658.2009.06908.x
15. Moelling K, Broecker F, Russo G, Sunagawa S (2017) RNase H as gene modifier, driver of evolution and antiviral defense. Front Microbiol 8:1745. https://doi.org/10.3389/fmicb.2017.01745
16. Kojima K, Baba M, Tsukiashi M et al (2018) RNA/DNA structures recognized by RNase H2. Brief Funct Genomics 18(3):169–173. https://doi.org/10.1093/bfgp/ely024
17. Stein H, Hausen P (1969) Enzyme from calf thymus degrading the RNA moiety of DNA-RNA hybrids: effect on DNA-dependent RNA polymerase. Science 166:393–395
18. Busen W, Hausen P (1975) Distinct Ribonuclease H activities in calf thymus. Eur J Biochem 52:179–190. https://doi.org/10.1111/j.1432-1033.1975.tb03985.x
19. Cerritelli SM, Frolova EG, Feng C et al (2003) Failure to produce mitochondrial DNA results in embryonic lethality in Rnaseh1 null mice. Mol Cell 11:807–815
20. Hiller B, Achleitner M, Glage S et al (2012) Mammalian RNase H2 removes ribonucleotides from DNA to maintain genome integrity. J Exp Med 209:1419–1426. https://doi.org/10.1084/jem.20120876
21. Fazzi E, Cattalini M, Orcesi S et al (2013) Aicardi–Goutieres syndrome, a rare neurological disease in children: a new autoimmune disorder? Autoimmun Rev 12:506–509. https://doi.org/10.1016/j.autrev.2012.08.012
22. Frank P, Braunshofer-Reiter C, Wintersberger U et al (1998) Cloning of the cDNA encoding the large subunit of human RNase HI, a homologue of the prokaryotic RNase HII. Proc Natl Acad Sci 95:12872–12877. https://doi.org/10.1073/pnas.95.22.12872
23. Eder PS, Walder RY, Walder JA (1993) Substrate specificity of human RNase H1 and its role in excision repair of ribose residues misincorporated in DNA. Biochimie 75:123–126
24. Reijns MAM, Bubeck D, Gibson LCD et al (2011) The structure of the human RNase H2 complex defines key interaction interfaces relevant to enzyme function and human disease. J Biol Chem 286:10530–10539. https://doi.org/10.1074/jbc.M110.177394
25. Zamaratski E, Pradeepkumar PI, Chattopadhyaya J (2001) A critical survey of the structure-function of the antisense oligo/RNA heteroduplex as substrate for RNase H. J Biochem Biophys Methods 48:189–208
26. Herdewijn P (1999) Conformationally restricted carbohydrate-modified nucleic acids and antisense technology. Biochim Biophys Acta 1489:167–179
27. Paterson BM, Roberts BE, Kuff EL (1977) Structural gene identification and mapping by DNA-mRNA hybrid-arrested cell-free translation. Proc Natl Acad Sci U S A 74:4370–4374. https://doi.org/10.1073/pnas.74.10.4370
28. Donis-Keller H (1979) Site specific enzymatic cleavage of RNA. Nucleic Acids Res 7:179–192. https://doi.org/10.1093/nar/7.1.179

29. Minshull J, Hunt T (1986) The use of single-stranded DNA and RNase H to promote quantitative "hybrid arrest of translation" of mRNA/DNA hybrids in reticulocyte lysate cell-free translations. Nucleic Acids Res 14:6433–6451. https://doi.org/10.1093/nar/14.16.6433
30. Walder RY, Walder JA (1988) Role of RNase H in hybrid-arrested translation by antisense oligonucleotides. Proc Natl Acad Sci U S A 85:5011–5015. https://doi.org/10.1073/pnas.85.14.5011
31. Blake KR, Murakami A, Miller PS (1985) Inhibition of rabbit globin mRNA translation by sequence-specific oligodeoxyribonucleotides. Biochemistry 24:6132–6138
32. Cazenave C, Loreau N, Toulmé JJ, Hélène C (1986) Anti-messenger oligodeoxynucleotides: specific inhibition of rabbit beta-globin synthesis in wheat germ extracts and Xenopus oocytes. Biochimie 68:1063–1069
33. Cazenave C, Loreau N, Thuong NT et al (1987) Enzymatic amplification of translation inhibition of rabbit β-globin mRNA mediated by anti-messenger oligodeoxynucleotides covalently linked to intercalating agents. Nucleic Acids Res 15:4717–4736. https://doi.org/10.1093/nar/15.12.4717
34. Lawson TG, Ray BK, Dodds JT et al (1986) Influence of 5′ proximal secondary structure on the translational efficiency of eukaryotic mRNAs and on their interaction with initiation factors. J Biol Chem 261:13979–13989
35. Kawasaki ES (1985) Quantitative hybridization-arrest of mRNA in Xenopus oocytes using single-stranded complementary DNA or oligonucleotide probes. Nucleic Acids Res 13:4991–5004. https://doi.org/10.1093/nar/13.13.4991
36. Dash P, Lotan I, Knapp M et al (1987) Selective elimination of mRNAs in vivo: complementary oligodeoxynucleotides promote RNA degradation by an RNase H-like activity. Proc Natl Acad Sci U S A 84:7896–7900. https://doi.org/10.1073/pnas.84.22.7896
37. Zamecnik PC, Goodchild J, Taguchi Y, Sarin PS (1986) Inhibition of replication and expression of human T-cell lymphotropic virus type III in cultured cells by exogenous synthetic oligonucleotides complementary to viral RNA. Proc Natl Acad Sci 83:4143–4146. https://doi.org/10.1073/pnas.83.12.4143
38. Smith CC, Aurelian L, Reddy MP et al (1986) Antiviral effect of an oligo(nucleoside methylphosphonate) complementary to the splice junction of herpes simplex virus type 1 immediate early pre-mRNAs 4 and 5. Proc Natl Acad Sci 83:2787–2791. https://doi.org/10.1073/pnas.83.9.2787
39. Gupta KC (1987) Antisense oligodeoxynucleotides provide insight into mechanism of translation initiation of two Sendai virus mRNAs. J Biol Chem 262:7492–7496
40. Cornelissen AWCA, Verspieren MP, Toulmé J-J et al (1986) The common 5′ terminal sequence on trypanosome mRNAs: a target for anti-messenger oligodeoxynucleotides. Nucleic Acids Res 14:5605–5614. https://doi.org/10.1093/nar/14.14.5605
41. Walder J, Eder P, Engman D et al (1986) The 35-nucleotide spliced leader sequence is common to all trypanosome messenger RNA's. Science 233:569–571. https://doi.org/10.1126/science.3523758
42. Verspieren P, Cornelissen AW, Thuong NT et al (1987) An acridine-linked oligodeoxynucleotide targeted to the common 5′ end of trypanosome mRNAs kills cultured parasites. Gene 61:307–315
43. Rebagliati MR, Melton DA (1987) Antisense RNA injections in fertilized frog eggs reveal an RNA duplex unwinding activity. Cell 48:599–605
44. Wickstrom E (1986) Oligodeoxynucleotide stability in subcellular extracts and culture media. J Biochem Biophys Methods 13:97–102. https://doi.org/10.1016/0165-022X(86)90021-7
45. Oberemok V, Laikova K, Repetskaya A et al (2018) A half-century history of applications of antisense oligonucleotides in medicine, agriculture and forestry: we should continue the journey. Molecules 23:1302. https://doi.org/10.3390/molecules23061302
46. Khvorova A, Watts JK (2017) The chemical evolution of oligonucleotide therapies of clinical utility. Nat Biotechnol 35:238–248. https://doi.org/10.1038/nbt.3765
47. Lundin KE, Gissberg O, Smith CIE (2015) Oligonucleotide therapies: the past and the present. Hum Gene Ther 26:475–485. https://doi.org/10.1089/hum.2015.070
48. Miller PS, Yano J, Yano E et al (1979) Nonionic nucleic acid analogues. Synthesis and characterization of dideoxyribonucleoside methylphosphonates. Biochemistry 18:5134–5143
49. Eckstein F (1966) Nucleoside Phosphorothioates. J Am Chem Soc 88:4292–4294. https://doi.org/10.1021/ja00970a054
50. Bobst AM, Cerutti PA, Rottman F (1969) Structure of poly(2'-O-methyladenylic acid) at acidic and neutral pH. J Am Chem Soc

91:1246–1248. https://doi.org/10.1021/ja01033a054

51. Blake KR, Murakami A, Spitz SA et al (1985) Hybridization arrest of globin synthesis in rabbit reticulocyte lysates and cells by oligodeoxyribonucleoside methylphosphonates. Biochemistry 24:6139–6145
52. Maher LJ, Dolnick BJ (1988) Comparative hybrid arrest by tandem antisense oligodeoxyribonucleotides or oligodeoxyribonucleoside methylpbosphonates in a cell-free system. Nucleic Acids Res 16:3341–3358. https://doi.org/10.1093/nar/16.8.3341
53. Akhtar S, Kole R, Juliano RL (1991) Stability of antisense DNA oligodeoxynucleotide analogs in cellular extracts and sera. Life Sci 49:1793–1801
54. Inoue H, Hayase Y, Iwai S, Ohtsuka E (1987) Sequence-dependent hydrolysis of RNA using modified oligonucleotide splints and RNase H. FEBS Lett 215:327–330
55. Campbell JM, Bacon TA, Wickstrom E (1990) Oligodeoxynucleoside phosphorothioate stability in subcellular extracts, culture media, sera and cerebrospinal fluid. J Biochem Biophys Methods 20:259–267. https://doi.org/10.1016/0165-022X(90)90084-P
56. Reed JC, Stein C, Subasinghe C et al (1990) Antisense-mediated inhibition of BCL2 protooncogene expression and leukemic cell growth and survival: comparisons of phosphodiester and phosphorothioate oligodeoxynucleotides. Cancer Res 50:6565–6570
57. Baker C, Holland D, Edge M, Colman A (1990) Effects of oligo sequence and chemistry on the efficiency of oligodeoxyribonucleotide-mediated mRNA cleavage. Nucleic Acids Res 18:3537–3543. https://doi.org/10.1093/nar/18.12.3537
58. Walder J, Walder R (1995) Nucleic acid hybridization and amplification method for detection of specific sequences in which a complementary labeled nucleic acid probe is cleaved. US Patent US5403711A, 4 April 1995
59. Quartin RS, Brakel CL, Wetmur G (1989) Number and distribution of methyiphosphonate linkages in oligodeoxynucleotides affect exo- and endonuclease sensitivity and ability to form RNase H substrates. Nucleic Acids Res 17:7253–7262. https://doi.org/10.1093/nar/17.18.7253
60. Furdon PJ, Dominski Z, Kole R (1989) RNase H cleavage of RNA hybridized to oligonucleotides containing methylphosphonate, phosphorothioate and phosphodiester bonds. Nucleic Acids Res 17:9193–9204. https://doi.org/10.1093/nar/17.22.9193
61. Hayase Y, Inoue H, Ohtsuka E (1990) Secondary structure in formylmethionine tRNA influences the site-directed cleavage of ribonuclease H using chimeric 2′-O-methyl oligodeoxyribonucleotides. Biochemistry 29:8793–8797
62. Agrawal S, Mayrand SH, Zamecnik PC, Pederson T (1990) Site-specific excision from RNA by RNase H and mixed-phosphate-backbone oligodeoxynucleotides. Proc Natl Acad Sci U S A 87:1401–1405. https://doi.org/10.1073/pnas.87.4.1401
63. Dagle JM, Walder JA, Weeks DL (1990) Targeted degradation of mRNA in Xenopus oocytes and embryos directed by modified oligonucleotides: studies of An2 and cyclin in embryogenesis. Nucleic Acids Res 18:4751–4757. https://doi.org/10.1093/nar/18.16.4751
64. Potts JD, Dagle JM, Walder JA et al (1991) Epithelial-mesenchymal transformation of embryonic cardiac endothelial cells is inhibited by a modified antisense oligodeoxynucleotide to transforming growth factor beta 3. Proc Natl Acad Sci 88:1516–1520. https://doi.org/10.1073/pnas.88.4.1516
65. Heikkila R, Schwab G, Wickstrom E et al (1987) A c-myc antisense oligodeoxynucleotide inhibits entry into S phase but not progress from G0 to G1. Nature 328:445–449. https://doi.org/10.1038/328445a0
66. Loke SL, Stein C, Zhang X et al (1988) Delivery of c-myc antisense phosphorothioate oligodeoxynucleotides to hematopoietic cells in culture by liposome fusion: specific reduction in c-myc protein expression correlates with inhibition of cell growth and DNA synthesis. Curr Top Microbiol Immunol 141:282–289
67. Tidd DM, Warenius HM (1989) Partial protection of oncogene, anti-sense oligodeoxynucleotides against serum nuclease degradation using terminal methylphosphonate groups. Br J Cancer 60:343–350. https://doi.org/10.1038/bjc.1989.283
68. Tidd DM, Hawley P, Warenius HM, Gibson I (1988) Evaluation of N-ras oncogene antisense, sense and nonsense sequence methylphosphonate oligonucleotide analogues. Anticancer Drug Des 3:117–127
69. Giles RV, Tidd DM (1992) Enhanced RNase H activity with methylphosphonodiester/phosphodiester chimeric antisense oligodeoxynucleotides. Anticancer Drug Des 7:37–48
70. Monia BP, Lesnik EA, Gonzalez C et al (1993) Evaluation of 2′-modified

oligonucleotides containing 2′-deoxy gaps as antisense inhibitors of gene expression. J Biol Chem 268:14514–14522

71. Giles RV, Tidd DM (1992) Increased specificity for antisense oligodeoxynucleotide targeting of RNA cleavage by RNase H using chimeric methylphosphonodiester/phosphodiester structures. Nucleic Acids Res 20:763–770. https://doi.org/10.1093/nar/20.4.763

72. Crooke ST, Lemonidis KM, Neilson L et al (1995) Kinetic characteristics of *Escherichia coli* RNase H1: cleavage of various antisense oligonucleotide-RNA duplexes. Biochem J 312:599–608. https://doi.org/10.1042/bj3120599

73. Braasch DA, Corey DR (2001) Locked nucleic acid (LNA): fine-tuning the recognition of DNA and RNA. Chem Biol 8:1–7

74. Obika S, Nanbu D, Hari Y et al (1998) Stability and structural features of the duplexes containing nucleoside analogues with a fixed N-type conformation, 2′-O,4′-C-methyleneribonucleosides. Tetrahedron Lett 39:5401–5404. https://doi.org/10.1016/S0040-4039(98)01084-3

75. Koshkin AA, Singh SK, Nielsen P et al (1998) LNA (locked nucleic acids): synthesis of the adenine, cytosine, guanine, 5-methylcytosine, thymine and uracil bicyclonucleoside monomers, oligomerisation, and unprecedented nucleic acid recognition. Tetrahedron 54:3607–3630. https://doi.org/10.1016/S0040-4020(98)00094-5

76. Wahlestedt C, Salmi P, Good L et al (2000) Potent and nontoxic antisense oligonucleotides containing locked nucleic acids. Proc Natl Acad Sci U S A 97:5633–5638. https://doi.org/10.1073/pnas.97.10.5633

77. Kasuya T, Hori S, Watanabe A et al (2016) Ribonuclease H1-dependent hepatotoxicity caused by locked nucleic acid-modified gapmer antisense oligonucleotides. Sci Rep 6:30377. https://doi.org/10.1038/srep30377

78. Swayze EE, Siwkowski AM, Wancewicz EV et al (2007) Antisense oligonucleotides containing locked nucleic acid improve potency but cause significant hepatotoxicity in animals. Nucleic Acids Res 35:687–700. https://doi.org/10.1093/nar/gkl1071

79. Kakiuchi-Kiyota S, Koza-Taylor PH, Mantena SR et al (2014) Comparison of hepatic transcription profiles of locked ribonucleic acid antisense oligonucleotides: evidence of distinct pathways contributing to non-target mediated toxicity in mice. Toxicol Sci 138:234–248. https://doi.org/10.1093/toxsci/kft278

80. Burel SA, Hart CE, Cauntay P et al (2016) Hepatotoxicity of high affinity gapmer antisense oligonucleotides is mediated by RNase H1 dependent promiscuous reduction of very long pre-mRNA transcripts. Nucleic Acids Res 44:2093–2109. https://doi.org/10.1093/nar/gkv1210

81. Martin P (1995) Ein neuer Zugang zu 2'-O-Alkylribonucleosiden und Eigenschaften deren Oligonucleotide. Helv Chim Acta 78:486–504. https://doi.org/10.1002/hlca.19950780219

82. Garay M, Gaarde W, Monia BP et al (2000) Inhibition of hypoxia/reoxygenation-induced apoptosis by an antisense oligonucleotide targeted to JNK1 in human kidney cells. Biochem Pharmacol 59:1033–1043. https://doi.org/10.1016/S0006-2952(99)00412-8

83. Levesque L, Dean NM, Sasmor H, Crooke ST (1997) Antisense oligonucleotides targeting human protein kinase C-alpha inhibit phorbol ester-induced reduction of bradykinin-evoked calcium mobilization in A549 cells. Mol Pharmacol 51:209–216

84. Moreno PMD, Pêgo AP (2014) Therapeutic antisense oligonucleotides against cancer: hurdling to the clinic. Front Chem 2:87. https://doi.org/10.3389/fchem.2014.00087

85. Jauvin D, Chrétien J, Pandey SK et al (2017) Targeting DMPK with antisense oligonucleotide improves muscle strength in Myotonic dystrophy type 1 mice. Mol Ther Nucleic Acids 7:465–474. https://doi.org/10.1016/j.omtn.2017.05.007

86. Yamamoto T, Wada F, Harada-Shiba M (2016) Development of antisense drugs for dyslipidemia. J Atheroscler Thromb 23:1011–1025. https://doi.org/10.5551/jat.RV16001

87. Lucas T, Bonauer A, Dimmeler S (2018) RNA therapeutics in cardiovascular disease. Circ Res 123:205–220. https://doi.org/10.1161/CIRCRESAHA.117.311311

88. Amodio N, Stamato MA, Juli G et al (2018) Drugging the lncRNA MALAT1 via LNA gapmeR ASO inhibits gene expression of proteasome subunits and triggers anti-multiple myeloma activity. Leukemia 32:1948–1957. https://doi.org/10.1038/s41375-018-0067-3

89. Salehi M, Sharifi M, Bagheri M (2019) Knockdown of long noncoding RNA Plasmacytoma variant translocation 1 with antisense locked nucleic acid GapmeRs exerts tumor-

suppressive functions in human acute Erythroleukemia cells through Downregulation of C-MYC expression. Cancer Biother Radiopharm 34(6):371–379. https://doi.org/10.1089/cbr.2018.2510
90. Mirtschink P, Bischof C, Pham M-D et al (2019) Inhibition of the HIF1α-induced Cardiospecific HERNA1 eRNA protects from heart disease. Circulation 139 (24):2778–2792. https://doi.org/10.1161/CIRCULATIONAHA.118.036769
91. Yu RZ, Kim T-W, Hong A et al (2006) Cross-species pharmacokinetic comparison from mouse to man of a second-generation antisense oligonucleotide, ISIS 301012, targeting human Apolipoprotein B-100. Drug Metab Dispos 35:460–468. https://doi.org/10.1124/dmd.106.012401
92. Seth PP, Siwkowski A, Allerson CR et al (2008) Design, synthesis and evaluation of constrained methoxyethyl (cMOE) and constrained ethyl (cEt) nucleoside analogs. Nucleic Acids Symp Ser (Oxf) 52:553–554. https://doi.org/10.1093/nass/nrn280
93. Shen W, De Hoyos CL, Migawa MT et al (2019) Chemical modification of PS-ASO therapeutics reduces cellular protein-binding and improves the therapeutic index. Nat Biotechnol 37(6):640–650. https://doi.org/10.1038/s41587-019-0106-2
94. Hagedorn PH, Yakimov V, Ottosen S et al (2013) Hepatotoxic potential of therapeutic oligonucleotides can be predicted from their sequence and modification pattern. Nucleic Acid Ther 23:302–310. https://doi.org/10.1089/nat.2013.0436
95. Burdick AD, Sciabola S, Mantena SR et al (2014) Sequence motifs associated with hepatotoxicity of locked nucleic acid—modified antisense oligonucleotides. Nucleic Acids Res 42:4882–4891. https://doi.org/10.1093/nar/gku142
96. Kasuya T, Kugimiya A (2018) Role of computationally evaluated target specificity in the hepatotoxicity of Gapmer antisense oligonucleotides. Nucleic Acid Ther 28:312–317. https://doi.org/10.1089/nat.2018.0724
97. Uppuladinne MVN, Sonavane UB, Deka RC, Joshi RR (2019) Structural insight into antisense gapmer-RNA oligomer duplexes through molecular dynamics simulations. J Biomol Struct Dyn 37:2823–2836. https://doi.org/10.1080/07391102.2018.1498390
98. Fazil MHUT, Ong ST, Chalasani MLS et al (2016) GapmeR cellular internalization by macropinocytosis induces sequence-specific gene silencing in human primary T-cells. Sci Rep 6:37721. https://doi.org/10.1038/srep37721
99. Hvam ML, Cai Y, Dagnæs-Hansen F et al (2017) Fatty acid-modified Gapmer antisense oligonucleotide and serum albumin constructs for pharmacokinetic modulation. Mol Ther 25:1710–1717. https://doi.org/10.1016/j.ymthe.2017.05.009
100. Crooke ST, Wang S, Vickers TA et al (2017) Cellular uptake and trafficking of antisense oligonucleotides. Nat Biotechnol 35:230–237. https://doi.org/10.1038/nbt.3779
101. Liang X-H, Sun H, Nichols JG, Crooke ST (2017) RNase H1-dependent antisense oligonucleotides are robustly active in directing RNA cleavage in both the cytoplasm and the nucleus. Mol Ther 25:2075–2092. https://doi.org/10.1016/j.ymthe.2017.06.002
102. Cheng X, Liu Q, Li H et al (2017) Lipid nanoparticles loaded with an antisense oligonucleotide Gapmer against Bcl-2 for treatment of lung cancer. Pharm Res 34:310–320

Chapter 2

Development and Clinical Applications of Antisense Oligonucleotide Gapmers

Leanna Chan and Toshifumi Yokota

Abstract

DNA-like molecules called antisense oligonucleotides have opened new treatment possibilities for genetic diseases by offering a method of regulating gene expression. Antisense oligonucleotides are often used to suppress the expression of mutated genes which may interfere with essential downstream pathways. Since antisense oligonucleotides have been introduced for clinical use, different chemistries have been developed to further improve efficacy, potency, and safety. One such chemistry is a chimeric structure of a central block of deoxyribonucleotides flanked by sequences of modified nucleotides. Referred to as a gapmer, this chemistry produced promising results in the treatment of genetic diseases. Mipomersen and inotersen are examples of recent FDA-approved antisense oligonucleotide gapmers used for the treatment of familial hypercholesterolemia and hereditary transthyretin amyloidosis, respectively. In addition, volanesorsen was conditionally approved in the EU for the treatment of adult patients with familial chylomicronemia syndrome (FCS) in 2019. Many others are being tested in clinical trials or under preclinical development. This chapter will cover the development of mipomersen and inotersen in clinical trials, along with advancement in gapmer treatments for cancer, triglyceride-elevating genetic diseases, Huntington's disease, myotonic dystrophy, and prion diseases.

Key words Mipomersen (trade name Kynamro), Inotersen (trade name Tegsedi), Apatorsen, Volanesorsen (trade name Waylivra), Hypertriglyceridemia, WVE-120101 /WVE-120102, Familial hypercholesterolemia, Hereditary transthyretin amyloidosis, familial chylomicronemia syndrome, Familial partial lipodystrophy

1 Introduction

In many genetic diseases, genes with disease-inducing mutations produce abnormal mRNA, leading to the translation of nonfunctional protein that can disrupt surrounding and downstream pathways [1]. An emerging therapeutic option for the treatment of genetic diseases is antisense-mediated therapy [2]. Antisense oligonucleotides (AONs) are a sequence of 8–50 synthetic nucleotides that complementarily bind to RNA by Watson–Crick pairing to regulate gene expression [3–5]. Once bound to its RNA complementary sequence, AONs can knock down gene expression

Toshifumi Yokota and Rika Maruyama (eds.), *Gapmers: Methods and Protocols*, Methods in Molecular Biology, vol. 2176, https://doi.org/10.1007/978-1-0716-0771-8_2, © Springer Science+Business Media, LLC, part of Springer Nature 2020

through either enzyme-mediated cleavage, such as RNase H or through steric blocking, which will prevent the binding of ribosomes to the mRNA and therefore inhibit protein production [2–4]. RNase H is a ubiquitous enzyme that normally removes RNA in an RNA/DNA substrate [6, 7]. It targets the RNA strand of RNA–DNA duplexes and is commonly used as a laboratory reagent in molecular biology and for therapeutic AON design as it can inhibit gene expression when bound to any part of the mRNA of interest [4, 8]. In contrast, AONs that operate through steric blocking, such as phosphorodiamidate morpholino oligomers (PMOs), must bind generally to the 5′ or AUG initiation codon region to disrupt the translational machinery [4]. Figure 2 illustrates these two mechanisms of action of AONs.

Gapmers are a type of AON chemistry that consists of a central block of deoxyribonucleotides (DNA), flanked by modified nucleotides such as phosphorothioates (PS), 2′-*O*-methyl (2′OMe), or 2′-*O*-methoxyethyl (2′MOE) on both sides [8]. The DNA central block binds to mRNA, making it a target for RNase H cleavage [8]. In recent years, AON therapies with gapmer chemistry have received approval by the Food and Drug Administration (FDA) for the treatment of several genetic diseases. Notably, AON gapmers mipomersen and inotersen were FDA-approved in 2013 and 2018, respectively. In this chapter, the clinical applications and recent advancements in AON gapmer therapy will be covered. Specifically, treatments and developing therapies for the following diseases are mentioned: familial hypercholesterolemia (FH), familial chylomicronemia syndrome (FCS), cancer, familial partial lipodystrophy (FPLD), familial hypertriglyceridemia, hereditary transthyretin amyloidosis (hATTR), myotonic dystrophy, and Huntington's disease.

2 AON Chemistries

An important aspect of gapmers to be taken into account is the chemistry of the AON [9]. Unmodified AONs, such as DNA, are easily degraded by intracellular nucleases before they reach their target receptor, so evasion of these mechanisms is required to prolong the efficacy of the treatment [3]. Phosphorothioates (PS) are the first generation of modifications developed for AONs, where a sulfur atom replaces a nonbridging-phosphate oxygen. This enables PS to evade degradation, enact RNase H cleavage of the target RNA, and prevents rapid renal excretion by increasing the binding affinity of the PS to serum proteins [3, 8, 10]. PS is the most widely used and successful chemical modification for loss of function studies [3]. The second-generation AONs have modifications to the 2′ribose sugar molecule of synthetic nucleotides. The most common modifications are 2′-*O*-methyl (2′OMe) and 2′-*O*-

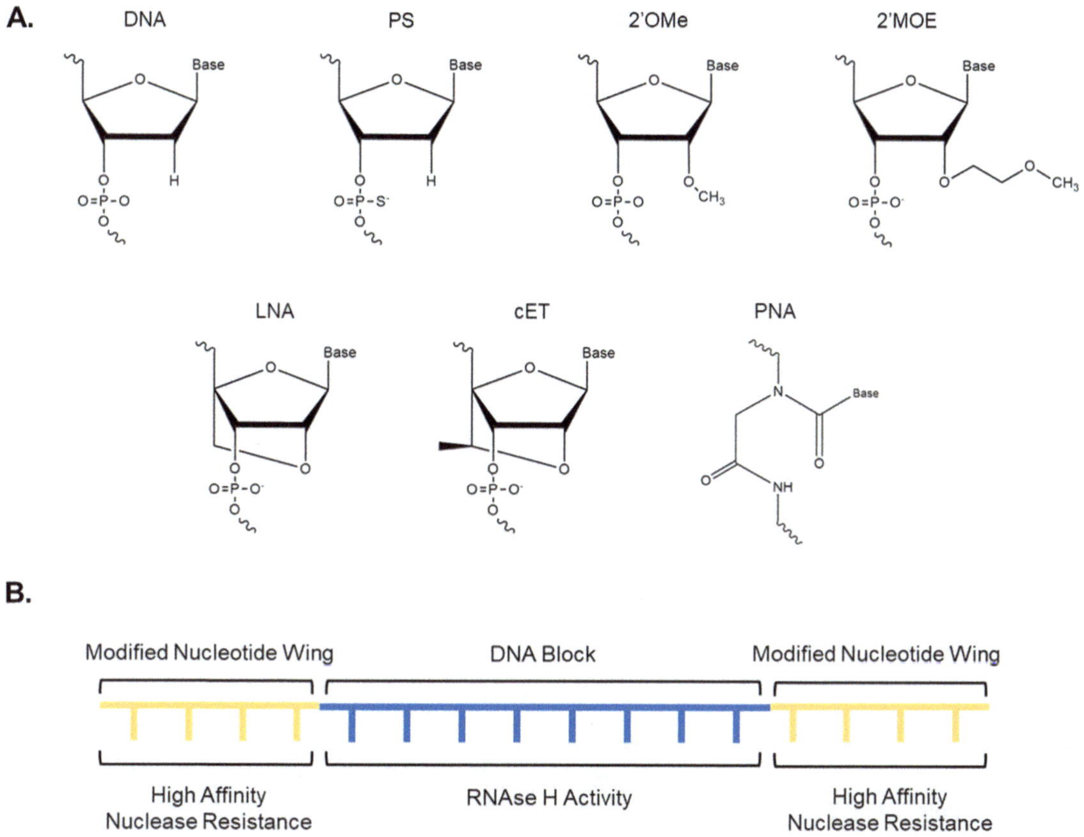

Fig. 1 Antisense oligonucleotide chemistry. (**a**) *DNA* deoxyribonucleic acid, *PS* phosphorothioate, *2′OMe* 2′-OMe, 2′-*O*-methyl, *2′-MOE* 2′-*O*-methoxyethyl, *LNA* locked nucleic acid, *cEt* constrained ethyl, *PNA* peptide nucleic acid. (**b**) Gapmers contain a central "gap" of DNA, flanked by "wings" of modified nucleotides

methoxyethyl (2′MOE) [10]. Like PS, these modifications help stabilize the AON against degradation by nucleases. Shorter AONs may be used with 2′ribose modification as a tighter bond can be achieved by increasing the thermal stability of the hybridization [8]. This 2′ribose AONs are unable to enact RNase H cleavage as they resemble RNA more closely than DNA [8]. As a result, 2′MOE and 2′OMe are often used for steric blocking instead, which preserves target mRNA integrity [8]. Figure 1a. illustrates these different AON chemistries.

Gapmers are a chimeric second-generation AON that contains a central DNA 'block', flanked by modified nucleotide "wings" [10]. Wings are usually composed of 2′OMe or 2′MOE; however gapmer designs using other modified nucleotides are being explored [2, 10]. The inclusion of a deoxyribonucleotide center allows RNase H to enact cleavage, while the modified "wings" increase hybridization strength and stability, which in turn prevents excessive cleavage to unwanted regions [10, 11]. Gapmer design circumvents some of the shortcomings of AONs by minimizing

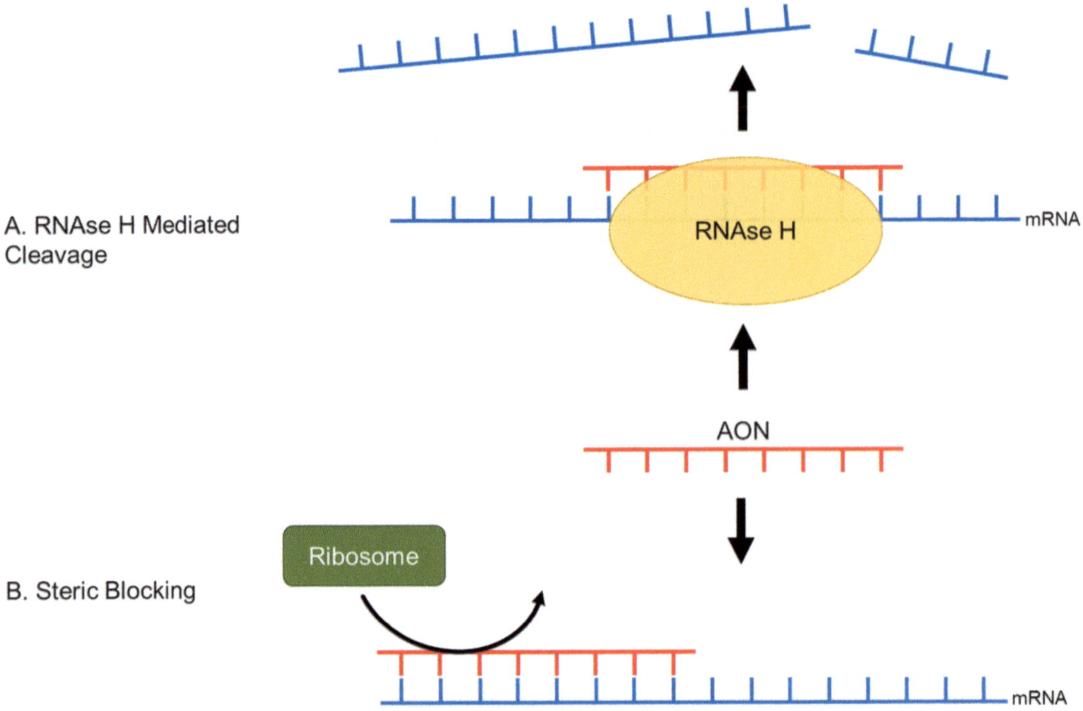

Fig. 2 Antisense oligonucleotide (AON) mechanism of action. (**a**) Enzyme-mediated cleavage by RNase H causes degradation of mRNA, while (**b**) steric blocking of AON to mRNA inhibits the action of translational machinery. Both methods lead to reduced translation of protein

toxicity while maintaining RNase H activity and evasion of nucleases [2]. The "wings" can be substituted with other AON chemistries, such as PS, or third-generation AONs like locked nucleic acids (LNA), constrained ethyl residues (cET), and peptide nucleic acids (PNA) [2, 12]. The ability to customize gapmer AONs has great potential for designing highly specific treatments.

3 Familial Hypercholesterolemia (FH)

Familial hypercholesterolemia (FH) is a genetic disorder that results in elevated levels of plasma low-density lipoprotein cholesterol (LDL-C) levels [13]. FH is one of the most frequent monogenic hereditary disorders [14]. Most FH cases are due to mutations in the LDL receptor gene *LDLR*, the apolipoprotein gene *APOB*, and the proprotein convertase subtilisin/kexin type 9 gene *PCSK9* [15]. Inheriting a mutated one of three genes, including *LDLR*, *APOB*, and *PCSK9*, is responsible for an estimated 70–95% of FH cases [16]. FH is most often caused by a mutated *LDLR* gene [17]. Specifically, LDL receptors are found on hepatocytes that are responsible for clearing out LDL-C from the bloodstream [18]. On the other hand, mutations in the *APOB*

Table 1
Gapmer AON Therapies

Treatment	Disease	Target	Status	Clinical Trial Identifiers
Mipomersen	Familial hypercholesterolemia (FH)	ApoB-100 expression	FDA approved for homozygous FH (HoFH)	NCT00706849 NCT00607373 NCT00694109
Inotersen	Hereditary transthyretin amyloidosis (hATTR)	TTR expression	FDA approved for stage 1 and 2 hATTR	NCT01737398 (NEURO-TTR) NCT03431896
Apatorsen	Cancer • Castration-resistant prostate cancer • Bladder cancer • Lung cancer • Pancreatic cancer	Hsp27 expression	Phase II clinical trials	NCT01120470 NCT01780545 (Borealis-2) NCT01829113 (The Spruce Clinical Trial) NCT01844817 (The Rainier Trial)
Volanesorsen	Familial chylomicronemia syndrome Familial partial lipodystrophy Familial hypertriglyceridemia	ApoC-III expression	EMA orphan drug	NCT02211209 (APPROACH) NCT02300233 (COMPASS) NCT02527343 (BROADEN)
IONIS-HTTRx	Huntington's disease (HD)	All HTT expression	Phase 3 clinical trials EMA orphan drug	NCT02519036 NCT03842969
WVE-120101 WVE-120102	Huntington's disease (HD)	Mutant HTT expression	Phase 1b/2a recruiting	NCT03225833 (PRECISION-HD1) NCT03225846 (PRECISION-HD2)
IONIS-DMPKRx	Myotonic dystrophy	DMPK expression	Discontinued	NCT02312011

gene inhibits the production of apolipoprotein B-100 (apoB-100), which is a structural component of lipoprotein particles, including LDL [19]. The reduction of apoB-100 decreases the binding affinity of LDL receptors to LDL-C and contributes to 2–5% of FH cases [13]. The protein PCSK9 is responsible for the degradation of LDLRs inside lysosomes [20]. Although loss-of-function mutations in the *PCSK9* gene result in increased levels of LDL-C clearance, gain of function mutations of the *PCSK9* gene decrease the amount of cellular LDLRs, resulting in elevated levels of plasma LDL-C [13]. Inheritance of a mutated *LDLR*, *APOB*, or *PCSK9* from one parent results in heterozygous familial hypercholesterolemia (HeFH), while inheriting a mutated *LDLR*, *APOB*, or *PCSK9* from both parents results in homozygous familial hypercholesterolemia (HoFH), which is a rare but life-threatening, leading to aggressive and premature heart disease [17, 21]. Conventional treatments for FH typically include diet management and the use of statin therapies to dramatically reduce LDL-C by at least 50% [13, 16]. Statins inhibit a rate-limiting enzyme involved in cholesterol synthesis, resulting in the reduction of hepatic cholesterol and upregulation of hepatic LDLRs, overall reducing the circulation of LDL-C [13]. However, a limitation of statin therapy is the dependence on the availability of LDLRs in the liver, which for HoFH patients have little to none LDLRs available [13, 16, 22]. In contrast, HeFH patients respond well to statin treatment as they retain one fully functional *LDLR* gene [13]. High-dose statin therapies greatly reduce the chance of cardiovascular disease (CVD), but adverse effects such as elevated serum transaminase and myalgias may occur [13]. Other conventional therapies available for statin-intolerant individuals are ezetimibe, niacin, and bile acid sequestrants, which all result in some reduction of LDL-C levels [13, 16, 23].

3.1 Mipomersen

Mipomersen (brand name Kynamro) is a second-generation AON used for the treatment of HoFH in conjunction with lifestyle management and other cholesterol-lowering agents [13]. Developed by IONIS (formerly ISIS) Pharmaceuticals and subsequently licensed to Genzyme, mipomersen was approved in 2013 by the Food and Drug Administration (FDA) as a nonstatin treatment for lowering LDL-C levels in patients with HoFH [24]. Mipomersen is a gapmer that targets the mRNA of apoB-100, an important structural component of LDL-C and very-low-density lipids (VLDL) [23]. The hybridization of mipomersen to apoB-100 mRNA enacts RNase H cleavage and inhibits the production of apoB-100, therefore reducing LDL-C levels [23].

3.1.1 Mipomersen Clinical Trials

1. Clinicaltrials.gov. Unique identifier: NCT00706849.

   In a double-blind, placebo-controlled, phase 3 trial completed in March 2013, randomized patients with HeFH and coronary heart disease were given a subcutaneous injection of 200 mg mipomersen or a placebo along with a maximally tolerated statin. The primary endpoint for this study was the percent change in baseline LDL-C after 28 weeks of treatment. In 124 patients, LDL-C levels were significantly reduced in the mipomersen group (−28%, baseline 152.9 mg/dl) than the placebo group (5.2%, baseline 142.9 mg/dl) [25]. As well, mipomersen reduced apolipoprotein B levels by 26.5%, total cholesterol by 19.4%, and lipoprotein by 21.1%. Adverse events were primarily injection site reactions (41.5% placebo, 92.8% mipomersen) and flu-like symptoms (31.7% placebo, 49.4% mipomersen) [25].

2. ClinicalTrials.gov Identifier: NCT00607373.

   A randomized, placebo-controlled phase 3 clinical trial was also conducted on HoFH patients in a multicenter study. Along with a maximally tolerated statin treatment, patients were given a 200 mg mipomersen subcutaneous injection weekly or a placebo for 26 weeks. Of 45 total participants, the mipomersen group had a significant reduction of LDL-C levels (−24.7%, baseline 11.4 mmol/L) compared to the placebo group (−3.3%, baseline 10.4 mmol/L) [22]. Four patients in the mipomersen group also had significantly elevated alanine aminotransferase concentrations [22].

3. Clinicaltrials.gov. Identifier: NCT00694109.

   An open-labeled phase 3 clinical trial conducted by Kastle Therapeutics in collaboration with Ionis Pharmaceuticals completed in December 2015 was an extension study of previous clinical trials. This multicenter clinical trial recruited 144 participants who had familial hypercholesterolemia or severe hypercholesterolemia and are currently taking other lipid-lowering therapies in conjunction with mipomersen. Participants have also completed NCT00607373, NCT00706849, NCT00477594, or NCT00794664 clinical drug trials. Each participant was given a 200 mg injection of mipomersen subcutaneously weekly for up to 4 years. Primary outcome measures include changes in baseline levels of LDL-C, apoB, total cholesterol, and non-high-density lipoprotein cholesterol (non-HDL-C). The secondary outcome measures include changes in triglyceride levels. At week 234 a mean 22.5% reduction from baseline LDL-C levels was measured (Clinicaltrials.gov Identifier: NCT00694109).

   Despite the effectiveness of mipomersen, it performed poorly as a marketable lipid-lowering therapy [8, 26]. Mipomersen was only approved for the treatment of HoFH, limiting its use to a small subset of patients [26]. The FDA simultaneously approved lomitapide (brand name Juxtapid) along with mipomersen in 2013, which is also used to treat HoFH [8]. Lomitapide is taken orally while mipomersen is administered by subcutaneous injection [8]. Injection site reactions along with safety concerns over hepatotoxicity and long-term effects of drug-induced hepatic fat accumulation may have dissuaded patients from continuing with mipomersen and favor lomitapide instead [8, 26].

4 Hereditary Transthyretin Amyloidosis (hATTR)

Amyloid transthyretin (ATTR) amyloidosis is a disorder caused by the misfolding of the transthyretin protein (TTR) produced by the liver [27]. TTR is used in the transport of thyroxine and retinol in blood circulation [27]. A misfolded TTR dissociates to form amyloids, which conglomerate to form amyloid fibrils which deposit in the peripheral nervous system, heart, and gastrointestinal system [28]. As a result, the symptoms of ATTR vary greatly and pose a diagnostic challenge. Hereditary ATTR (hATTR) has an autosomal dominant inheritance pattern and is caused by mutations to the *TTR* gene [27, 28]. Most hATTR patients have one mutant and one wild type copy of the *TTR* gene, resulting in unstable heterotetramer TTR protein that dissociates more readily [29]. Common mutations to the *TTR* gene are Val30Met and Val122Ile, the former found predominantly in people of European and Japanese descent and the latter found almost exclusively in those of male African descent [28]. Liver transplantations are the standard treatment for hATTR, as 95% of TTR is produced by the liver [29]. Patients with the Val30Met mutation have a greater success rate with liver transplantations, with survival improving significantly and have about 80–90% stabilization of the disease [29]. However, patients with other mutations than Val30Met have less favorable results, with only 60% stabilization of the disease after liver transplantation [29]. Other limitations to liver transplantation are the shortage of donors, the need for surgery, and the use of posttransplantation immunosuppressive drugs [30]. Other therapies like diflunisal (brand name Dolobid) and tafamidis (brand name Vyndaqel) aim to stabilize the TTR heterotetramer [27, 30]. Diflunisal potentially prevents the dissociation of the heterotetramer into amyloid fibrils, while tafamidis help stabilize TTR by binding to the thyroxine-binding site and have been approved in Europe and Japan for the treatment of hATTR with polyneuropathy [30, 31].

4.1 Inotersen

Inotersen (brand name Tegsedi) is a second-generation AON inhibitor developed by IONIS (formerly ISIS) Pharmaceuticals and Akcea Therapeutics [28]. In 2018, inotersen was approved by the FDA as a subcutaneous injection of 284 mg weekly for the treatment of stage 1 and stage 2 hATTR patients [31]. Inotersen has a 5-10-5 gapmer structure, with five 2′MOE ribonucleotides flanking ten central 2′deoxyribonucleotide residues [32]. Complementary to the 3′UTR region of the TTR mRNA, inotersen inhibits TTR production, causing degradation of both wildtype and mutant mRNA through RNase H cleavage [28, 32]. This prevents the formation of the heterotrimer, therefore preventing amyloid fibril deposits from accumulating and thus halting disease progression [32].

4.1.1 Inotersen Clinical Trials

1. ClinicalTrials.gov Identifier: NCT01737398 NEURO-TTR.

   In an international, randomized, double-blind, placebo-controlled phase 3 clinical trial, inotersen was administered to stage 1 and stage 2 hATTR patients to evaluate the efficacy and safety of this treatment. Subcutaneous injections of 300 mg or placebo were administered weekly and patients were monitored for 65 weeks. Primary endpoints of this study were changes in the baseline scores of The Modified Neuropathy Impairment Score+7 (mNIS+7) composite score at Week 66 and the total score of the Norfolk Quality of Life-Diabetic Neuropathy (QoL-DN) questionnaire. The mNIS+7 score (range: −22.32 to 346.32) measures muscle fatigue and nervous condition, where a higher score indicates poorer function. The QoL-DN questionnaire (range: −4 to 136) is patient-reported and measures physical condition, symptoms, and daily activities, where a higher Norfolk QoL-DN score indicates poorer quality of life. Of 172 patients, 112 received inotersen treatment and 60 received a placebo [33]. 139 (81%) of participants completed the trial [33]. At Week 66, a significant difference was measured between the two groups (inotersen minus placebo) of −19.7 points for mNIS+7 and −11.7 points for QoL-DN, indicating an improvement in symptoms independent of disease stage or mutation type [33]. The inotersen group saw an average 5.8-point increase in baseline mNIS+7 score, and a 1.0-point increase in baseline QoL-DN score, whereas the placebo group had a 25.5-point increase in baseline mNIS+7 score and a 12.7-point increase in baseline QoL-DN score [33]. Within the inotersen group, 36% of patients had no increase from baseline in mNIS+7 scores and 50% improvement in QOL-DN scores [33]. Adverse events include glomerulonephritis (3 patients, 3%) and thrombocytopenia (3 patients, 3%) [33]. There were five deaths in the inotersen group, four of which are consistent with underlying disease progression, while one patient suffered a fatal intracranial hemorrhage due to reduced platelet count (<10,000 per cubic millimeter) [33]. The imbalance of deaths between the inotersen group and the placebo group is unclear whether they are due to disease progression, chance, or other causes [33]. From this clinical trial, the FDA approved inotersen in October 2018 and it is being marketed with a Risk Evaluation and Mitigation Strategy (REMS) due to its adverse events [31].

2. ClinicalTrials.gov Identifier: NCT03431896 (MED-hATTR).

    There is currently a recruiting phase 3 clinical trial to monitor the early disease progression of hATTR using a novel peptide-based probe that labels misfolded TTR oligomers (ClinicalTrials.gov Identifier: NCT03431896). One of the primary endpoints is assessing the degree in which different ATTR-medications like diflunisal, tafamidis, inotersen, etc. can reduce probe signals of serum misfolded TTR oligomers. The estimated completion date of this study is February 2024.

5 Heat Shock Protein 27 (Hsp27) in Cancer Progression

Heat shock proteins (HSP) are released in environmental stressors and help cells survive otherwise lethal conditions [34, 35]. Most act as chaperones that help facilitate the proper folding, transport, and assembly of proteins, as well as their removal when a protein is beyond repair [35]. Cancers typically have an upregulation of signaling pathways to facilitate increased proliferation, so HSPs are often a therapeutic target to inhibit the folding of oncogenic proteins [34]. In cancers, HSPs are highly recruited to meet the demands of promoting cancer cell growth and proliferation [34].

Douglas Hanahan and Robert Weinburg characterized six features of cancer cells: (1) sustained proliferative signaling, (2) evasion of growth suppressors, (3) cell invasion and metastasis, (4) replicative immortality, (5) angiogenesis, and (6) evasion of apoptosis [34, 36]. Heat shock protein 27 (Hsp27) is one of the many HSPs in which upregulation can aggravate and promote the development of malignant tumor formation [34, 35]. Hsp27 is unlike other HSPs in that it forms large oligomers that store protein, can function without cochaperones, and is ATP-independent [34]. Cell proliferation may be sustained by the action of many HSPs relaying signals to oncogenic proteins [34]. Hsp27 is used to evade anti-growth signals by inactivating a p53-mediated pathway, ultimately suppressing cell senescence and achieving replicative immortality [34, 37]. Cell invasion and metastasis can be aggravated by Hsp27, which promotes the epithelial-mesenchymal transition (EMT) in cancer cells and enhance the ability to establish new sites for growth [34, 38]. Hsp27 is also a regulator of apoptotic mechanisms such as the apoptosome, caspase activation complex, and degradation of apoptotic regulatory proteins [39]. Overall Hsp27 is highly involved in cancer cell progression and a suitable therapeutic target for slowing malignant tumor formation.

5.1 Apatorsen

Apatorsen (OGX-427) is an AON gapmer complementary to Hsp27 mRNA, leading to degradation and inhibition of protein production [40]. It follows a 4-12-4 gapmer structure with the wings composing of 2′MOE modified nucleotides [41]. Apatorsen has so far undergone several Phase 2 clinical trials for the treatment of castration-resistant prostate cancer (NCT01120470, NCT01681433), bladder cancer (NCT01780545), lung cancer (NCT01829113), and pancreatic cancer (NCT01844817), in which it was referred to as OGX-427.

5.1.1 Apatorsen Clinical Trials

1. ClinicalTrials.gov Identifier: NCT01120470.

   In a randomized phase II clinical trial, OGX-427 was studied in castration-resistant prostate cancer patients who have not previously received chemotherapy. This study was completed in

October 2012 and sponsored by the British Columbia Cancer Agency. Specifically, 74 participants randomly received 3 doses of 600 mg intravenously administered OGX-427 over 5–9 days in conjunction with 5 mg of prednisone, an anti-inflammatory drug, or prednisone alone as the control. Afterward, a 1000 mg weekly dose of OGX-427 was administered over 24 weeks. Primary endpoints measured disease progression at 12 weeks, which can primarily be characterized by a 25% greater increase of prostate-specific antigen (PSA) from baseline. At 12 weeks into treatment, 50% of apatorsen plus prednisone patients and 42% of prednisone control patients did not have disease progression ($P = 0.33$) [42]. Therefore, the primary endpoint was not met [42]; however, 47% of the apatorsen plus prednisone group experienced at least 50% reduction of PSA, versus 24% in the prednisone group [42]. The most common adverse event was injection site reactions in 77% of apatorsen treated patients [42]. Some limitations of this study were the open-label design and low enrollment.

Another phase 2 clinical trial also conducted for the treatment of castration-resistant prostate cancer in December 2012 (ClinicalTrials.gov Identifier: NCT01681433). However, this trial was terminated due to lack of patient accrual (ClinicalTrials.gov Identifier: NCT01681433).

2. ClinicalTrials.gov Identifier: NCT01780545 (Borealis-2) [43].

   In a multicenter, open-label, randomized phase 2 clinical trial, OGX-427 was evaluated for the treatment of locally advanced or metastatic urothelial carcinoma (UC) with docetaxel, a chemotherapy drug. This trial, also known as Borealis-2, was conducted between August 2013 and September 2015, where three doses of 600 mg of apatorsen were administered intravenously over a 9-day period before a weekly 600 mg dose of apatorsen with 75 mg/m^2 dose of docetaxel every 21 days. The primary endpoint of this study was overall survival in a 36-month treatment period of apatorsen plus docetaxel versus the control of docetaxel alone. The median overall survival was 6.4 months for the apatorsen plus docetaxel group, versus 5.9 months for the docetaxel only group [43]. Patients with a baseline Hsp27 level of less than 5.7 ng/ml had improved overall survival than patients with higher Hsp27 levels [43]. Overall, despite a 2-week improvement of median survival in the apatorsen plus docetaxel group over docetaxel alone, secondary endpoints measuring hazard of disease progression or death were improved for the apatorsen treated group, although this result was not statistically significant (one-sided $P = 0.1069$) [43].

3. ClinicalTrials.gov Identifier: NCT01829113 (The Spruce Clinical Trial).

   A double-blind, randomized phase 2 study was conducted for the treatment of previously untreated Stage IV nonsquamous non–small-cell lung cancer with apatorsen and carboplatin. Also known as the Spruce Clinical Trial, 155 participants received carboplatin with 600 mg dose of apatorsen or placebo on days 1, 8, and 15 of each 21-day cycle for a maximum of four treatment cycles. If their condition was stable, then patients can continue to receive 600 mg of apatorsen weekly until toxicity or disease progression. Primary outcome measures are the median progression-free survival measured every 6 weeks up to 24 months. This was measured by the Response Evaluation Criteria in Solid Tumors (RECIST) v 1.1, or date of death. Results were a median progression-free survival of 6 months for the apatorsen treated group, and 4.9 months for the placebo-treated group (ClinicalTrials.gov Identifier: NCT01829113).

4. ClinicalTrials.gov Identifier: NCT01844817 (The Rainier Trial) [44].

   A randomized, double-blinded phase 2 trial evaluated the treatment of metastatic pancreatic cancer with gemcitabine with either apatorsen or a placebo, where132 participants were randomized such that 66 received apatorsen and gemcitabine treatment, and 66 received placebo and gemcitabine treatment in 28-day cycles. The primary endpoint of this study was overall survival for up to 2 years. The addition of apatorsen did not result in significant improvement, as the objective response rate was the same for both treatment groups (18%) [44]. Overall survival was not significant between the two groups, although apatorsen-treated patients had slightly worse numbers in progression-free survival (PFS) and overall survival (OS) than the placebo group (median PFS, 2.7 vs. 3.8 months, respectively [$p = 5.92$]; median OS, 5.3 vs. 6.9 months, respectively [$p = 5.62$]) [44]. The subgroup that did benefit from apatorsen treatment was patients with high serum Hsp27 concentrations, which saw a reduction in hazard ratio (HR) for death (median PFS for patients with high baseline serum Hsp27 levels, 3.3 vs. 0.9 months for apatorsen vs. placebo, respectively [HR 0.38], median OS 3.3 vs. 1.0 months [HR 0.59]) [44]. However, this subgroup represents only 14% of the study population, which makes it difficult to draw definitive conclusions [44].

6 Familial Chylomicronemia Syndrome (FCS)

Familial chylomicronemia syndrome (FCS) is a rare autosomal recessive genetic disease that is characterized by elevated serum triglyceride (TG) levels [45]. Healthy individuals can clear triglyceride from plasma 3–4 h after eating, maintaining a TG level of <150 mg/dl, whereas FCS patients sustain levels greater than 885 mg/dl for 12–14 h after a meal [45, 46]. FCS patients have an increased risk of developing complications such as pancreatitis that can result in fatality [47]. Approximately 95% of monogenic FCS cases are due to mutations on *LPL*, a gene encoding for lipoprotein lipase (LPL), resulting in symptoms appearing in infancy and childhood [48]. LPL plays an important role in promoting the clearance of TG, which is transported by chylomicrons or very-low-density lipoprotein (VLDL) [49]. Loss of function mutations into *LPL* cause inefficient removal of serum TG [46]. A small subset of FCS cases is caused by mutations to *APOC2* (encoding apolipoprotein C-II), *GPIHBP1* (encoding glycosylphosphatidylinositol-anchored high-density lipoprotein-binding protein 1), *APOA5* (encoding apolipoprotein A-V), and *LMF1* (encoding lipase maturation factor 1), all of which encode chaperone proteins or cofactors of LPL [48]. There is currently no effective treatment for FCS except for restriction of dietary fats, avoiding alcohol consumption, and certain medications [46]. One target for antisense therapy in the treatment of FCS is *APOC3*, which encodes APOC3 lipoprotein. APOC3 has an inhibitory effect in the LPL pathways, such that it reduces the clearance of TG-rich lipoproteins [46]. However, the mechanism of how this is done is under study [45, 46]. Targeting an inhibitor of LPL such as apoC-III will subsequently lower plasma TG by facilitating the action of LPL, which increases the clearance of TG [46].

6.1 Familial Partial Lipodystrophy (FPLD)

Familial partial lipodystrophy is a rare autosomal dominant genetic condition that results in abnormal distribution of fatty tissue [50]. Normally, fatty tissue is distributed among many areas of the body, but FPLD patients have reduced fatty tissue in their arms, legs, and hips [50]. Complications like diabetes and hypertriglyceridemia have increased risk, along with infertility in females due to the formation of cysts in the ovaries [50]. FPDL cases are classified into five categories according to their mutated gene, however the most common are mutations to *LMNA*, causing FPDL Type 2 (FPDL2) [51]. *LMNA* encodes for lamin A and lamin C, which are important structural components of cell scaffolding [50].

6.2 Familial Hypertriglyceridemia

Familial hypertriglyceridemia is characterized as high concentrations of the triglyceride-rich lipoproteins VLDL in the blood [52]. In severe cases, patients have plasma TG concentrations

>10 mmol/L [53]. Familial hypertriglyceridemia may have monogenic origins or multigenic, making it difficult for clinicians to diagnose and treat [53]. Several environmental factors like lifestyle contribute to the cause of familial hypertriglyceridemia, while external environmental factors may exacerbate the symptoms [53]. Loss of function mutations in genes responsible for lipoprotein production, such as *LDL*, *APOC2*, *APOA5*, *LMF1*, *GPIHBP1*, and *GPD1* affects the clearance of TG from circulation [53].

6.3 Volanesorsen

Volanesorsen (brand name Waylivra) is an AON with a gapmer structure developed by Ionis Pharmaceuticals through Akcea Therapeutics [47]. It is used for the treatment of FCS, FPLD, and hypertriglyceridemia by binding to the 3′ end of apoCIII mRNA and degrading it through RNase H cleavage [47]. The breakdown of apoCIII can increase the action of LDL; therefore promoting TG clearance and reducing plasma TG levels [47]. Volanesorsen is administered through subcutaneous injection at a dosage of 285 mg/dl weekly for the first 3 months and then administered at the same dosage biweekly [47]. Although the FDA Advisory Committee had voted in favor of the approval of volanesorsen in May 2018, its New Drug Application was ultimately rejected in September 2018 [54, 55]. Akcea Therapeutics announced in response that they aim to work with the FDA for volanesorsen approval [55]. In November 2018, Canada had rejected volanesorsen's New Drug Submission [56]. Volanesorsen is under conditional marketing by the European Commission as of May 7, 2019, for the treatment of FCS alongside other triglyceride-lowering therapies, although postmarketing studies are still needed for full approval [47].

6.3.1 Volanesorsen Clinical Trials

1. ClinicalTrials.gov Identifier: NCT02211209 (Phase 3 APPROACH, FCS).

   Volanesorsen has undergone phase III clinical trials for the treatment of three disorders; FCS, hypertriglyceridemia, and FPLD. For FCS, a randomized, double-blind, placebo-controlled APPROACH study was conducted over a 52-week period on the reduction of TG in FCS and hypertriglyceridemia patients [47]. Participants were given a 300 mg/dl subcutaneous dose of volanesorsen or placebo weekly, with the primary outcome was the percent change in fasting TG levels at 13 weeks [57]. In comparison to the placebo, the volanesorsen group had a 77% decrease of TG levels by 3 months, whereas the placebo group had an 18% increase [47]. By 52 weeks, volanesorsen treated patients maintained a decreased level of TG (−50.1%) [47]. During the 52-week period, volanesorsen treated patients who had experienced at least two pancreatitis attacks in the past 5 years had no attacks [58]. The most common adverse events were injection site reactions (11.8%

of all participants), and five early terminations were made due to reduced platelet counts (<25,000/μl) [58]. Currently, there is an open-label extension study of volanesorsen on the treatment of FCS using participants from the APPROACH study and COMPASS study to examine long term effects (ClinicalTrials.gov Identifier: NCT02658175).

2. ClinicalTrials.gov Identifier: NCT02300233 (Phase 3 COMPASS, Hypertriglyceridemia).

   COMPASS was a phase 3 randomized, multinational, double-blind, placebo-controlled trial of volanesorsen on the treatment for hypertriglyceridemia (ClinicalTrials.gov Identifier: NCT02300233). Over a 26-week trial, 114 adults with hypertriglyceridemia were given a 300 mg subcutaneous dose of volanesorsen or placebo injection. Primary outcomes measured the percent change in fasting TG from baseline over 13 weeks. Results indicate a 71.8% decrease of TG in the volanesorsen group versus 0.8% in the placebo [47]. A subgroup of FCS patients who receive volanesorsen treatment and placebo was also monitored for pancreatitis events, and the volanesorsen group had no attacks, while in placebo there were 5 in 3 patients [47].

3. ClinicalTrials.gov Identifier: NCT02527343 (Phase 3 BROADEN, FPLD).

   For the treatment of Familial partial lipodystrophy, a phase 2/3 randomized, double-blind, placebo-controlled trial is underway. Participants will undergo treatment with volanesorsen for 52 weeks, and then be offered to continue treatment in a 2-year open-label study. Primary outcomes measure the percent change in fasting TG from baseline. The estimated completion date of this study is September 2021.

7 Huntington's Disease (HD)

Huntington's disease (HD) is an autosomal dominant genetic disorder causing progressive degeneration of motor and cognitive abilities [59]. Symptoms of late-stage HD are the inability to maintain a voluntary muscle contraction, cognitive impairment, dementia, and chorea, eventually leading to fatality 15–18 years after disease onset [59]. The genetic cause of HD is attributed to mutations to the *huntingtin (HTT)* gene, which encodes huntingtin protein (HTT) [60]. The role of wild-type HTT is only partially understood, but most research has elucidated its function in early embryonic development, protein scaffolding, synaptic connectivity, cell trafficking, and as a transcriptional regulator [61, 62]. HD severity is characterised by the number of CAG trinucleotide repeats in exon 1 of *HTT*. Wild-type HTT has approximately

27 CAG trinucleotide repeats, while asymptomatic HD individuals with 27–35 repeats have a higher risk of producing offspring with an HD allele containing ≤36 repeats [60]. Individuals with ≥40 CAG repeats manifest HD, with greater CAG repeat numbers resulting in earlier disease onset [60, 61]. Excessive CAG repeats result in unstable HTT protein which leads to the formation of abnormal nuclear and cytoplasmic aggregates [60]. There are currently no disease-modifying treatments to slow or stop the progression of HD [62, 63]. Drugs like tetrabenazine and antipsychotics may be used to alleviate the symptoms of HD [61].

7.1 Current Developments in Therapeutic Gapmers for HD

The primary cause of HD is the toxic properties of mutant HTT, which results in disruptive effects such as impairment of protein degradation, transcriptional dysregulation, and altered synaptic plasticity [61, 62]. Therefore lowering mutant HTT production is the therapeutic goal for developing AON treatments for HD. Two approaches have been explored thus far in attempts to lower mutant HTT levels; non–allele-specific AONs and allele-specific AONs.

Non-allele-specific AONs target both mutant and wild-type HTT, resulting in total HTT lowering in HD patients and a net beneficial effect [62]. Although this approach has lower development costs and is easier to deploy, there are several concerns surrounding the long-term impact of reducing wild-type HTT [62]. Specifically, wild-type HTT is involved in the production and transport of brain-derived neurotrophic factor (BDNF) and BDNF mRNA [62, 64]. BDNF has a significant role in the survival of medium spiny neurons of the striatum, such that the long-term reduction of wild-type HTT by non–allele-specific AONs may induce adverse effects and further contribute to HD pathogenesis [62, 65, 66]. The reduction of wild-type HTT may also affect vesicle packaging and transportation processes [62, 67].

Allele-specific AONs aim to target mutant HTT exclusively. Wild-type HTT is known to have an essential role in embryogenesis, early development, BDNF production, and potentially more functions as the function of HTT is only partially understood [61, 62]. Using an allele-specific therapy removes the concern of the long-term impacts of wild-type HTT lowering, but there are many challenges for its development. Common targets of allele-specific therapies for HD are the CAG tract and single-nucleotide polymorphisms (SNPs) to overall reduce the production of mutant HTT [68]. However, developing AONs or RNAi for the expanded CAG tract may result in off-target effects for other genes that contain a polyCAG stretch [68]. Using SNPs to achieve allele-specificity also has its disadvantages as there is no commonly shared SNP among all HD patients [62, 68]. This means that different agents must be developed depending on the SNP the patient has, along with diagnostic kits to determine what SNPs are present [62, 68]. This creates a significant barrier for development and deployment as each agent must reach approval [62].

7.2 IONIS-HTTRx

A non-allele-specific AON gapmer called IONIS-HTTrx (also known as ISIS 443139 and RG6042) was developed by Ionis Pharmaceuticals in collaboration with Hoffman-La Roche. It is complementary to a 20 nucleotide stretch on HTT mRNA in HD patients and has a 5-10-5 gapmer structure, where 10 deoxyribonucleotides are flanked by five 2′MOE nucleosides [69]. RNase H cleavage is activated to reduce the production of both wild-type and mutant HTT mRNA [62, 69]. IONIS-HTTrx has completed phase 1/2a clinical trails and is currently recruiting for phase 3 clinical trials (NCT02519036, NCT03842969). The results of the phase1/2a trial of IONIS-HTTrx met safety primary endpoints despite concerns about wildtype-HTT lowering [70, 71].

7.2.1 IONIS-HTTRx Clinical Trials

1. ClinicalTrials.gov Identifier: NCT02519036.

   In a phase 1/2a randomized, double-blind, placebo-controlled clinical trial was conducted to evaluate the safety and tolerability of an ascending dosage of IONIS-HTTrx (ClinicalTrials.gov Identifier: NCT02519036). 52 participants had early stages of HD and received four injections of a dose category (10 mg, 30 mg, 60 mg, 90 mg, 120 mg) of IONIS-HTTrx or placebo through intrathecal injection. Primary outcomes were the number of patients that experienced unfavorable treatment-related adverse events over 28 weeks of treatment. 98% of patients reported adverse events; however, 83% of cases were mild and 17% were moderate in severity (ClinicalTrials.gov Identifier: NCT02519036). Most adverse events were related to pain and headaches associated with intrathecal injection, which were resolved after 2 days (ClinicalTrials.gov Identifier: NCT02519036). An exploratory endpoint of this trial was measuring the concentration of mutant HTT protein in the cerebral spinal fluid (CSF). They found that a mean percent change of −20%, −25%, −28%, −42%, and −38% decrease of mutant HTT protein in the 10 mg, 30 mg, 60 mg, 90 mg, and 120 mg cohorts, respectively (vs. 10% increase in placebo group) (ClinicalTrials.gov Identifier: NCT02519036). This trial was sponsored by Ionis and now led by Roche Pharmaceuticals [71].

2. ClinicalTrials.gov Identifier: NCT03842969.

   Currently, there is a phase 3 open-label extension study of RO7234292 (formerly IONIS-HTTrx) that has started recruiting. Participants for this study must have previously been involved in a Roche-sponsored or Genentech-sponsored study in HD for RO7234292, and 950 participants have been enrolled thus far. The primary outcome measures are the percentage of patients with adverse events, changes from baseline in the Columbia-Suicide Severity Rating Scale (C-SSRS), which assess suicidal ideation behavior, and change in baseline in cognition using Montreal Cognitive Assessment (MoCA) for up to 5 years. The expected completion date is November 2023.

7.3 WVE-120101 and WVE-120102

WVE-120101 and WVE-120102 are allele-specific AONs developed by Wave Life Sciences Ltd. [62, 70]. These are the first allele-specific therapeutic AONs to advance to clinical trials, although the chemical composition of these AONs, including whether these are gapmers or not, has not been disclosed to date [62]. Unlike IONIS-HTTrx, a non-allele specific AON, WVE-120101 and WVE-120102 target only mutant HTTrx [62, 70]. Three SNPs are attributed to the CAG expansion in exon 1 of *HTT* and are found in 80% of gene expansion carriers of European ancestry [70, 72]. WVE-120101 and WVE-120102 target two of the three SNPs: rs362307 (SNP1) and rs362331 (SNP2), respectively (ClinicalTrials.gov Identifier: NCT03225833, NCT03225846).

7.3.1 WVE-120101 and WVE-120102 Clinical Trials

For both WVE-120101 and WVE-120102, phase 1b/2a multicenter, randomized, double-blind, and placebo-controlled clinical trials that will be conducted in parallel (ClinicalTrials.gov Identifier: NCT03225833, NCT03225846). WVE-120101 or WVE-120102 will be administered in a combined single ascending dose and multiple ascending dose design [70]. Both clinical trials are currently recruiting in Australia, Canada, Denmark, Poland, and the United Kingdom, with an estimated enrollment of 48 participants for each trial [70]. Participants are pre-screened for targeted SNPs on the same allele as the pathogenic CAG expansion [70]. The primary outcome of these studies is safety and tolerability over 210 days of treatment, which determined by the number of patients with adverse events and the severity of those adverse events [70]. The secondary outcomes are the pharmacokinetics of the plasma and cerebral spinal fluid, as well as changes in total functional capacity (TCF) as measured by the Unified Huntington's Disease Rating Scale (UHDRS) [70]. The estimated completion date of both studies is March 2020.

8 Myotonic Dystrophy

Myotonic dystrophy is an autosomal dominant genetic disorder with a prevalence of 1 in 8000 [73, 74]. There are two types of myotonic dystrophy, with myotonic dystrophy 2 (DM2) being milder in symptoms than myotonic dystrophy 1 (DM1) [75]. Symptoms of DM1 include myotonia, cataract formation, muscle wasting and weakness, and mental retardation [75]. The DM1 causing mutation was discovered in 1992 as a CTG trinucleotide expansion in the 3′ untranslated region of the *dystrophia myotonica protein kinase* (*DMPK*) gene [74, 76, 77]. DM2 is caused by a CCTG repeat in *Zinc Finger 9 (ZMF9)* (also known as *CNBP*) [78]. *DMPK* in unaffected individuals generally has 5–38 CTG repeats, while DM1 individuals have 50–3000 CTG repeats [76]. DMPK transcripts that have large unstable regions of CUG repeats remain in the nucleus and form nuclear foci with splicing factors like muscleblind-like 1 (MBNL1) [3, 75, 76]. MBNL1, a key protein for alternative splicing regulation, becomes absent in the nucleoplasm, contributing to multiple symptoms of DM1 [3, 75, 76]. Along with the phosphorylation of a splicing factor called CUP-binding protein 1 (CUGBP1), incorrect mRNA splicing events occur more frequently [76, 78]. Currently, there is no cure for myotonic dystrophy or treatments for slowing the disease progression [78]. Disease management is through genetic counseling and treating the symptoms of myotonic dystrophy [78].

8.1 IONIS-DMPKRx Clinical Trial

In a phase 1/2a blinded, placebo-controlled clinical trial, the safety and tolerability of IONIS-DMPKRx, a 2′ MOE gapmer, was assessed with 48 participants (ClinicalTrials.gov Identifier: NCT02312011). IONIS-DMPKRx targets *DMPK* RNA to reduce its accumulation. Primary outcomes measured the safety and tolerability over 24–32 weeks of treatment of subcutaneous injection weekly. The clinical trial completed in August 2016. In a letter by Ionis Pharmaceuticals and Biogen dated January 2017, RNA splicing and drug presence were detected in muscle tissue; however the degree of which IONIS-DMPKRx enacted clinical benefits was weak [79]. Ionis Pharmaceuticals and Biogen decided to not advance IONIS-DMPKRx in favor of using ligand-conjugated antisense (LICA) technology [79]. Different gapmer designs are still being investigated for the treatment of myotonic dystrophy [73, 80].

9 Prion Diseases

Prion diseases are fatal neurodegenerative illnesses that have either sporadic, acquired, or genetic causes [81]. Familial Cruetzfeldt–Jakob disease (CJD), fatal familial insomnia, and Gerstmann–Sträussler–Scheinker syndrome are some examples of prion diseases caused by an autosomal dominant mutation to the *prion protein gene* (*PRNP*) [82, 83]. Despite presentation, all prion diseases are caused by a misfolding of a prion protein (PrP) from its normal conformation (cellular prion protein or PrP^{C}) to a pathogenic and self-propagating form (scrapie prion protein or PrP^{Sc})

[82, 83]. PrP^C is normally located on cell membranes and is highly abundant in neurons [84]. In the conversion from PrP^C to the pathogenic PrP^{Sc}, alpha-helical structures are misfolded in beta-sheets [84]. Changes in the secondary and tertiary structures make the PrP^{Sc} form resistant to proteases and normal means of degradation [83]. There is currently no treatment to cure prion diseases in animals and humans, although several options have been explored for genetic knockdown of PrP transcript, which arrests the PrP^C to PrP^{Sc} conversion [82, 84, 85]. Previous studies have shown that depletion of PrP^C in prion-infected mice prevents early spongiform pathology, neuronal loss, and disease progression [85, 86].

9.1 Recent Advances in PrP Transcript Lowering Therapies

Several methods have been devised to knock down *PRNP* activity, including RNA interference [82, 86]. A previous study by Nazor Friberg et al. was first to investigate the efficacy of gapmer AONs in the treatment of prion-infected mice [87]. However, the study had inconclusive results about the mechanism of action of the gapmer AONs they tested. PRNP lowering was observed, which was the intended effect of the AONs administered, but it was uncertain whether AONs inhibits PrP^{Sc} formation through sequence-specific degradation of the PRNP mRNA or if they increase the clearance of PrP^{Sc} through sequence-independent pathways. AONs were delivered directly to the CSF via intracerebral administration through surgically implanted osmotic pumps in the left lateral ventricle. Half of the cannulated mice died after surgery due to complications in administration, and the clinical translation of this delivery method in humans was unclear. Therefore, gapmer AONs for prion diseases were not pursued extensively until a recent study by Raymond et al., which utilized recent advancements in other AON therapies [82].

In this study, Raymond et al. identified two gapmer AONs that target different regions of the *PRNP* gene [82]. Intracerebroventricular (ICV) injections, which previously have shown better tolerance in AON delivery for rodents, were used for drug administration. These AONs were used to test drug efficacy in prion-infected wild type mice in prophylactic treatment. No treatment, saline, or 300 ug of either AON1 or AON2 were given to mice 14 days before infection, and then 46 and 106 days postinfection. AON1 and AON2 in comparison to saline treatment delayed prion disease clinical signs by 82% and 99%, respectively, and delayed all-cause mortality by 81% and 98%, respectively. The two gapmer AONs were also used to treat prion disease in postinfected mice. Wild type mice were administered treatment 120 days after infection, in which AON2 was not tolerated, but AON1 resulted in a slower disease progression. AON1 administered 120 days after infection delayed prion disease clinical signs by 33% and increased survival with respect to all-cause mortality by 55% in comparison to the saline control. Neuropathology for both prophylactic and post-infection treatments showed reduced pathological changes, and although all mice succumbed to disease-related causes, their neuropathology resembled those of untreated animals at the study's endpoint. In comparison, control mice had granular PrP deposits and spongiform vacuolation; pathology that is typical of advanced prion disease progression. These results suggest that AON therapy delay disease progression by reducing the accumulation of PrP^{Sc} and changes to neuropathology.

From this study, the mechanism of action has been identified as the lowering of PrP RNA, due to amelioration of prion disease progression was demonstrated by PrP-targeting AONs and not control AON [82]. These AONs were effective in both prophylactic treatments and postinfection treatments, while administration through ICV injections effectively delivered gapmer AONs with reduced adverse effects. These results may suggest a potential benefit for treating individuals who are at risk for familial prion disease onset and a method of treating prion disease in affected populations. Improved diagnostics technology would further contribute to this direction, as prion diseases rapidly progress over 5 months. Overall, further investigation into gapmer AONs for prion disease has a promising outlook.

10 Future Perspectives

In clinical treatments, AON gapmers have had some success in controlling the expression of mutated RNA transcripts. However, major challenges such as toxicity and delivery methods require further optimization of the design of AON gapmers.

Toxicity due to the AON gapmer is hypothesized to be caused by (1) hybridization dependent binding to unintended transcripts and (2) hybridization independent binding to essential proteins. The former can enact off-target cleavage and downregulation of proteins or regulatory RNAs, while the latter can disrupt the function of otherwise normal proteins [88]. In a study by Dieckmann et al., toxicity caused by hybridization dependent binding was explored [88]. The optimal melting temperature (T_m) of the LNA-AON gapmer is between 50 and 55 °C, in which hybridization to the intended target is balanced with off-target hybridization [88]. Introduction of 2′OMe nucleotides to the "wing" region of LNA-AON gapmers may also reduce hepatoxicity by disrupting the action of RNase H cleavage. The addition of 2′OMe increases T_m, but with the disruption to RNase H action, off-target effects can be minimized [88]. An in vitro assay was also developed to determine the hepatoxicity of LNA-AONs by measuring apoptosis induction [88]. These findings may contribute to designing safer LNA-AON gapmers for clinical use.

Hybridization independent toxicity was examined in a study by Shen et al. [89]. "Wings" of gapmer AONs are typically 2′MOE modified nucleotides, but recently LNA and cET have been used to stabilize hybridization to the RNA–DNA duplex. However, LNA and cET gapmers in clinical use have a greater incidence of hepatoxicity. Shen et al. examined the molecular effects of different AON gapmer chemistries and found that reportedly toxic cET gapmers induce p53-regulated transcript upregulation. Toxic LNA, cET, and 2′MOE gapmers also cause hepatic apoptosis, as indicated by caspase 3 activity. The rapid response rate after administration of AON treatment suggests that protein delocalization and binding to gapmers play a major role in the toxicity. Proteins tend to associate with toxic LNA-AONs, as well as nuclear paraspeckle proteins can become delocalized from the nucleoplasm. As a strategy to mitigate toxic response, the introduction of a 2′OMe substantially reduced protein interactions [89]. cET gapmers upregulate liver transaminases after a single dose, but with the 2′OMe addition, this was not observed. This modification to gapmer design may pave the way for a more clinically effective AON treatment.

AON therapy also has great difficulty crossing cell membranes and the blood-brain barrier (BBB), presenting an ongoing challenge in developing treatments [90]. In terms of administering an effective concentration, drug delivery to tissues, cells, and organs requires some creative solutions. Along with the second and third generation modifications created to improve stability and ribonuclease resistance, nanovectors such as dendrimers, peptides, antibodies, and ligand conjugated antisense (LICA) technology are options for genetic material delivery, although these approaches are also associated with technical challenges [90–93].

11 Conclusion

Over recent years antisense therapies have made great strides in mitigating the symptoms of genetic disorders that otherwise have limited treatment options. With recent FDA approval of mipomersen and inotersen for familial hypercholesterolemia and hereditary transthyretin amyloidosis, respectively, and many gapmer AONs in clinical trials, we may see additional gapmer AONs introduced to the market in the near future. Apatorsen may be a potential treatment option for preventing cancer cell progression by inhibiting the expression of Hsp27, a heat shock protein in which upregulation aggravates malignant tumor formation. In addition, volanesorsen can promote clearance of plasma triglyceride levels for genetic diseases like familial chylomicronemia syndrome, familial partial lipodystrophy, and familial hypertriglyceridemia by suppressing *apoC-III* expression. IONIS-HTTRx has produced favorable safety outcomes and exploratory endpoint results in phase 1/2a trials (ClinicalTrials Identifier: NCT02519036) and is currently recruiting for a phase 3 trials (ClinicalTrials Identifier: NCT03842969). Although IONIS-DMPKRx has been discontinued, alternative gapmer designs are being explored for the treatment of myotonic dystrophy. Finally, recent advancements in gapmer AON design and delivery offer a promising treatment direction for addressing familial prion diseases by targeting PrP RNA.

The principle challenge for gapmer AONs to reach widespread clinical use is addressing the concerns over toxicity and delivery. Toxicity due to hybridization independent and dependent effects can disrupt the translation of essential transcripts or proteins, while insufficient delivery of gapmer AONs to the target can reduce the efficacy of the treatment. Despite these challenges, gapmer AONs allow for a great amount of customization due to their design. Modifications have promising results in reducing the toxic effects of gapmer AONs [89]. Exploring other designs, such as different antisense chemistries or altering the "wings" at different positions, may offer greater success for gapmer AONs in terms of reducing toxicity, as well as using in silico tools or finding accessible protocols to test toxicity in preclinical exploration. With these developments and the present success in mitigating some of the symptoms of several severe genetic disorders, gapmer AONs provide a viable treatment option that may also encourage alternative therapies for other genetic diseases.

References

1. Lee TI, Young RA (2013) Transcriptional regulation and its misregulation in disease. Cell 152(6):1237–1251. https://doi.org/10.1016/j.cell.2013.02.014
2. Scoles DR, Minikel EV, Pulst SM (2019) Antisense oligonucleotides. Neurol Genet 5(2): e323. https://doi.org/10.1212/NXG.0000000000000323
3. Bennett CF, Swayze EE (2010) RNA targeting therapeutics: molecular mechanisms of antisense oligonucleotides as a therapeutic platform. Annu Rev Pharmacol Toxicol 50 (1):259–293. https://doi.org/10.1146/annurev.pharmtox.010909.105654
4. Dias N, Stein CA (2002) Antisense oligonucleotides: basic concepts and mechanisms. Mol Cancer Ther 1(5):347–355
5. Rinaldi C, Wood MJA (2017) Antisense oligonucleotides: the next frontier for treatment of neurological disorders. Nat Rev Neurol 14 (1):9–21. https://doi.org/10.1038/nrneurol.2017.148
6. Miller HI, Riggs AD, Gill GN (1973) Ribonuclease H (hybrid) in *Escherichia coli*. Identification and characterization. J Biol Chem 248 (7):2621–2624
7. Tadokoro T, Kanaya S (2009) Ribonuclease H: molecular diversities, substrate binding domains, and catalytic mechanism of the prokaryotic enzymes. FEBS J 276(6):1482–1493. https://doi.org/10.1111/j.1742-4658.2009.06907.x
8. Shen X, Corey DR (2018) Chemistry, mechanism and clinical status of antisense oligonucleotides and duplex RNAs. Nucleic Acids Res 46(4):1584–1600. https://doi.org/10.1093/nar/gkx1239
9. Grunweller A, Hartmann RK (2016) Chemical modification of nucleic acids as a key technology for the development of RNA-based therapeutics. Pharmazie 71(1):8–16
10. Chan JHP, Lim S, Wong WSF (2006) Antisense oligonucleotides: from design to therapeutic application. Clin Exp Pharmacol Physiol

33(5-6):533–540. https://doi.org/10.1111/j.1440-1681.2006.04403.x
11. Lundin KE, Gissberg O, Smith CIE (2015) Oligonucleotide therapies: the past and the present. Hum Gene Ther 26(8):475–485. https://doi.org/10.1089/hum.2015.070
12. Uppuladinne MVN, Sonavane UB, Deka RC et al (2018) Structural insight into antisense gapmer-RNA oligomer duplexes through molecular dynamics simulations. J Biomol Struct Dyn 37(11):2823–2836. https://doi.org/10.1080/07391102.2018.1498390
13. Patel RS, Scopelliti EM, Savelloni J (2015) Therapeutic Management of Familial Hypercholesterolemia: current and emerging drug therapies. Pharmacotherapy 35 (12):1189–1203. https://doi.org/10.1002/phar.1672
14. M-A K, Rader DJ (2019) Familial hypercholesterolemia. In: Human Pathobiochemistry. Springer, New York, pp 43–56
15. Bruikman CS, Hovingh GK, Kastelein JJ (2017) Molecular basis of familial hypercholesterolemia. Curr Opin Cardiol. https://doi.org/10.1097/HCO.0000000000000385
16. Youngblom E, Pariani M, Knowles JW (1993) Familial hypercholesterolemia. In: Adam MP, Ardinger HH, Pagon RA et al (eds) GeneReviews®. University of Washington, Seattle (WA)
17. De Castro-Oros I, Pocovi M, Civeira F (2010) The genetic basis of familial hypercholesterolemia: inheritance, linkage, and mutations. Appl Clin Genet 3:53–64. https://doi.org/10.2147/tacg.s8285
18. Santini A, Novellino E (2017) Nutraceuticals in hypercholesterolaemia: an overview. Br J Pharmacol 174(11):1450–1463. https://doi.org/10.1111/bph.13636
19. Innerarity TL, Mahley RW, Weisgraber KH et al (1990) Familial defective apolipoprotein B-100: a mutation of apolipoprotein B that causes hypercholesterolemia. J Lipid Res 31 (8):1337–1349
20. Vogt A (2015) The genetics of familial hypercholesterolemia and emerging therapies. Appl Clin Genet 8:27–36. https://doi.org/10.2147/TACG.S44315
21. Huang CH, Chiu PC, Liu HC et al (2015) Clinical observations and treatment of pediatric homozygous familial hypercholesterolemia due to a low-density lipoprotein receptor defect. J Clin Lipidol 9(2):234–240. https://doi.org/10.1016/j.jacl.2014.11.011
22. Raal FJ, Santos RD, Blom DJ et al (2010) Mipomersen, an apolipoprotein B synthesis inhibitor, for lowering of LDL cholesterol concentrations in patients with homozygous familial hypercholesterolaemia: a randomised, double-blind, placebo-controlled trial. Lancet 375(9719):998–1006. https://doi.org/10.1016/S0140-6736(10)60284-X
23. Hair P, Cameron F, McKeage K (2013) Mipomersen sodium: first global approval. Drugs 73 (5):487–493. https://doi.org/10.1007/s40265-013-0042-2
24. Robinson JG (2013) Management of familial hypercholesterolemia: a review of the recommendations from the National Lipid Association Expert Panel on familial hypercholesterolemia. J Manag Care Pharm 19(2):139–149. https://doi.org/10.18553/jmcp.2013.19.2.139
25. Stein EA, Dufour R, Gagne C et al (2012) Apolipoprotein B synthesis inhibition with mipomersen in heterozygous familial hypercholesterolemia: results of a randomized, double-blind, placebo-controlled trial to assess efficacy and safety as add-on therapy in patients with coronary artery disease. Circulation 126 (19):2283–2292
26. Parham JS, Goldberg AC (2019) Mipomersen and its use in familial hypercholesterolemia. Expert Opin Pharmacother 20(2):127–131. https://doi.org/10.1080/14656566.2018.1550071
27. Gertz M (2016) Hereditary ATTR amyloidosis: burden of illness and diagnostic challenges. J Peripher Nerv Syst 23(7):5–9. https://doi.org/10.1111/jns.12153
28. Keam SJ (2018) Inotersen: first global approval. Drugs 78(13):1371–1376. https://doi.org/10.1007/s40265-018-0968-5
29. Ankarcrona M, Winblad B, Monteiro C et al (2016) Current and future treatment of amyloid diseases. J Intern Med 280(2):177–202. https://doi.org/10.1111/joim.12506
30. Sekijima Y (2014) Recent progress in the understanding and treatment of transthyretin amyloidosis. J Clin Pharm Ther 39 (3):225–233. https://doi.org/10.1111/jcpt.12145
31. Gertz MA, Scheinberg M, Waddington-Cruz M et al (2019) Inotersen for the treatment of adults with polyneuropathy caused by hereditary transthyretin-mediated amyloidosis. Expert Rev Clin Pharmacol 12:701–711. https://doi.org/10.1080/17512433.2019.1635008
32. Mathew V, Wang AK (2019) Inotersen: new promise for the treatment of hereditary transthyretin amyloidosis. Drug Des Devel Ther 13:1515–1525. https://doi.org/10.2147/DDDT.S162913

33. Benson MD, Waddington-Cruz M, Berk JL et al (2018) Inotersen treatment for patients with hereditary Transthyretin amyloidosis. N Engl J Med 379(1):22–31. https://doi.org/10.1056/NEJMoa1716793
34. Calderwood SK, Gong J (2016) Heat shock proteins promote cancer: It's a protection racket. Trends Biochem Sci 41(4):311–323. https://doi.org/10.1016/j.tibs.2016.01.003
35. Wang X, Chen M, Zhou J et al (2014) HSP27, 70 and 90, anti-apoptotic proteins, in clinical cancer therapy (review). Int J Oncol 45 (1):18–30. https://doi.org/10.3892/ijo.2014.2399
36. Hanahan D, Weinberg RA (2000) The hallmarks of cancer. Cell 100(1):57–70. https://doi.org/10.1016/S0092-8674(00)81683-9
37. O'Callaghan-Sunol C, Gabai VL, Sherman MY (2007) Hsp27 modulates p53 Signaling and suppresses cellular senescence. Cancer Res 67 (24):11779–11788. https://doi.org/10.1158/0008-5472.CAN-07-2441
38. Shiota M, Bishop JL, Nip KM et al (2013) Hsp27 regulates epithelial Mesenchymal transition, metastasis, and circulating tumor cells in prostate cancer. Cancer Res 73 (10):3109–3119. https://doi.org/10.1158/0008-5472.CAN-12-3979
39. Zoubeidi A, Zardan A, Beraldi E et al (2007) Cooperative interactions between androgen receptor (AR) and heat-shock protein 27 facilitate AR transcriptional activity. Cancer Res 67 (21):10455–10465. https://doi.org/10.1158/0008-5472.CAN-07-2057
40. Dellis AE, Papatsoris AG (2019) Perspectives on the current and emerging chemical androgen receptor antagonists for the treatment of prostate cancer. Expert Opin Pharmacother 20 (2):163–172. https://doi.org/10.1080/14656566.2018.1548611
41. Chi KN, Yu EY, Jacobs C et al (2016) A phase I dose-escalation study of apatorsen (OGX-427), an antisense inhibitor targeting heat shock protein 27 (Hsp27), in patients with castration-resistant prostate cancer and other advanced cancers. Ann Oncol 27 (6):1116–1122. https://doi.org/10.1093/annonc/mdw068
42. Yu EY, Ellard SL, Hotte SJ et al (2018) A randomized phase 2 study of a HSP27 targeting antisense, apatorsen with prednisone versus prednisone alone, in patients with metastatic castration resistant prostate cancer. Investig New Drugs 36(2):278–287. https://doi.org/10.1007/s10637-017-0553-x
43. Rosenberg JE, Hahn NM, Regan MM et al (2018) Apatorsen plus docetaxel versus docetaxel alone in platinum-resistant metastatic urothelial carcinoma (Borealis-2). Br J Cancer 118(11):1434–1441. https://doi.org/10.1038/s41416-018-0087-9
44. Ko AH, Murphy PB, Peyton JD et al (2017) A randomized, double-blinded, phase II trial of gemcitabine and nab-paclitaxel plus Apatorsen or placebo in patients with metastatic pancreatic cancer: the RAINIER trial. Oncologist 22 (12):1427–e1129. https://doi.org/10.1634/theoncologist.2017-0066
45. Chyzhyk V, Brown AS (2019) Familial chylomicronemia syndrome: A rare but devastating autosomal recessive disorder characterized by refractory hypertriglyceridemia and recurrent pancreatitis. Trends Cardiovasc Med 30 (2):80–85. https://doi.org/10.1016/j.tcm.2019.03.001
46. Gaudet D, Brisson D, Tremblay K et al (2014) Targeting APOC3 in the familial Chylomicronemia syndrome. N Engl J Med 371 (23):2200–2206. https://doi.org/10.1056/NEJMoa1400284
47. Paik J, Duggan S (2019) Volanesorsen: first global approval. Drugs 79(12):1349–1354. https://doi.org/10.1007/s40265-019-01168-z
48. Brahm AJ, Hegele RA (2015) Chylomicronaemia—current diagnosis and future therapies. Nat Rev Endocrinol 11(6):352–362. https://doi.org/10.1038/nrendo.2015.26
49. Reference GH LPL gene. https://ghr.nlm.nih.gov/gene/LPL
50. Reference GH Familial partial lipodystrophy. https://ghr.nlm.nih.gov/condition/familial-partial-lipodystrophy
51. Jéru I, Vatier C, Araujo-Vilar D et al (2017) Clinical utility gene card for: familial partial lipodystrophy. Eur J Hum Genet 25 (2):271–271. https://doi.org/10.1038/ejhg.2016.102
52. Ferns G, Keti V, Griffin B (2008) Investigation and management of hypertriglyceridaemia. J Clin Pathol 61(11):1174–1183. https://doi.org/10.1136/jcp.2008.055756
53. Hegele RA, Ginsberg HN, Chapman MJ et al (2014) The polygenic nature of hypertriglyceridaemia: implications for definition, diagnosis, and management. Lancet Diabetes Endocrinol 2(8):655–666. https://doi.org/10.1016/S2213-8587(13)70191-8
54. Akcea Therapeutics I (2018) FDA Advisory Committee Votes in Favor of WAYLIVRA for Treatment of Familial Chylomicronemia Syndrome. https://ir.akceatx.com/news-releases/news-release-details/fda-advisory-committee-

votes-favor-waylivra-treatment-familial. Accessed October 27, 2019
55. Akcea Therapeutics I (2018) Akcea and Ionis Receive Complete Response Letter for WAYLIVRA from FDA. https://ir.akceatx.com/news-releases/news-release-details/akcea-and-ionis-receive-complete-response-letter-waylivra-fda. Accessed October 29, 2019
56. Government of Canada HC (2014) Search Page - Drug and Health Product Register. https://hpr-rps.hres.ca/reg-content/regulatory-decision-summary-detail.php?lang=en&linkID=RDS00490
57. Hegele RA, Berberich AJ, Ban MR et al (2018) Clinical and biochemical features of different molecular etiologies of familial chylomicronemia. J Clin Lipidol 12(4):920–927. e924. https://doi.org/10.1016/j.jacl.2018.03.093
58. Gaudet D, Digenio A, Alexander VJ et al (2017) The APPROACH study: A randomized, double-blind, placebo-controlled, phase 3 study of Volanesorsen administered subcutaneously to patients with familial Chylomicronemia syndrome (FCS). J Clin Lipidol 11 (3):814–815. https://doi.org/10.1016/j.jacl.2017.04.071
59. Caron NS, Wright GEB, Hayden MR (1993) Huntington disease. In: Adam MP, Ardinger HH, Pagon RA et al (eds) GeneReviews®. University of Washington, Seattle (WA)
60. Mahalingam S, Levy LM (2014) Genetics of Huntington disease. Am J Neuroradiol 35 (6):1070–1072. https://doi.org/10.3174/ajnr.A3772
61. Jimenez-Sanchez M, Licitra F, Underwood BR et al (2017) Huntington's disease: mechanisms of pathogenesis and therapeutic strategies. Cold Spring Harb Perspect Med 7(7). https://doi.org/10.1101/cshperspect.a024240
62. Kaemmerer WF, Grondin RC (2019) The effects of huntingtin-lowering: what do we know so far? Degener Neurol Neuromuscul Dis 9:3–17. https://doi.org/10.2147/DNND.S163808
63. Errington SJ, Mann CJ, Fletcher S et al (2003) Target selection for antisense oligonucleotide induced exon skipping in the dystrophin gene. J Gene Med 5(6):518–527
64. Ma B, Culver BP, Baj G et al (2010) Localization of BDNF mRNA with the Huntington's disease protein in rat brain. Mol Neurodegener 5:22. https://doi.org/10.1186/1750-1326-5-22
65. Baquet ZC, Gorski JA, Jones KR (2004) Early striatal dendrite deficits followed by neuron loss with advanced age in the absence of anterograde cortical brain-derived neurotrophic factor. J Neurosci 24(17):4250–4258. https://doi.org/10.1523/JNEUROSCI.3920-03.2004
66. Zhang Y, Engelman J, Friedlander RM (2009) Allele-specific silencing of mutant Huntington's disease gene. J Neurochem 108 (1):82–90. https://doi.org/10.1111/j.1471-4159.2008.05734.x
67. Gauthier LR, Charrin BC, Borrell-Pages M et al (2004) Huntingtin controls neurotrophic support and survival of neurons by enhancing BDNF vesicular transport along microtubules. Cell 118(1):127–138. https://doi.org/10.1016/j.cell.2004.06.018
68. Wild EJ, Tabrizi SJ (2017) Therapies targeting DNA and RNA in Huntington's disease. Lancet Neurol 16(10):837–847. https://doi.org/10.1016/S1474-4422(17)30280-6
69. Tabrizi SJ, Leavitt BR, Landwehrmeyer GB et al (2019) Targeting Huntingtin expression in patients with Huntington's disease. N Engl J Med 381(14):1398. https://doi.org/10.1056/NEJMoa1900907
70. Rodrigues FB, Wild EJ (2018) Huntington's disease clinical trials corner: February 2018. J Huntingtons Dis 7(1):89–98. https://doi.org/10.3233/JHD-189001
71. Ionis Pharmaceuticals I (2017) Ionis Pharmaceuticals Licenses IONIS-HTT Rx to Partner Following Successful Phase 1/2a Study inPatients with Huntington's Disease. https://ir.ionispharma.com/news-releases/news-release-details/ionis-pharmaceuticals-licenses-ionis-htt-rx-partner-following
72. Kay C, Collins JA, Skotte NH et al (2015) Huntingtin haplotypes provide prioritized target panels for allele-specific silencing in Huntington disease patients of European ancestry. Mol Ther 23(11):1759–1771. https://doi.org/10.1038/mt.2015.128
73. Lee JE, Bennett CF, Cooper TA (2012) RNase H-mediated degradation of toxic RNA in myotonic dystrophy type 1. Proc Natl Acad Sci U S A 109(11):4221–4226. https://doi.org/10.1073/pnas.1117019109
74. Brook JD, McCurrach ME, Harley HG et al (1992) Molecular basis of myotonic dystrophy: expansion of a trinucleotide (CTG) repeat at the 3′ end of a transcript encoding a protein kinase family member. Cell 68(4):799–808. https://doi.org/10.1016/0092-8674(92)90154-5
75. Lee Johanna E, Cooper Thomas A (2009) Pathogenic mechanisms of myotonic dystrophy. Biochem Soc Trans 37(6):1281–1286. https://doi.org/10.1042/BST0371281

76. Thornton CA, Wang E, Carrell E (2017) Myotonic dystrophy: approach to therapy. Curr Opin Genet Dev 44:135–140. https://doi.org/10.1016/j.gde.2017.03.007
77. Ranum LPW, Day JW (2004) Myotonic dystrophy: RNA pathogenesis comes into focus. Am J Hum Genet 74(5):793–804. https://doi.org/10.1086/383590
78. Thornton CA (2014) Myotonic dystrophy. Neurol Clin 32(3):705–719. https://doi.org/10.1016/j.ncl.2014.04.011
79. Ionis, Biogen (2017) Letter from Ionis Pharmaceuticals & Biogen to the MDF Community. https://us8.campaign-archive.com/?u=8f5969cac3271759ce78c8354&id=1109538bcf&e=[UNIQID].
80. Jauvin D, Chrétien J, Pandey SK et al (2017) Targeting DMPK with antisense oligonucleotide improves muscle strength in Myotonic dystrophy type 1 mice. Mol Ther Nucleic Acids 7:465–474. https://doi.org/10.1016/j.omtn.2017.05.007
81. Prusiner SB (1998) Prions. PNAS 95 (23):13363–13383. https://doi.org/10.1073/pnas.95.23.13363
82. Raymond GJ, Zhao HT, Race B et al (2019) Antisense oligonucleotides extend survival of prion-infected mice. JCI Insight 5:131175
83. Imran M, Mahmood S (2011) An overview of human prion diseases. Virol J 8(1):1–9. https://doi.org/10.1186/1743-422X-8-559
84. Atkinson CJ, Zhang K, Munn AL et al (2016) Prion protein scrapie and the normal cellular prion protein. Prion 10(1):63–82. https://doi.org/10.1080/19336896.2015.1110293
85. Mallucci G, Dickinson A, Linehan J et al (2003) Depleting neuronal PrP in prion infection prevents disease and reverses Spongiosis. Science 302(5646):871–874. https://doi.org/10.1126/science.1090187
86. White MD, Farmer M, Mirabile I et al (2008) Single treatment with RNAi against prion protein rescues early neuronal dysfunction and prolongs survival in mice with prion disease. PNAS 105(29):10238–10243. https://doi.org/10.1073/pnas.0802759105
87. Nazor Friberg K, Hung G, Wancewicz E et al (2012) Intracerebral infusion of antisense oligonucleotides into prion-infected mice. Mol Ther Nucleic Acids 1(2):e9. https://doi.org/10.1038/mtna.2011.6
88. Dieckmann A, Hagedorn PH, Burki Y et al (2018) A sensitive in vitro approach to assess the hybridization-dependent toxic potential of high affinity Gapmer oligonucleotides. Mol Ther Nucleic Acids 10:45–54. https://doi.org/10.1016/j.omtn.2017.11.004
89. Shen W, De Hoyos CL, Migawa MT et al (2019) Chemical modification of PS-ASO therapeutics reduces cellular protein-binding and improves the therapeutic index. Nat Biotechnol 37(6):640–650. https://doi.org/10.1038/s41587-019-0106-2
90. Mignani S, Shi X, Zablocka M et al (2019) Dendrimer-enabled therapeutic antisense delivery systems as innovation in medicine. Bioconjug Chem 30(7):1938–1950. https://doi.org/10.1021/acs.bioconjchem.9b00385
91. Alexander VJ, Digenio A, Xia ST et al (2018) Inhibition of Apolipoprotein C-iii with Galnac conjugated antisense drug potently lowers fasting serum Apolipoprotein C-iii and triglyceride levels in healthy volunteers with elevated triglycerides. J Am Coll Cardiol 71(11):1724–1724. https://doi.org/10.1016/S0735-1097(18)32265-4
92. Marshall NB, Oda SK, London CA et al (2007) Arginine-rich cell-penetrating peptides facilitate delivery of antisense oligomers into murine leukocytes and alter pre-mRNA splicing. J Immunol Methods 325(1-2):114–126. https://doi.org/10.1016/j.jim.2007.06.009
93. Walker I, Irwin WJ, Akhtar S (1995) Improved cellular delivery of antisense oligonucleotides using transferrin receptor antibody-oligonucleotide conjugates. Pharm Res 12 (10):1548–1553. https://doi.org/10.1023/a:1016260110049

Chapter 3

Knocking Down Long Noncoding RNAs Using Antisense Oligonucleotide Gapmers

Rika Maruyama and Toshifumi Yokota

Abstract

Long noncoding RNAs (lncRNAs) are a class of RNA with 200 nucleotides or longer that are not translated into protein. lncRNAs are highly abundant; a study estimates that at least four times more lncRNAs are typically present than coding RNAs in humans. However, function of more than 95% of human lncRNAs are still unknown. Synthetic antisense oligonucleotides called gapmers are powerful tools for lncRNA loss-of-function studies. Gapmers contain a central DNA part, which activates RNase H-mediated RNA degradation, flanked by modified oligonucleotides, such as 2′-*O*-methyl RNA (2′OMe), 2′-*O*-methoxyethyl RNA (2′MOE), constrained ethyl nucleosides (cEt), and locked nucleic acids (LNAs). In contrast to siRNA or RNAi-based methods, antisense oligonucleotide gapmer-based knockdown is often more effective against nuclear-localized lncRNA targets, since RNase H is mainly localized in nuclei. As such, gapmers are also potentially a powerful tool for therapeutics targeting lncRNAs in various diseases, including cancer, cardiovascular diseases, lung fibrosis, and neurological/neuromuscular diseases. This chapter will discuss the development and applications of gapmers for lncRNA loss-of-function studies and tips to design effective antisense oligonucleotides.

Key words MIAT (myocardial infarction associated transcript) also known as RNCR2 (retinal noncoding RNA 2) or Gomafu, CRISPR/Cas9, PRNCR1 (prostate cancer noncoding RNA 1), PVT1 (Pvt1 Oncogene), HOTAIR (HOX transcript antisense RNA), BC200 (Brain cytoplasmic 200), HULC (Highly Upregulated in Liver Cancer), NEAT1 (Nuclear Enriched Abundant Transcript 1), Metastasis-associated lung adenocarcinoma transcript 1 (MALAT1), DNM3OS (DNM3 opposite strand/antisense RNA), microRNAs (miRNAs)

1 Introduction

Identification of noncoding RNAs (ncRNAs) with different sizes and structures has made the most significant shift in our understanding of the role of RNA and the classical central dogma theory [1]. There are many classes of ncRNAs described such as microRNAs (miRNA), snoRNA, and long noncoding RNAs (lncRNA) [2]. Specifically, ncRNAs are classified based on the length, that is, (1) small noncoding RNAs (sncRNAs), which are less than 200 nt, including microRNAs (miRNAs, 21–22 nt), piwi-interacting RNAs

Toshifumi Yokota and Rika Maruyama (eds.), *Gapmers: Methods and Protocols*, Methods in Molecular Biology, vol. 2176, https://doi.org/10.1007/978-1-0716-0771-8_3, © Springer Science+Business Media, LLC, part of Springer Nature 2020

(piRNAs, 24–32 nt), small nucleolar RNAs (snoRNAs, 70–200 nt), and transfer RNAs (tRNAs, 76–90 nt); and (2) long noncoding RNAs (lncRNAs), which are longer than 200 nt [3–6]. lncRNAs are the longest member of the ncRNA family with 200 nucleotides or longer sequence [7]. lncRNAs, transcribed either intergenic or intronic regions, have diverse function in gene expression, including structural components for shaping the nuclear organization, chromatin organization, and as decoys for other RNAs or proteins [8]. lncRNAs are highly abundant; a study estimates that at least four times more lncRNAs are typically present than coding RNAs in human cells [9]. Currently, more than 50,000 lncRNAs are listed in LNCipedia 5 [10]. Although thousands of lncRNAs are now listed, most of these molecules' function is still poorly understood [11]. lncRNAs also have diverse localization [12]. Some lncRNAs localize in the nucleus and others localize in the cytoplasm [13]. Interestingly, some lncRNAs localize evenly between the nucleus and cytoplasm [14].

To understand the role of lncRNAs by knocking down the expression, three strategies are commonly used, including RNA interference (RNAi), CRISPR-Cas9 genome editing, and antisense oligonucleotide gapmers [15]. Depending on the subcellular localization of the lncRNA, different techniques are employed [16]. Since RNAi-mediated mRNA degradation occurs mainly in the cytoplasm, this method is used for lncRNAs localized in the cytoplasm [16]. CRISPR-Cas9 genome editing is a more universal and genetic approach; however, using this method for lncRNA is tricky [17]. Making a small indel in the targeted lncRNA sequence by nonhomologous end joining (NHEJ) is most likely not enough since no "open reading frame" exists in ncRNAs [18]. As such, researchers need to use dual guide RNAs to excise a large fragment or target an RNA destabilizing element [19–21]. On the other hand, a more simple method to knockdown lncRNAs is to use synthetic antisense oligonucleotides called gapmers [22]. In this chapter, we will summarize the development and application of gapmers to knockdown lncRNAs.

2 Gapmers

In 1987, Inoue et al. developed a technique to knockdown RNAs at specific sites using synthetic antisense oligonucleotides [23]. These oligonucleotides, now often called gapmers, are synthetic and chimeric single-stranded oligonucleotides which bind to and cleavage a complementary RNA through the recruitment and activation of an endonuclease, RNase H. Gapmers typically contain a central DNA part, which activates RNase H, flanked by modified oligonucleotides, such as 2′-*O*-methyl RNA (2′OMe), 2′-*O*-methoxyethyl RNA (2′MOE), constrained ethyl nucleosides (cEt), and locked nucleic acids (LNAs) [24, 25]. Currently, all of these modified oligonucleotides are commercially available (e.g., QIAGEN/Exiqon, Eurogentec, Integrated DNA Technologies (IDT)). Since unmodified DNAs are highly unstable, it is important to protect them by chemically modified oligonucleotides. The DNA–RNA heteroduplex is recognized and cleaved by RNase H [26]. In addition, gapmers commonly have the phosphorothioate backbone as opposed to the phosphodiester backbone seen in DNAs and RNAs [27]. This modification increases nuclease resistance while maintaining RNase recognition and activity [27]. To date, gapmers have found widespread use as research reagents for inducing RNA degradation, as well as for therapeutics [28, 29]. Since RNase H is most abundant in the nucleus, gapmers are now also considered powerful tools to knockdown lncRNAs [30]. According to a study, nuclear lncRNAs are more easily degraded by gapmers, including LNA gapmers and 2′-*O*-methyl RNA gapmers, while cytoplasmic targets are more easily knocked down by RNAi [16]. In addition, gapmers are also often effective at degrading cytoplasmic RNAs [16]. As such, when the localization of lncRNAs is unclear, selecting gapmers is likely the safest choice for successful knockdown experiments.

3 Design

Designing good gapmers is a critical step toward successful knockdown of lncRNAs [15]. Gapmers generally have 15–20 nucleotides [31]. Specifically, LNA and cEt gapmers are typically 15–16 nucleotides while 2′OMe and 2′MOE gapmers are 18–20 nucleotides since LNA and cEt have higher RNA binding affinity [32]. Typical design of gapmers includes 3× LNAs –10× DNAs– 3× LNAs, or 5× 2′MOE-10× DNAs– 5× 2′MOE; that is, the central core has 10 nucleotides of DNAs, flanked by 3 nucleotides of LNAs (or cEt) or 5 nucleotides of 2′MOEs (or 2′OMe) on both sides. Several factors should be taken into account to choose the target site. First, GC content is one of the most important parameters. The GC content ranged between 40% and 60% is typically chosen [33]. Second, the self-complementarity of the antisense oligonucleotides (AONs) should be taken into account, which is typically above −9 kcal/mol [33]. These parameters can be analyzed using online software such as the OligoAnalyzer Tool (IDT). Third, AONs should be analyzed for potential off-target effects, as all techniques that suppress gene expression have a risk of such effects. This can be done using Bowtie, Bowtie2 (Ensembl release 85) or GGGenome, typically allowing for a maximum of two mismatches [34]. Forth, RNA target accessibility should be considered. This can be analyzed by software such as UNAFold [35]. Fifth, protein binding motifs should be avoided. RBPmap is a useful tool to predict it [36]. Lastly, the melting temperature of AONs potentially has a significant impact on the efficiency [37]. "TM"."LNA oligo Tm prediction tool", a program developed by Exiqon (QIAGEN), can predict the melting temperature of LNA and DNA-containing mixmer AONs [38]. The T_m of 2′,4′-BNA/LNA/DNA mixmers and 2′-*O*-Me AONs can be also estimated by the OligoAnalyzer 3.1 (IDT). LNA shows remarkably high RNA-binding affinity, with melting temperature being increased by 3–9 °C/modification. Optimal Tm is not clear; however, 55 °C is typically chosen as the cut-off, using the default conditions against an RNA target, from IDT OligoAnalyzer [39]. To our knowledge, there is no freely available program to estimate T_m of other AON chemistries. Overall, these tips are helpful to design gapmers; however, the effects of gapmers are still highly unpredictable. As such, 5–10 gapmers are typically designed and tested to select the most efficient one experimentally. Alternatively, Integrated DNA Technologies customer care services and QIAGEN on their website or their support team can design gapmers for a fee.

4 Long Noncoding RNA Knockdown and Therapeutics

More than 200 diseases are known to be associated with dysregulated or dysfunctional lncRNAs to date [40]. The list is growing dramatically in the literature. Examples of these lncRNAs include PRNCR1 (prostate cancer noncoding RNA 1), PVT1 (Pvt1 Oncogene) HOTAIR (HOX transcript antisense RNA), BC200 (Brain cytoplasmic 200), HULC (Highly Upregulated in Liver Cancer), NEAT1 (Nuclear Enriched Abundant Transcript 1), and MIAT (myocardial infarction associated transcript) also known as RNCR2 (retinal noncoding RNA 2) or Gomafu. These lncRNAs are dysregulated in cancers, Alzheimer's disease, and human immunodeficiency virus infection, among others [41–49]. Since lncRNAs are dysregulated in many diseases, these could be promising targets for gapmer therapeutics [7].

4.1 Cancers

Metastasis-associated lung adenocarcinoma transcript 1 (MALAT1), also known as nuclear-enriched transcript 2 (NEAT2), is a bona fide lncRNA discovered as a prognostic marker for lung cancer metastasis [50]. MALAT1 is associated with many pathological processes, ranging from diabetes complications to cancers [51]. As such, gapmer-mediated knockdown of MALAT1 is a promising therapeutic strategy [52]. In fact, MALAT-1 targeting LNA-gapmers are being developed to treat tumors, including multiple myeloma and others [53]. Researchers also use gapmers to treat ovarian cancer. A lncRNA called RPSAP52 (Ribosomal Protein SA Pseudogene 52) plays an important role in ovarian cancer. Specifically, RPSAP52 is an antisense transcript to the well-described oncogene HMGA2 (High Mobility Group A2).

As such, the use of gapmers for the depletion of RPSAP52 can be a potential therapeutic strategy in ovarian cancer [54].

4.2 Spinal Muscular Atrophy

Another interesting example of using gapmers for treating a disease is one for spinal muscular atrophy (SMA). Woo and colleagues identified gapmers targeting a lncRNA that recruits the repressive complex of *SMN2*, a duplicated gene that is almost identical to *SMN1* but can typically produce only 10% of normal protein [55]. Specifically, a lncRNA called SMN-antisense 1 (SMN-AS1), which suppresses SMN2, was targeted by gapmers that increase *SMN2* expression in primary neuronal cultures. This study is particularly interesting because this approach increases SMN expression through a distinct mechanism from that of a splice switching AON already approved by the FDA called nusinersen [56]. As such, the combination therapy of these two AONs might be more effective than a single AON approach.

4.3 Cardiovascular Disease

Cardiovascular diseases are also among the most promising targets for therapeutic applications using gapmers against lncRNAs. For example, LNA-gapmer-mediated silencing of CHAST (cardiac hypertrophy-associated transcript) efficiently attenuated transverse aortic constriction-induced cardiac remodeling in mice [57].

4.4 Lung Fibrosis

Researchers also use gapmers to treat lung fibrosis. The lncRNA DNM3OS (DNM3 opposite strand/antisense RNA) is an miRNA reservoir in TGF-β signaling and identified as a promising therapeutic target. Jian and colleagues demonstrated using gapmers that DNM3OS is a crucial downstream effector of TGF-β signaling in lung fibroblasts in vitro, and in bleomycin-induced lung fibrosis in mice in vivo [58].

5 Conclusions

In this chapter, we described a gapmer-based approach and applications to efficiently knockdown lncRNAs. Overall, gapmers targeting lncRNAs hold a great promise. Our understanding of lncRNAs has been dramatically improved in recent years. However, the role of most lncRNAs is still poorly understood. Reliable knockdown techniques, such as gapmers, are important tools to help better determine their individual functions.

The field of gapmer therapeutics targeting lncRNAs is still in its infancy but represents a promising and novel avenue for targeted treatments in many diseases. Within the next several years, the function of lncRNAs will be better understood and gapmers targeting lncRNAs might enter clinical trials. However, there are still several hurdles. First, the function of individual lncRNAs needs to be further elucidated, as transcriptional changes may or may not be causative. Second, delivery of AONs is a key challenge. Development of tissue-specific delivery of AONs would greatly advance the field. Such delivery methods can include cell-penetrating peptides/moieties, polymers, antibodies, and/or nanoparticles [59–63]. In addition, the route of delivery, low toxicity, duration of treatment, dosage adjustment, and off-target effects are all important factors to be considered.

Acknowledgments

This work is supported by Muscular Dystrophy Canada, the Friends of Garrett Cumming Research Fund, HM Toupin Neurological Science Research Fund, Canadian Institutes of Health Research (CIHR), Alberta Innovates: Health Solutions (AIHS), Canada Foundation for Innovation (CFI), Alberta Advanced Education and Technology, and Women and Children's Health Research Institute (WCHRI).

References

1. Costa FF (2010) Non-coding RNAs: meet thy masters. BioEssays 32(7):599–608
2. Beermann J, Piccoli M-T, Viereck J et al (2016) Non-coding RNAs in development and disease: background, mechanisms, and therapeutic approaches. Physiol Rev 96(4):1297–1325
3. Gomes AQ, Nolasco S, Soares H (2013) Non-coding RNAs: multi-tasking molecules in the cell. Int J Mol Sci 14(8):16010–16039
4. Krishnan P, Ghosh S, Graham K et al (2016) Piwi-interacting RNAs and PIWI genes as novel prognostic markers for breast cancer. Oncotarget 7(25):37944
5. Yin Q-F, Yang L, Zhang Y et al (2012) Long noncoding RNAs with snoRNA ends. Mol Cell 48(2):219–230
6. Ho PY, Yu AM (2016) Bioengineering of non-coding RNAs for research agents and therapeutics. Wiley Interdiscip Rev RNA 7(2):186–197
7. Esteller M (2011) Non-coding RNAs in human disease. Nat Rev Genet 12(12):861
8. Engreitz JM, Ollikainen N, Guttman M (2016) Long non-coding RNAs: spatial amplifiers that control nuclear structure and gene expression. Nat Rev Mol Cell Biol 17(12):756
9. Mercer TR, Dinger ME, Mattick JS (2009) Long non-coding RNAs: insights into functions. Nat Rev Genet 10(3):155
10. Volders PJ, Anckaert J, Verheggen K et al (2019) LNCipedia 5: towards a reference set of human long non-coding RNAs. Nucleic Acids Res 47(D1):D135–D139. https://doi.org/10.1093/nar/gky1031
11. Moran VA, Perera RJ, Khalil AM (2012) Emerging functional and mechanistic paradigms of mammalian long non-coding RNAs. Nucleic Acids Res 40(14):6391–6400
12. Chen L-L (2016) Linking long noncoding RNA localization and function. Trends Biochem Sci 41(9):761–772
13. Cabili MN, Dunagin MC, McClanahan PD et al (2015) Localization and abundance analysis of human lncRNAs at single-cell and single-molecule resolution. Genome Biol 16(1):20
14. van Heesch S, van Iterson M, Jacobi J et al (2014) Extensive localization of long noncoding RNAs to the cytosol and mono-and polyribosomal complexes. Genome Biol 15(1):R6
15. Lennox KA, Behlke MA (2016) Mini-review: current strategies to knockdown long non-coding RNAs. J Rare Dis Res Treat 1:66–70
16. Lennox KA, Behlke MA (2015) Cellular localization of long non-coding RNAs affects silencing by RNAi more than by antisense oligonucleotides. Nucleic Acids Res 44 (2):863–877
17. Szafranski P, Karolak JA, Lanza D et al (2017) CRISPR/Cas9-mediated deletion of lncRNA Gm26878 in the distant Foxf1 enhancer region. Mamm Genome 28(7–8):275–282

18. Yang J, Meng X, Pan J et al (2018) CRISPR/Cas9-mediated noncoding RNA editing in human cancers. RNA Biol 15(1):35–43
19. Aparicio-Prat E, Arnan C, Sala I et al (2015) DECKO: single-oligo, dual-CRISPR deletion of genomic elements including long non-coding RNAs. BMC Genomics 16(1):846
20. Gutschner T (2015) Silencing long noncoding RNAs with genome-editing tools. In: Chromosomal mutagenesis. Springer, New York, pp 241–250
21. Koirala P, Zou D-H, Mo Y-Y (2016) Long non-coding RNAs as key regulators of cancer metastasis. J Cancer Metastasis Treat 2:1–10
22. Salehi M, Sharifi M, Bagheri M (2019) Knockdown of long noncoding RNA plasmacytoma variant translocation 1 with antisense locked nucleic acid GapmeRs exerts tumor-suppressive functions in human acute erythroleukemia cells through downregulation of C-MYC expression. Cancer Biother Radiopharm 34(6):371–379
23. Inoue H, Hayase Y, Iwai S et al (1987) Sequence-dependent hydrolysis of RNA using modified oligonucleotide splints and RNase H. FEBS Lett 215(2):327–330. https://doi.org/10.1016/0014-5793(87)80171-0
24. GrÜnweiler A, Hartmann RK (2007) Locked nucleic acid oligonucleotides. BioDrugs 21 (4):235–243
25. Gait MJ, Agrawal S (2019) Advances in nucleic acid therapeutics, vol 68. Royal Society of Chemistry, London
26. Nakamura H, Oda Y, Iwai S et al (1991) How does RNase H recognize a DNA. RNA hybrid? Proc Natl Acad Sci U S A 88 (24):11535–11539
27. Seth PP, Jazayeri A, Yu J et al (2012) Structure activity relationships of α-L-LNA modified phosphorothioate gapmer antisense oligonucleotides in animals. Mol Ther Nucleic Acids 1:e47
28. Lundin KE, Gissberg O, Smith CI (2015) Oligonucleotide therapies: the past and the present. Hum Gene Ther 26(8):475–485. https://doi.org/10.1089/hum.2015.070
29. Lee JJ, Yokota T (2013) Antisense therapy in neurology. J Pers Med 3(3):144–176. https://doi.org/10.3390/jpm3030144
30. Matsui M, Corey DR (2017) Non-coding RNAs as drug targets. Nat Rev Drug Discov 16(3):167
31. Khan H, Missailidis S (2008) Antisense Agents. In: Anticancer therapeutics, vol 317. Wiley, Chichester
32. Shen W, De Hoyos CL, Migawa MT et al (2019) Chemical modification of PS-ASO therapeutics reduces cellular protein-binding and improves the therapeutic index. Nat Biotechnol 37(6):640–650
33. Naessens S, Ruysschaert L, Lefever S et al (2019) Antisense oligonucleotide-based downregulation of the G56R pathogenic variant causing NR2E3-associated autosomal dominant retinitis Pigmentosa. Genes 10 (5):363
34. Yoshida T, Naito Y, Yasuhara H et al (2019) Evaluation of off-target effects of gapmer antisense oligonucleotides using human cells. Genes Cells 24(12):827–835
35. Markham NR, Zuker M (2008) UNAFold: software for nucleic acid folding and hybridization. Methods Mol Biol 453:3–31. https://doi.org/10.1007/978-1-60327-429-6_1
36. Paz I, Kosti I, Ares M Jr et al (2014) RBPmap: a web server for mapping binding sites of RNA-binding proteins. Nucleic Acids Res 42: W361–W367. https://doi.org/10.1093/nar/gku406
37. Pedersen L, Hagedorn PH, Lindholm MW et al (2014) A kinetic model explains why shorter and less affine enzyme-recruiting oligonucleotides can be more potent. Mol Ther Nucleic Acids 3:e149. https://doi.org/10.1038/mtna.2013.72
38. Tolstrup N, Nielsen PS, Kolberg JG et al (2003) OligoDesign: optimal design of LNA (locked nucleic acid) oligonucleotide capture probes for gene expression profiling. Nucleic Acids Res 31(13):3758–3762
39. Moreno PM, Ferreira AR, Salvador D et al (2018) Hydrogel-assisted antisense LNA Gapmer delivery for in situ gene silencing in spinal cord injury. Mol Ther Nucleic Acids 11:393–406
40. Leti F, DiStefano J (2017) Long noncoding RNAs as diagnostic and therapeutic targets in type 2 diabetes and related complications. Genes 8(8):207
41. Chung S, Nakagawa H, Uemura M et al (2011) Association of a novel long non-coding RNA in 8q24 with prostate cancer susceptibility. Cancer Sci 102(1):245–252
42. Guan Y, Kuo W-L, Stilwell JL et al (2007) Amplification of PVT1 contributes to the pathophysiology of ovarian and breast cancer. Clin Cancer Res 13(19):5745–5755
43. Panzitt K, Tschernatsch MM, Guelly C et al (2007) Characterization of HULC, a novel gene with striking up-regulation in hepatocellular carcinoma, as noncoding RNA. Gastroenterology 132(1):330–342
44. Pibouin L, Villaudy J, Ferbus D et al (2002) Cloning of the mRNA of overexpression in

colon carcinoma-1: a sequence overexpressed in a subset of colon carcinomas. Cancer Genet Cytogenet 133(1):55–60
45. Bussire T, Gold G, Kvari E et al (2003) Stereologic analysis of neurofibrillary tangle formation in prefrontal cortex area 9 in aging and Alzheimer's disease. Neuroscience 117 (3):577–592
46. Lukiw W, Handley P, Wong L et al (1992) BC200 RNA in normal human neocortex, non-Alzheimer dementia (NAD), and senile dementia of the Alzheimer type (AD). Neurochem Res 17(6):591–597
47. Nair M, Sagar V, Pilakka-Kanthikeel S (2016) Gene-expression reversal of lncRNAs and associated mRNAs expression in active vs latent HIV infection. Sci Rep 6:34862
48. Zhang Q, Chen C-Y, Yedavalli VS et al (2013) NEAT1 long noncoding RNA and paraspeckle bodies modulate HIV-1 posttranscriptional expression. MBio 4(1):e00596–e00512
49. Ishii N, Ozaki K, Sato H et al (2006) Identification of a novel non-coding RNA, MIAT, that confers risk of myocardial infarction. J Hum Genet 51(12):1087
50. Gutschner T, Hämmerle M, Diederichs S (2013) MALAT1—a paradigm for long noncoding RNA function in cancer. J Mol Med 91 (7):791–801
51. Liu J, Yao J, Li X et al (2014) Pathogenic role of lncRNA-MALAT1 in endothelial cell dysfunction in diabetes mellitus. Cell Death Dis 5 (10):e1506
52. Arun G, Diermeier SD, Spector DL (2018) Therapeutic targeting of long non-coding RNAs in cancer. Trends Mol Med 24 (3):257–277
53. Amodio N, Stamato MA, Juli G et al (2018) Drugging the lncRNA MALAT1 via LNA gapmeR ASO inhibits gene expression of proteasome subunits and triggers anti-multiple myeloma activity. Leukemia 32(9):1948
54. Castillo AS, Mateos CO, Riera MS et al (2019) Non-coding RNAs as key players for understanding oncogenic pathways in ovarian cancer: studying new therapeutic strategies. AACR, Philadelphia
55. Woo CJ, Maier VK, Davey R et al (2017) Gene activation of SMN by selective disruption of lncRNA-mediated recruitment of PRC2 for the treatment of spinal muscular atrophy. Proc Natl Acad Sci U S A 114(8):E1509–E1518
56. Goodkey K, Aslesh T, Maruyama R et al (2018) Nusinersen in the treatment of spinal muscular atrophy. In: Exon skipping and inclusion therapies. Springer, New York, pp 69–76
57. Viereck J, Kumarswamy R, Foinquinos A et al (2016) Long noncoding RNA Chast promotes cardiac remodeling. Sci Transl Med 8 (326):326ra322
58. Jiang D, Liang J (2019) A lncRNA Links TGF-β Signaling in Lung Fibrosis. Am J Respir Crit Care Med 200(2):123–125
59. Echigoya Y, Nakamura A, Nagata T et al (2017) Effects of systemic multiexon skipping with peptide-conjugated morpholinos in the heart of a dog model of Duchenne muscular dystrophy. Proc Natl Acad Sci U S A 114 (16):4213–4218. https://doi.org/10.1073/pnas.1613203114
60. Lu QL, Rabinowitz A, Chen YC et al (2005) Systemic delivery of antisense oligoribonucleotide restores dystrophin expression in body-wide skeletal muscles. Proc Natl Acad Sci U S A 102(1):198–203. https://doi.org/10.1073/pnas.0406700102
61. Ezzat K, Aoki Y, Koo T et al (2015) Self-assembly into nanoparticles is essential for receptor mediated uptake of therapeutic antisense oligonucleotides. Nano Lett 15 (7):4364–4373. https://doi.org/10.1021/acs.nanolett.5b00490
62. Walker I, Irwin WJ, Akhtar S (1995) Improved cellular delivery of antisense oligonucleotides using transferrin receptor antibody-oligonucleotide conjugates. Pharm Res 12 (10):1548–1553
63. Echigoya Y, Aoki Y, Miskew B et al (2015) Long-term efficacy of systemic multiexon skipping targeting dystrophin exons 45-55 with a cocktail of vivo-morpholinos in mdx52 mice. Mol Ther Nucleic Acids 4:e225. https://doi.org/10.1038/mtna.2014.76

Chapter 4

Development of Antisense Oligonucleotide Gapmers for the Treatment of Huntington's Disease

Tejal Aslesh and Toshifumi Yokota

Abstract

The field of neuromuscular and neurodegenerative diseases has been revolutionized by the advent of genetics and molecular biology to evaluate the pathogenicity, thereby providing considerable insight to develop suitable therapies. With the successful translation of antisense oligonucleotides (AOs) from in vitro into animal models and clinical practice, modifications are being continuously made to the AOs to improve the pharmacokinetics and pharmacodynamics. In order to activate RNase H-mediated cleavage of the target mRNA, as well as to increase the binding affinity and specificity, gapmer AOs are designed to have a phosphorothioate (PS) backbone flanked with the modified AOs on both sides. Antisense-mediated knockdown of mutated huntingtin is a promising therapeutic approach for Huntington's disease (HD), a devastating disorder affecting the motor and cognitive abilities. This chapter focuses on the modified gapmer AOs for the treatment of HD.

Key words Huntingtin (HTT), Trinucleotide repeat disorders, 2′-*O*-methoxyethyl (MOE), Gapmer, HTT_{RX} (Alternative Names: HTT ASO; IONIS-HTTRx; ISIS-443139; ISIS-HTT Rx; RG 6042; RO7234292), WVE-120101/WVE-120102, Nusinersen (brand name Spinraza), Eteplirsen (brand name Exondys 51), Zinc-finger proteins (ZFPs), Short interfering RNA (siRNA)

1 Introduction

To date, significant technological advancements have been made in understanding the genetic causes underlying various neurodegenerative disorders. There is a shift from simply treating the symptoms to addressing the root cause of the disease, based on the progress made in the genetic diagnosis field and tools to directly address genetic defects [1]. Over the past decade, efforts are being implemented to develop antisense oligonucleotide (AO)-based therapies to treat neuromuscular diseases based on the genetic insights associated with the disease pathophysiology [1]. With the approval of AO therapy with nusinersen and eteplirsen for the treatment of spinal muscular atrophy (SMA) and Duchenne muscular dystrophy

Toshifumi Yokota and Rika Maruyama (eds.), *Gapmers: Methods and Protocols*, Methods in Molecular Biology, vol. 2176,
https://doi.org/10.1007/978-1-0716-0771-8_4,

(DMD), several modified AOs are now being evaluated in clinical trials for the treatment of a number of neurological disorders [2].

Huntington's disease (HD) is a devastating neurodegenerative disorder characterized by the decline in the cognitive and motor abilities due to the expansion of a polyglutamine-encoding cytosine adenine guanine (CAG) tract in exon 1 of the *HTT* gene [3]. Several non-AO therapies, such as the use of zinc-finger proteins (ZFPs) and short interfering RNA (siRNA), are being pursued and tested to reduce the mutant Huntingtin (mHTT) expression as well as to attenuate the HTT toxic gain-of-function [4, 5]. Wild-type HTT is ubiquitously expressed and plays an important role in transcriptional regulation [6]. Hence, a strategy deploying antisense therapy targeting the mHTT alone seems to be an ideal candidate for the treatment of HD.

This chapter aims to highlight the advancements in AO therapy for HD that aims to ameliorate the devastating effects of the neurocognitive deterioration and improve the quality of life of the patients.

2 Huntington's Disease

2.1 Pathology

HD is an autosomal neurodegenerative disorder, which is monogenetic in nature and is clinically characterized by dystonia, hyperkinetic involuntary movements (chorea), and a decline in cognitive functions [3]. The gradual loss of spiny neurons in the striatum, coupled with progressive atrophy of the caudate nucleus, the putamen, and the globus pallidus is the hallmark of HD [3]. As the disease worsens, the atrophy rapidly progresses to other parts of the brain such as the cerebrum and hippocampus [3]. The spiny neurons present in the striatum of the brain are the most affected and considered to be a harbinger of HD [3]. Ultimately, the HD brain becomes much smaller than a normal brain due to the deterioration as well as the reduced volume of the striatum and other parts of the brain [3].

HD is caused by the expansion of a polyglutamine-encoding cytosine adenine guanine (CAG) tract in exon 1 of the *HTT* gene [3]. The protein HTT is expressed ubiquitously and plays an active role in transcriptional regulation, suppression of apoptosis, oxidative stress, and synaptic transmission [6]. Mutant huntingtin (mHTT) arises from this CAG expansion and gains toxic functions such as intraneuronal aggregation and abnormal interactions with other proteins, depending on the number of CAG repeats present [3, 6]. In the normal population, the repeat number is 6–35; alleles with 36–39 CAG repeats may exhibit reduced penetrance [6]. The juvenile form of HD shows the presence of 56–120 CAG repeats, while HD with adult-onset typically has around 39–55 CAG repeats [6].

2.2 Standard Treatment

Ever since the pathology of HD was described in 1872, tremendous effort has been spent in developing a successful disease-modifying treatment. However, the only form of therapy available is in the form of supportive and palliative care [2]. Several groups have made painstaking efforts to develop RNA silencing therapies such as short hairpin RNAs (shRNAs) or microRNAs (miRNAs) to suppress the expression of *HTT* [7–9]. Since wild-type *HTT* is ubiquitously expressed and has many functions, a strategy to inhibit the expression of mHTT only would be of more clinical interest.

3 Modification of Antisense Oligonucleotides

AOs are single-stranded sequences made up of modified nucleic acids (8–50 bases) [10]. AOs can be classified as those that promote RNA degradation by recruiting RNase H enzyme (i.e., gapmers) and those that do not promote RNA degradation (e.g., splice-switching AOs) [2]. An illustration of this classification is seen in Fig. 1 [11]. Splice-switching AOs bind to the target mRNA and can redirect the splicing machinery to either repress or activate the expression of a particular gene [2]. For clinical applications, the use of unmodified oligonucleotides (e.g., DNA) was difficult since they were rapidly degraded by the endonucleases and activate immune response [12]. The creation of a phosphorothioate (PS) backbone by the addition of a sulphur atom to the phosphate backbone of the deoxyribose ring enhanced the nuclease resistance and efficient protein-binding in the first generation of AO modification as seen in Fig. 2a [13, 14]. The second-generation AOs with 2′ modifications such as 2′-*O*-methyl (2′-OMe) and 2′-*O*-methoxyethyl (2′-MOE) are resistant to nucleases and demonstrate increased binding affinity, but do not activate RNase H, as seen in Fig. 2a [2, 14, 15]. This was the driving force for the development of chimeric AOs in which the central region of a DNA with PS backbone is flanked on both sides by nuclease-resistant arms consisting of the modified AO [2, 16]. The resulting "gapmer" ensures the RNase H-mediated cleavage of the target mRNA at the central region, whereas the flanking ends provide nuclease resistance and binding affinity. An illustration of a gapmer is provided in Fig. 2b [2, 15, 17, 18].

4 Antisense Therapy for Huntington's Disease

A breakthrough to treat HD occurred with the development of 20-mer MOE, an AO gapmer with a phosphorothioate backbone modification that binds to the target human huntingtin mRNA [7]. When infused continuously through the central nervous system (CNS) for 2 weeks (10, 25, or 50 μg/day) through the right

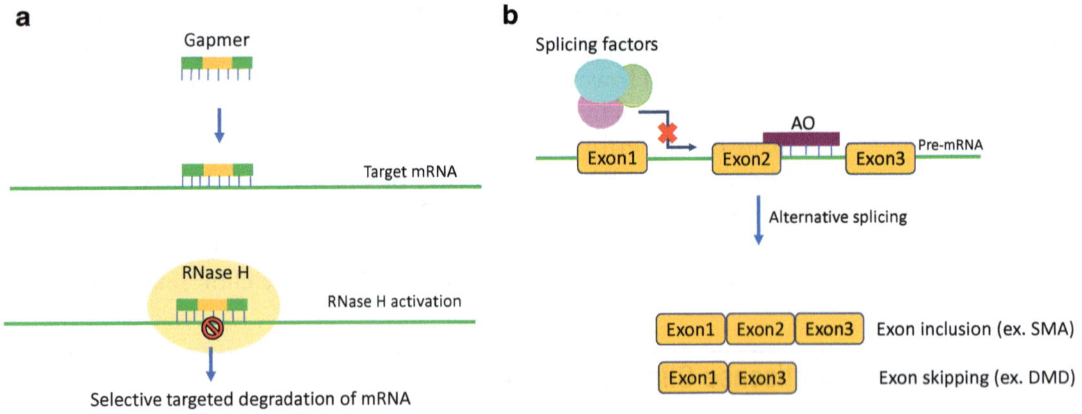

Fig. 1 Mechanism of action of AOs. Antisense oligonucleotides (AOs) interact with the target RNA in either of the 2 mechanisms. (**a**) Target degradation of the target by activating ribonuclease H (RNase H)-mediated activity (e.g., gapmers). (**b**) Steric blocking of splicing leading to splice switching leading to either inclusion or skipping of exons (e.g., MOE)

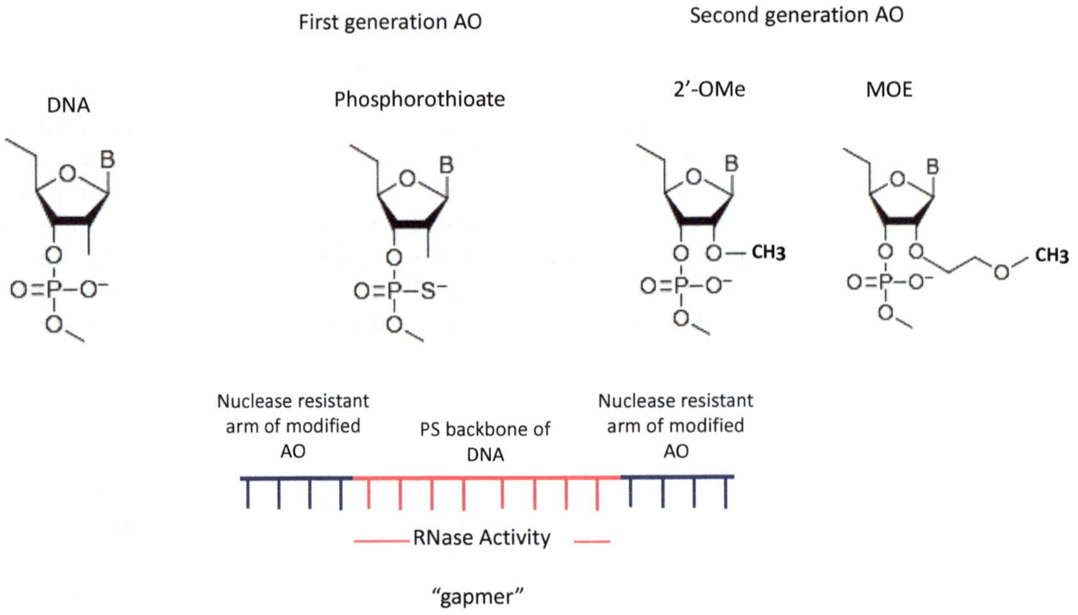

Fig. 2 Evolution of AOs. (**Upper Panel**) The first-generation AO was created by the addition of a sulphur atom to the phosphate backbone of the deoxyribose ring, leading to the formation of a phosphorothioate (PS) backbone. The second-generation AOs had an addition of a methyl (2′-OMe) and a methoxy group (MOE) to increase nuclease resistance and binding affinity, but do not promote RNase H activity. (**Lower Panel**) A chimeric molecule (gapmer) created by a central region of a DNA with PS backbone, flanked on both the sides by nuclease-resistant arms consisting of the modified AO

ventricle of the BACHD mouse model expressing a neuropathogenic, full-length human mutant Huntingtin (*mHTT*), it effectively suppressed the accumulated HTT in a dose-dependent manner and reversed the HD phenotype [7]. Remarkably, they also found that simultaneous suppression of both wtHTT and mHTT did not attenuate the benefits of AO-mediated mHTT suppression [7]. In addition to this, 2-month-old YAC128 mice expressing a full-length *mHTT* transgene also responded positively to the 2-week ASO infusion with improved motor function. ASO treatment also provided sustained suppression of mHTT in older YAC128 mice treated at 6 months of age that reversed some behavioral characteristics evaluated 3 months after stopping the treatment in three different mouse models of HD with varying severity of the phenotype [7]. They also assessed the efficacy of AO treatment in a fatal HD model called R6/2 that survives only 16 weeks [7, 19]. ASO treatment with 50 μg/day for 4 weeks suppressed the human HTT protein expression [7]. In addition to this, MOE infusion ensured a widespread distribution in a nonhuman primate (NHP) brain when administered intrathecally, providing sustained suppression of *HTT* mRNA in different regions associated with HD pathology [7]. In spite of the successful story underlying the first HD preclinical studies, questions concerning the suppression of wtHTT and its underlying effects in the adult nervous system still remained.

In order to preserve the wtHTT, allele-selective mHTT suppression can be achieved by targeting the CAG expansion tract. However, this expansion also occurs in several other transcripts and their downregulation would have deleterious effects [20]. Previous studies have suggested that genetic inactivation of the mHTT in a mouse model of HD reversed the pathology and also improved motor performance [21, 22]. The mutant allele selectivity can be achieved by targeting single nucleotide polymorphisms (SNPs) associated with the CAG expansion. Population studies have identified three to five SNPs linked to the CAG expansion that would be ideal targets for allele-selective ASOs for 80–90% of HD patients globally [23–28]. The most common haplogroups, A and B, were found on CAG-expanded chromosomes in Latin American, European, Caucasian, and South African populations which led to the identification of two promising allele-specific ASOs targeting two SNPs that would suppress 50% of the mHTT [23–28]. A 16 nucleotide MOE, in particular, proved to be successful in ameliorating the cognitive and psychiatric deficits associated with HD pathology in a recently developed humanized mouse model of HD (Hu97/18) when infused into the CNS at a later stage of treatment (6 months of age) [21]. Studies carried out by Southwell et al. emphasized on mHTT suppression, which managed to have benefits in several tests, proving its potency and the benefits of allelic specificity without significant suppression of wtHTT [21].

In addition to the nonselective gapmer approach developed by Ionis Pharmaceuticals and Roche, WAVE Life Sciences recently developed stereopure antisense oligonucleotides that selectively target the *mHTT* mRNA that led to the selective inhibition of mHTT production. As opposed to the conventional AOs with PS-backbone modifications comprise of nucleotides that differ from each other in terms of stereochemistry, which are thereby heterogeneous and exhibit uncontrolled pharmacologic properties [29], stereopure AOs have stable PS-linkages with optimized and desirable pharmacologic properties [30]). WAVE's candidates target 2 different single nucleotide morphisms (SNPs) common to mHTT [31]. About two-thirds of the HD patients carry either one of the two markers SNP1 or SNP2 that identify the *mHTT* gene [31, 32]. The candidate stereopure AO WVE-120101 targets patients with SNP1 (rs362307), while WVE-120102 is designed to treat patients with SNP2 (rs362331) [31]. The allele-selective AO WVE-120101 has received an orphan drug designation in the USA by the FDA for the treatment of HD [31]. This allele-selective approach enables the silencing of only the mHTT, leaving the healthy wtHTT to produce a functional protein. The sequence and chemistry of the stereopure AOs, including the composition and whether these are gapmers or not, has not been disclosed by WAVE Life Sciences.

5 Clinical Trials

An allele-nonselective gapmer AO to lower HTT called IONIS HTTRx, developed by Ionis Pharmaceuticals (Alternative Names: HTT ASO; IONIS-HTTRx; ISIS-443139; ISIS-HTT Rx; RG 6042; RO7234292) was approved for the human-phase clinical trial 1b/2a in December 2017 to assess the safety and tolerability of its ascending doses administered intrathecally [33]. The ASO, a 20-mer MOE with a PS backbone gapmer chemistry RG6042 (RO7234292; formerly known as IONIS-HTT_{RX}) reached its primary endpoint of safety that was assessed in the clinical trial NCT02519036 [33]. The safety evaluations comprised of physical and neurological examinations, the Columbia Suicide Severity Rating Scale, electrocardiograms, and neuroimaging [34]. The secondary endpoint focused on the pharmacokinetics of RG6042 in the cerebrospinal fluid (CSF) as well as the plasma. Forty-six patients that were enrolled in the study received ISIS 443139 at different doses (10 mg, 30 mg, 60 mg, 90 mg, and 120 mg) through intrathecal injections at 4-week intervals over the course of a 13-week treatment period [34]. According to the results posted by ClinicalTrials.gov, a dose-dependent reduction of mHTT was observed in HD patients with no serious adverse effects. Post-completion of this trial, the 46 participants were given the opportunity to enroll in a 15-month open-label extension study to evaluate the effects of the 120 mg intrathecal administration of HTT_{RX} on a monthly or bimonthly basis (NCT03342053) [2]. Additionally, a phase 3 open-label extension study (sponsored by Roche/Genentech) is currently active and has recruited 950 participants so far. The aim of this intervention study is to assess the long-term safety and tolerability of intrathecally administered RG6042 every 8 or 16 weeks in patients with HD. The anticipated completion date is November 18, 2023. These large-scale studies for an extended duration will be crucial in determining whether RG6042-mediated mHTT reduction in the CSF is associated with a treatment effect as the disease progresses over time.

PRECISION-HD1 (NCT03225833)-an interventional, multicenter, randomized, double-blind, placebo-controlled phase Ib/IIa trial was initiated in July 2017 to test the safety and tolerability of intrathecally administered WVE-120101 in patients with HD. The trial has enrolled 48 patients with the target SNP1 on the same allele so far and is currently recruiting participants. The secondary outcome of this study is to study the pharmacokinetics, pharmacodynamics, and clinical effects based on the Unified Huntington's Disease Rating Scale (UHDRS). The outcome following the administration of the stereopure allele-specific AO is expected to be more efficient than the allele nonselective approach. Table 1 summarizes the clinical trials for the AOs tested for HD.

6 Challenges and Future Perspectives

Gapmers have made significant advancements in research and therapeutics, as demonstrated by approval for the treatment of several disorders and those that are currently being tested in clinical trials, including HD. In spite of their success, challenges persist that can limit efficiency. The major obstacles include limited uptake in the target tissues and toxicity. Gapmer toxicity arises partly due to hybridization onto the nontargets, leading to off-target effects [35]. These off-target issues can be minimized by using bioinformatics and in silico tools that can improve the gapmer design. Renal and hepatotoxicity have been reported due to the accumulation of gapmer AOs in the kidney and liver, respectively [36]. This can be further accompanied by inflammation and degeneration of the tissues because of the activation of the inflammatory players like the T-cells, macrophages, and microglia.

To offset these potential drawbacks, efforts are being implemented to design novel AO modifications that can enhance the availability of the AO gapmers to the target tissues. Two approaches can be implemented: (1) Modification of the AO itself, or by conjugating the AOs to cell-penetrating peptides (CPPs). With one such conversion of a single nucleotide in the central gap from a PS-DNA to a PS-2′-OMe residue, gapmers were made less toxic [37]. (2) Incorporation of the AO into carrier moieties such as nanoparticle polymers or antibodies that can direct the AO in the

Table 1
Clinical status of gapmer AO RG6042 (RO7234292; formerly known as IONIS-HTT$_{RX}$) and stereopure AO WVE-120101

Clinical trials identifier	Summary of the study	Study design	Intervention/ Treatment	Status
NCT02519036	Tested safety, tolerability, pharmacokinetics, and pharmacodynamics of several doses	46 participants in an interventional study that was a randomized, double-blind, placebo-controlled study	Intrathecal administration of RG6042 at various doses −10 mg, 30 mg, 60 mg, and 120 mg at 4-week intervals for a 13-week period	Phase I and II completed as of May 31, 2019. A dose-dependent reduction of mHTT was observed in HD patients with no serious adverse effects
NCT03342053	Currently testing the safety, tolerability, pharmacokinetics, and pharmacodynamics of RG6042 administered intrathecally to adult patients with Huntington's disease	46 participants who previously participated in the RG6042 study were enrolled in a randomized, open-label extension study	Intrathecal administration of RG6042 every 4 weeks for 14 months	Active as of October 30, 2019
NCT03842969	Evaluating the long-term safety and tolerability of intrathecally administered RG6042 in patients with HD	A randomized open-label extension study (interventional study), with 950 participants recruited	Intrathecal bolus administration of RG6042 every 8 weeks or every 16 weeks	Recruiting (expected date of completion is November 18, 2023)
NCT03225833 (PRECISION-HD1)	Currently evaluating the safety, tolerability, pharmacokinetics, and pharmacodynamics of WVE-120101 administered intrathecally to adult patients with Huntington's disease	48 participants in an interventional study that was a randomized, double-blind, placebo-controlled study	Intrathecal administration of WVE-120101	Phase Ib/IIa currently active and recruiting participants (as of July 10, 2019)

target tissue and determine its localization within the cellular compartment [13].

To date, significant advancements have been made and continue to grow to make developments to the field of AO chemistry for the effective treatment of a number of genetic diseases and manifestations.

7 Conclusion

The field of antisense therapy is evolving, with clinical developments made for approximately 30 years. ASOs are promising for neurodegenerative and neuromuscular disorders, as evidenced by the approval of nusinersen and eteplirsen for the treatment of SMA and DMD, respectively. Numerous ASOs with modified chemistries, including those of gapmers are now available or are being tested in clinical trials for the treatment of many genetic diseases, beyond the scope of neurodegenerative disorders.

In spite of the debate over how well the HD mouse models recapitulate the neurodegenerative diseases, the HD studies in preclinical mouse models provided the rationale to pursue AO treatment for HD. The AO RG6042, a 20-mer gapmer MOE, has successfully reached its primary and secondary endpoints of safety and a dose-dependent reduction of mHTT with no associated severe adverse effects. With several allele-specific MOE gapmers being currently evaluated in preclinical studies, the field of AO therapy for the treatment of HD continues to expand.

Acknowledgments

We would like to thank our funding sources, namely the University of Alberta, the Women and Children's Health Research Institute, The Friends of Garrett Cumming Research Fund, Muscular Dystrophy Canada, the HM Toupin Neurological Science Research Fund, the Canadian Institutes of Health Research Foundation Grant 143251, and Jesse's Journey—Foundation for Gene and Cell Therapy.

References

1. Lane RM, Smith A, Baumann T et al (2018) Translating antisense technology into a treatment for Huntington's disease. Methods Mol Biol 1780:497–523. https://doi.org/10.1007/978-1-4939-7825-0_23
2. Wurster CD, Ludolph AC (2018) Antisense oligonucleotides in neurological disorders. Ther Adv Neurol Disord 11:1756286418776932. https://doi.org/10.1177/1756286418776932
3. Vonsattel JP, Keller C, Cortes Ramirez EP (2011) Huntington's disease—neuropathology. Handb Clin Neurol 100:83–100. https://doi.org/10.1016/B978-0-444-52014-2.00004-5

4. Zielonka D, Mielcarek M, Landwehrmeyer GB (2015) Update on Huntington's disease: advances in care and emerging therapeutic options. Parkinsonism Relat Disord 21 (3):169–178. https://doi.org/10.1016/j.parkreldis.2014.12.013
5. Wild EJ, Tabrizi SJ (2014) Targets for future clinical trials in Huntington's disease: what's in the pipeline? Mov Disord 29(11):1434–1445. https://doi.org/10.1002/mds.26007
6. Zuccato C, Valenza M, Cattaneo E (2010) Molecular mechanisms and potential therapeutical targets in Huntington's disease. Physiol Rev 90(3):905–981. https://doi.org/10.1152/physrev.00041.2009
7. Kordasiewicz HB, Stanek LM, Wancewicz EV et al (2012) Sustained therapeutic reversal of Huntington's disease by transient repression of huntingtin synthesis. Neuron 74 (6):1031–1044. https://doi.org/10.1016/j.neuron.2012.05.009
8. Franich NR, Fitzsimons HL, Fong DM et al (2008) AAV vector-mediated RNAi of mutant huntingtin expression is neuroprotective in a novel genetic rat model of Huntington's disease. Mol Ther 16(5):947–956. https://doi.org/10.1038/mt.2008.50
9. Harper SQ, Staber PD, He X et al (2005) RNA interference improves motor and neuropathological abnormalities in a Huntington's disease mouse model. Proc Natl Acad Sci U S A 102 (16):5820–5825. https://doi.org/10.1073/pnas.0501507102
10. Schoch KM, Miller TM (2017) Antisense oligonucleotides: translation from mouse models to human neurodegenerative diseases. Neuron 94(6):1056–1070. https://doi.org/10.1016/j.neuron.2017.04.010
11. Shen X, Corey DR (2018) Chemistry, mechanism and clinical status of antisense oligonucleotides and duplex RNAs. Nucleic Acids Res 46(4):1584–1600. https://doi.org/10.1093/nar/gkx1239
12. Dagle JM, Weeks DL, Walder JA (1991) Pathways of degradation and mechanism of action of antisense oligonucleotides in *Xenopus laevis* embryos. Antisense Res Dev 1(1):9
13. Juliano RL (2016) The delivery of therapeutic oligonucleotides. Nucleic Acids Res 44 (14):6518–6548. https://doi.org/10.1093/nar/gkw236
14. Rinaldi C, Wood MJA (2018) Antisense oligonucleotides: the next frontier for treatment of neurological disorders. Nat Rev Neurol 14 (1):9–21. https://doi.org/10.1038/nrneurol.2017.148
15. Bennett CF, Baker BF, Pham N et al (2017) Pharmacology of antisense drugs. Annu Rev Pharmacol Toxicol 57:81–105. https://doi.org/10.1146/annurev-pharmtox-010716-104846
16. Sardone V, Zhou H, Muntoni F et al (2017) Antisense oligonucleotide-based therapy for neuromuscular disease. Molecules 22(4): E563. https://doi.org/10.3390/molecules22040563
17. Eckstein F (1970) Nucleoside phosphorothioates. J Am Chem Soc 92(15):4718–4723
18. Stephenson ML, Zamecnik PC (1978) Inhibition of Rous sarcoma viral RNA translation by a specific oligodeoxyribonucleotide. Proc Natl Acad Sci U S A 75(1):285–288
19. Mangiarini L, Sathasivam K, Seller M et al (1996) Exon 1 of the HD gene with an expanded CAG repeat is sufficient to cause a progressive neurological phenotype in transgenic mice. Cell 87(3):493–506
20. Ostergaard ME, Southwell AL, Kordasiewicz H et al (2013) Rational design of antisense oligonucleotides targeting single nucleotide polymorphisms for potent and allele selective suppression of mutant huntingtin in the CNS. Nucleic Acids Res 41(21):9634–9650. https://doi.org/10.1093/nar/gkt725
21. Southwell AL, Kordasiewicz HB, Langbehn D et al (2018) Huntingtin suppression restores cognitive function in a mouse model of Huntington's disease. Sci Transl Med 10(461): eaar3959. https://doi.org/10.1126/scitranslmed.aar3959
22. Yamamoto A, Lucas JJ, Hen R (2000) Reversal of neuropathology and motor dysfunction in a conditional model of Huntington's disease. Cell 101(1):57–66. https://doi.org/10.1016/S0092-8674(00)80623-6
23. Warby SC, Visscher H, Collins JA et al (2011) HTT haplotypes contribute to differences in Huntington disease prevalence between Europe and East Asia. Eur J Hum Genet 19 (5):561–566. https://doi.org/10.1038/ejhg.2010.229
24. Pfister EL, Kennington L, Straubhaar J et al (2009) Five siRNAs targeting three SNPs may provide therapy for three-quarters of Huntington's disease patients. Curr Biol 19 (9):774–778. https://doi.org/10.1016/j.cub.2009.03.030
25. Kay C, Tirado-Hurtado I, Cornejo-Olivas M et al (2017) The targetable A1 Huntington disease haplotype has distinct Amerindian and European origins in Latin America. Eur J Hum Genet 25(3):332–340. https://doi.org/10.1038/ejhg.2016.169

26. Warby SC, Montpetit A, Hayden AR et al (2009) CAG expansion in the Huntington disease gene is associated with a specific and targetable predisposing haplogroup. Am J Hum Genet 84(3):351–366. https://doi.org/10.1016/j.ajhg.2009.02.003
27. Kay C, Collins JA, Skotte NH et al (2015) Huntingtin haplotypes provide prioritized target panels for allele-specific silencing in Huntington disease patients of European ancestry. Mol Ther 23(11):1759–1771. https://doi.org/10.1038/mt.2015.128
28. Baine FK, Kay C, Ketelaar ME et al (2013) Huntington disease in the south African population occurs on diverse and ethnically distinct genetic haplotypes. Eur J Hum Genet 21 (10):1120–1127. https://doi.org/10.1038/ejhg.2013.2
29. Wave Life Sciences. (2016) Wave Life Sciences-Platform. https://www.wavelifesciences.com/science/platform/. Accessed November 8 2019
30. Iwamoto N, Butler DCD, Svrzikapa N et al (2017) Control of phosphorothioate stereochemistry substantially increases the efficacy of antisense oligonucleotides. Nat Biotechnol 35 (9):845–851. https://doi.org/10.1038/nbt.3948
31. WAVE Life Sciences Receives Orphan Drug Designation from FDA for its Lead Candidate Designed to Treat Huntington's Disease. (2016). https://hdsa.org/news/wave-life-sciences-receives-orphan-drug-designation-from-fda-for-its-lead-candidate-designed-to-treat-huntingtons-disease/. Accessed November 8 2019
32. Achey MA (2018) Meet the Compounds: WVE-120101 and WVE-120102. https://huntingtonstudygroup.org/hd-insights/meet-the-compounds-wve-120101-and-wve-120102/. Accessed November 8 2019
33. Rodrigues FB, Wild EJ (2018) Huntington's disease clinical trials corner: august 2018. J Huntingtons Dis 7(3):279–286. https://doi.org/10.3233/JHD-189003
34. Tabrizi SJ, Leavitt BR, Landwehrmeyer GB et al (2019) Targeting huntingtin expression in patients with Huntington's disease. N Engl J Med 380(24):2307–2316. https://doi.org/10.1056/NEJMoa1900907
35. Bennett CF, Swayze EE (2010) RNA targeting therapeutics: molecular mechanisms of antisense oligonucleotides as a therapeutic platform. Annu Rev Pharmacol Toxicol 50:259–293. https://doi.org/10.1146/annurev.pharmtox.010909.105654
36. Yu RZ, Kim TW, Hong A et al (2007) Cross-species pharmacokinetic comparison from mouse to man of a second-generation antisense oligonucleotide, ISIS 301012, targeting human apolipoprotein B-100. Drug Metab Dispos 35(3):460–468. https://doi.org/10.1124/dmd.106.012401
37. Shen W, De Hoyos CL, Migawa MT et al (2019) Chemical modification of PS-ASO therapeutics reduces cellular protein-binding and improves the therapeutic index. Nat Biotechnol 37(6):640–650

Chapter 5

Development of Antisense Oligonucleotide Gapmers for the Treatment of Dyslipidemia and Lipodystrophy

Tejal Aslesh and Toshifumi Yokota

Abstract

Although technological advances in molecular genetics over the last few decades have greatly expedited the identification of mutations in many genetic diseases, the translation of the genetic mechanisms into a clinical setting has been quite challenging, with a minimum number of effective treatments available. The advancements in antisense therapy have revolutionized the field of neuromuscular disorders as well as lipid-mediated diseases. With the approval of splice-switching antisense oligonucleotide (AO) therapy for nusinersen and eteplirsen for the treatment of spinal muscular atrophy (SMA) and Duchenne muscular dystrophy (DMD), several modified AOs are now being evaluated in clinical trials for the treatment of a number of disorders. In order to activate RNase H-mediated cleavage of the target mRNA, as well as to increase the binding affinity and specificity, gapmer AOs are designed that have a PS backbone flanked with the modified AOs on both sides. Mipomersen (trade name Kynamro), a 2′-*O*-methoxyethyl (MOE) gapmer, was approved by the Food and Drug Administration (FDA) for the treatment of homozygous familial hypercholesterolemia (HoFH) in 2013. Volanesorsen, another 20-mer MOE gapmer has shown to be successful in lowering the levels of triglycerides (TGs) in several lipid disorders and has received conditional approval in the European Union for the treatment of Familial chylomicronemia syndrome (FCS) in May 2019 following successful results from phase II/III clinical trials. This chapter focuses on the clinical applications of gapmer AOs for genetic dyslipidemia and lipodystrophy.

Key words 2′-*O*-methoxyethyl (MOE), Low-density lipoprotein (LDL), Apolipoprotein B-100 (ApoB-100), Homozygous familial hypercholesterolemia, Hypertriglyceridemia, Familial chylomicronemia, Familial partial lipodystrophy, Nusinersen (brand name Spinraza), Eteplirsen (brand name Exondys 51), Cardiovascular diseases (CVD)

1 Introduction

Considerable advancements have been made toward understanding the genetic causes and pathogenicity of many rare diseases, including neurodegenerative disorders, muscular dystrophies, and lipid-mediated disorders [1]. However, the translation of this understanding into a clinical setting has been quite challenging and effective treatments are very few in number [1]. In 2016, eteplirsen (brand name Exondys 51), an antisense oligonucleotide

Toshifumi Yokota and Rika Maruyama (eds.), *Gapmers: Methods and Protocols*, Methods in Molecular Biology, vol. 2176, https://doi.org/10.1007/978-1-0716-0771-8_5, © Springer Science+Business Media, LLC, part of Springer Nature 2020

(AO) for the treatment of Duchenne muscular dystrophy (DMD), was approved by the Food and Drug Administration (FDA) [2]. In addition, nusinersen (brand name Spinraza) a 2'*O*-methoxyethyl (MOE) AO was approved by the FDA in December 2016 for the treatment of spinal muscular atrophy (SMA) [1, 3]. Antisense oligonucleotide technology would open a new and exciting era in the treatment of genetic diseases. AOs are small DNA or modified nucleotide sequences made up of 8-50 bases [4]. The structure facilitates the complementary binding to the target mRNA by Watson–Crick base pairing [4]. The most common modification made to the phosphate backbone is the replacement of the single-bonded oxygen moiety with a sulfur ion, creating a phosphorothioate (PS) linkage [4]. The PS backbone enhances the binding affinity, increases the cellular uptake, and provides resistance against degradation by nucleases such as RNase-H [4, 5]. The most common AOs that have a PS backbone are the 2′-*O*-methyl (2′-*O*-Me) and MOE [4, 6]. These AOs do not activate RNase H but support non-degrading AO mechanisms such as splice switching or translational inhibition [4, 6]. In order to mediate the activation of RNase H enzyme and subsequent cleavage of the target mRNA, the AOs with the PS backbones were modified, thereby creating "gapmers" [4]. These chimeric AOs comprise 2′ modified rings that are separated by a central region of unmodified nucleotides that can mediate the RNase degradation. On the other hand, the 2 flanking ends of the gapmer are resistant to the degradation and enhance the binding affinity as seen in Fig. 1 [1, 4, 7].

The MOE gapmers are now being developed and tested extensively in several clinical trials for several diseases such as Huntington's disease and lipid-related disorders [8] In this chapter, we will be looking at a comprehensive overview of the advancements of gapmer AOs to treat dyslipidemia and lipodystrophy. A list of the AOs mentioned in this chapter is provided in Table 1.

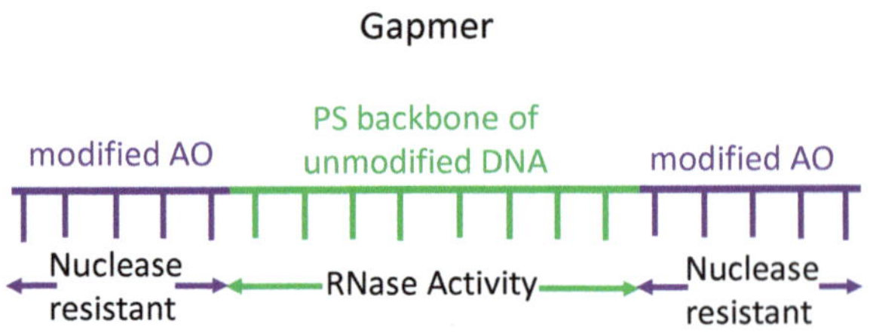

Fig. 1 Creation of a gapmer. A gapmer consists of a central region of deoxyribose rings with PS backbone, flanked on both sides by nuclease-resistant arms consisting of the modified AO

Table 1
List of gapmer AOs for the diseases mentioned in this study

Drug	Registration ID	Chemistry of AO	Indication	Sponsor and Partner	Status
Mipomersen (brand name Kynamro)	–	20mer MOE gapmer	HoFH	IONIS/ Kastle	Commercial, approved only in the USA
Volanesorsen (brand name Waylivra)	NCT02300233	20mer MOE gapmer	HTG	IONIS/ Akcea	Phase 3 completed as of May 9, 2018
Volanesorsen (brand name Waylivra)	NCT02211209	20mer MOE gapmer	FCH	IONIS/ Akcea	Phase 3 completed. Approved in the European Union on May 7, 2019
Volanesorsen (brand name Waylivra)	NCT02527343	20mer MOE gapmer	FLPD	IONIS/ Akcea	Phase 3 completed; active extension study. Expected to be completed by September 2021

2 Homozygous Familial Hypercholesterolemia (HoFH)

2.1 Pathology

Familial hypercholesterolemia (FH) is an autosomal codominant genetic disease, where two mutant LDL receptor alleles (FH Homozygotes; HoFH) lead to a much more severely affected phenotype than one mutant allele (FH Heterozygotes; HeFH) [9], characterized by an elevated level of LDL cholesterol in the blood, xanthomas, and premature coronary atherosclerosis [10]. HoFH affects 1 in 160,000 to 1 in 300,000 individuals, associated with an extremely high concentration of low-density lipoprotein (LDL) due to 2 defective alleles in at least one of several genes coding the LDL receptor (LDL-R) and its associated ligand Apo-B among others [9]. Apo-B is a crucial protein, mediating the clearance of lipoprotein particles by binding to the LDL-R present on hepatocytes [11]. Additionally, it takes part in the production of very-low-density lipoprotein (VLDL) in the liver, which, in turn, act as metabolic precursors of LDL particles [12]. This leads to reduced uptake of plasma LDL cholesterol (LDL-C). In addition to this, when the intracellular cholesterol decreases, the synthesis of endogenous LDL-C is activated by β-hydroxy β-methylglutaryl-CoA (HMG-CoA) reductase, further shooting the levels of LDL-C. This accumulation of LDL-C which is two to three times higher than in the normal population leads to the formation of benign cholesterol-laden tumors called xanthomas, in addition to several cardiovascular diseases (CVD) at a very young age [13]. Patients untreated for this condition cannot survive beyond the age of 30 [14]. Clinical trials have shown that lowering LDL-C levels can reduce the risk of CVD [15–20].

2.2 Standard Treatment

HMG-CoA reductase inhibitors (statins) are the commonly prescribed agents for HoFH. They inhibit the HMG-CoA reductase enzyme, which is the rate-limiting step in the biosynthetic cholesterol pathway [21]. In spite of being successful in dramatically reducing LDL-C levels, even a high-intensity dose of statins often fails to reach therapeutic goals of eliminating the risk of coronary heart disease [13].

2.3 Preclinical Studies

Gapmer oligonucleotides targeting the production of apoB (also known as apoB-100) is an emerging approach for lipid-lowering therapies [22]. Since the sequence of mRNAs for Apo-B is different in different species, species-specific AOs targeting ApoB were evaluated in multiple species including mice, hamsters, rabbits, and monkeys [lean and high fat (HF)-fed] [23, 24]. The first in vivo study to evaluate the effect of ISIS 147764 on the ApoB levels were carried out by Crooke et al. in HF-fed C57BL/6 mice that were injected intraperitoneally twice a week (5, 25, and 50 mg/kg) [8]. The liver ApoB levels reduced in a dose-dependent manner with the greatest inhibition of 87% at a dose of 50 mg/kg. The ApoB mRNA levels were unaffected by the mock AO used as a control in the experiment [25]. Six weeks posttreatment, dose-dependent reductions were observed in the ApoB-100 levels, and at the highest dose (50 mg/kg), the ApoB-100 levels were almost undetectable [25].

Human ApoB-expressing transgenic mice (C57BL/6 hemizygous for full-length human gene-Taconic, Hudson, NY) were administered with the 20-mer MOE gapmer (ISIS301012; also know as mipomersen) intraperitoneally at 2.5, 5, 10, and 25 mg/kg or control AO at 25 mg/kg twice a week for 6 weeks [23]. Lean cynomolgus monkeys were intravenously administered with ISIS301012 weekly at 4 mg/kg for 4 weeks [23]. Additionally, the cynomolgus monkeys were fed an HF-diet prior to the administration regime of monkey-specific apoB antisense inhibitor ISIS326358 (differs from ISIS301012 only by two bases) [23]. The regime consisted of weekly subcutaneous administration of 5, 10, or 35 mg/kg for 5 weeks [23]. The results derived from all these animal models (ISIS147764 in mice, ISIS301012 in transgenic mice ISIS301012 in lean cynomolgus monkeys and ISIS326358 in HF-fed cynomolgus monkeys) exhibited a dose-dependent reduction in the liver ApoB mRNA levels as well as the serum ApoB, LDL-C and total cholesterol levels [23]. These favorable dynamics provided guidance for clinical development of the ApoB AO for humans.

2.4 Clinical Trials

The first phase of the clinical trial in humans was carried out in patients with mild dyslipidemia. Weekly subcutaneous (SC) injections of mipomersen at 200 mg dose resulted in a 50% reduction in the circulating ApoB after 4 weeks of treatment [24]. The circulating ApoB levels remained low for several weeks, after which it returned to the baseline [23]. Additionally, four phase-I and II trials studied the pharmacodynamics of mipomersen in patients with mild dyslipidemia who were not on lipid-lowering therapy; patients suffering from moderate hyperlipidemia and patients with hypercholesterolemia treated with statins [26, 27]. Mipomersen lowered the levels of ApoB lipoproteins in a dose-dependent manner. In a pivotal multicenter phase III study, patients with clinically severe HoFH were subjected to weekly mipomersen or placebo for 26 weeks [23]. This study showcased a successful reduction in LDL by 36% in the mipomersen-treated cohort along with a statistical difference in the reduction of Apo-B and lipoprotein, without lowering the levels of high-density lipoprotein (HDL) [23].

A large double-blinded study evaluated the side effects of mipomersen in HoFH patients, which include injection site pain, inflammation, erythema, and pruritis, resulting in 5% of the patients discontinuing the therapy in pooled phase 3 trials [23]. In addition to this, 12% of the HoFH patients had elevated levels of liver enzymes, indicating a risk of hepatotoxicity. For the aforementioned reason, mipomersen is approved only for restricted use for physicians registered in a Risk Evaluation and Mitigation Strategy (REMS) [28].

Mipomersen has shown efficacy in lowering the LDL-C levels in patients with both HoFH and heterozygous FH (HeFH); however, it is associated with adverse effects, such as hepatotoxicity and injection site reactions, which can lead it to be discontinued by the patients or their physicians [27, 28]. Currently, mipomersen is approved only for patients with HoFH [28]. The long-term effects on the liver are currently unknown, creating a concern to treat younger patients. In addition to this, the drug is made available to patients as an alternative to other failed therapies [28]. The development of treatment for severe HoFH continues to move forward, with next-generation AOs and RNA silencing [28].

3 Hypertriglyceridemia

3.1 Pathology

Hypertriglyceridemia (HTG) is a prevalent disorder associated with elevated levels of triglyceride (TG), obesity, and type II diabetes (insulin resistance) [29]. HTG can be defined as fasting plasma concentration above a threshold value (e.g., >1.7 mmol/L [>150 mg/dL]) [30, 31]. HTG can be categorized as either primary or secondary disorders based on the cause of the condition. Inherited or familial HTG is classified as primary, while secondary HTG coexists with other conditions such as type II diabetes (T2D), metabolic syndrome, renal issues, and CVD [29]. In addition to the genetic factors, the lifestyle (obesity, alcohol consumption) exacerbates the condition, elevating the TG levels [29]. Severe monogenic HTG is usually autosomal recessive in nature and has a prevalence of 1 in 1,000,000 [31]. The patients usually are homozygous or heterozygous for loss-of-function mutations in TG-rich lipoproteins, such as *LPL*, *APOC2*, and *GPD1* [32–34].

Apolipoprotein III (ApoC-III), a 79 amino-acid glycoprotein synthesized in the liver, plays a crucial role in regulating the plasma TG levels [35–37]. It has a vital function in the conversion of VLDL to LDL [38]; thus the elevated levels of the enzyme can result in impaired clearance of TG-rich lipoproteins from the circulation, leading to a risk of CVD. It has also been shown that the loss-of-function nature of ApoC-III is associated with a reduced risk of coronary heart disease [39–41]. Considering the above factors, ApoC-III was identified as a suitable target to treat HTG using AOs. A number of lipid-lowering drugs such as atorvastatin [42], rosuvastatin [43], and omega-3 polyunsaturated fatty acids (PUFA) [44] lower the ApoC-III levels by 10–30%. However, a suitable promising therapeutic agent for robust reduction of APOC-III levels using an AO was not available until the generation of a promising 20-mer MOE gapmer volanesorsen [37]. This gapmer binds to *ApoC-III* mRNA and inhibits the expression when administered subcutaneously [37].

3.2 Preclinical and Clinical Trials

Reports suggest that volanesorsen selectively reduced both ApoC-III and TG levels in several animal models including rats, mice, human transgenic ApoC-III mice, and nonhuman primates (NHP) [45, 46]. A phase I double-blind, placebo-controlled, dose-escalation clinical study was performed by Graham et al. [26] where healthy volunteers with normal TG levels were given a total of 6 doses (50, 100, 200, or 400 mg) of volanesorsen (ISIS 304081) (N = 33, 25 received ISIS 304081, 8 placebo) over 22 days. Volanesorsen administration resulted in a dose-dependent and prolonged reduction in ApoC-III and TG levels that were sustained for at least 4 weeks after the sixth dose. No severe clinical adversities were reported [37].

Subsequently, a phase 2 double-blind placebo-controlled, dose-ranging study evaluated the effects of volanesorsen in patients whose TG levels between 225 mg/dL and 2000 mg/dL as an add-on to stable fibrate therapy, a group of amphipathic carboxylic acids that are used to lower the TG levels, or only on volanesorsen [45]. Fibrates are often prescribed to patients with moderate-to-severe HTG, and they lower TG levels by 30–50% and sometimes also increase HDL-C levels [47]. The cohort receiving volanesorsen alone showed a higher reduction in plasma ApoC-III levels than the cohort receiving volanesorsen as an add-on. Of note, 10% of the patients discontinued treatment because of adverse effects including injection site reactions, nausea, fatigue, and musculoskeletal pain. However, no signs and symptoms of renal or hepatic failure were observed [45]. In lines with the finding that HTG is associated with insulin resistance, Digenio et al., in another phase II study, showed that patients receiving weekly 300 mg SC of volanesorsen showed a significant reduction in TG and plasma APOC-III levels and an increase in HDL-C, along with an improvement in whole-body insulin sensitivity [48, 49].

The COMPASS study (NCT02300233) is a 26-week randomized, double-blind, placebo-controlled phase III trial that recruited patients with HTG (>700 mg/dL) and treated with 300 mg volanesorsen SC or placebo weekly [37]. The study monitored the percent change in fasting TG from baseline (completed as of April 2019). A mean reduction of 72.7% in the TG levels was observed in patients treated with volanesorsen [37]. However, larger studies are required to assess the effects on cardiovascular morbidity and mortality.

Overall, volanesorsen, the ApoC-III inhibitor, appears to be effective in reducing TG levels, as well as increasing the HDL-C levels and also at improving insulin sensitivity [45]. However, since it also augments the levels of LDL-C, the long-term effects of volanesorsen on cardiovascular risks are yet to be evaluated [45]. A few cases of thrombocytopenia were also observed in phase I and II trials, which could pose a worrisome threat if the frequency of cases increases [45]. The efficacy of volanesorsen coupled with statins must be evaluated since the primary target of the treatment is lowering the levels of LDL-C with statins, followed by lowering the levels of non-HDL-C which are the secondary targets [45]. Lastly, it is noteworthy to observe that volanesorsen has reduced the risk of pancreatitis in patients. Volanesorsen also seems to be a suitable option for patients with chronic kidney disease in which fenofibrate, a prodrug of fenofibric acid that is known to reduce the cholesterol and TG levels [50], is contraindicated and the cardiovascular risk is increased [49].

4 Familial Chylomicronemia Syndrome

4.1 Pathology

Familial chylomicronemia syndrome (FCS) is a rare monogenic autosomal lipid disorder affecting 1 to 2 per 1,000,000 people, that overlaps with HTG and exhibits recurrent pancreatitis [51]. Individuals are usually homozygous or compound heterozygous for loss-of-function mutations in the genes, mainly the *lipoprotein lipase (LPL)* gene and the *ApoC-II* gene that are responsible for the regulation of TG catabolism [31]. LPL facilitates the removal of TGs from chylomicrons and VLDL and breaks them down into free fatty acids. LPL binds to ApoC-II to become fully functional [52, 53]. Herewith, mutations in *LPL* and its associated cofactors such as *Apo-II* and *ApoA-V* can result in excess circulating chylomicrons leading to FCS. In healthy individuals, TGs are usually cleared from the plasma within 3–4 h after food consumption. However, there is a possibility of the onset of FCS if the TG levels are higher than 885 mg/dL, 12–14 h after the food intake [54]. Large quantities of chylomicrons make the blood in FCS patients appear milky and creamy [55].

4.2 Standard Treatment

FCS is resistant to TG lowering therapy, making it a difficult disease to treat [54]. To date, restriction of dietary fat intake (30–50 g/day) is considered to be effective to reduce the chylomicron formation, since FCS patients are resistant to the existing medical therapies [54, 55]. Infants diagnosed with FCS are given milk that contains medium-chain triglycerides (MCT) that can enter circulation without being incorporated into the chylomicrons. Pregnant women with FCS are at high risk for pancreatitis and fetal morbidity because of HTG during the pregnancy period. Treatments with statins, fibrates, niacin are not very effective in lowering the TG levels in FCS patients since they usually do not reduce chylomicrons (CM) but rather decrease VLDL production [31, 54, 56].

4.3 Clinical Trials

ApoC-III leads to the increment in the plasma TG levels, which was effectively lowered by volanesorsen, a 20-mer MOE gapmer, in patients with HTG of different etiology, other than FCS [54, 57]. It was suggested for the treatment of FCS because it targets the *ApoC-III* mRNA that, in turn, mediates the clearance of TG from the plasma via an *LPL*-independent pathway [58]. This led to an international phase III trial to evaluate the efficacy and safety of volanesorsen to treat patients with FCS.

In addition to the COMPASS study (NCT02300233) in patients with HTG, a parallel phase III randomized, multicenter, double-blind, placebo-controlled trial called the APPROACH study (ISIS 304801-CS6; NCT02211209) was initiated to verify the clinical hypothesis that volanesorsen (formerly IONIS-APOCIIIRx) reduces plasma TG levels in FCS patients compared to the placebo. Sixty-six patients over the age of 18 years with a fasting TG $\geq$ 750 mg/dL were administered with 300 mg volanesorsen SC weekly for 52 weeks. At the end of the study, patients could either enroll in an open-label extension (OLE) study (NCT02658175) or participate in a 13-week post-evaluation period [37]. In this study, the primary end point was met after 3 months, with a 94% reduction in the TG levels in the volanesorsen-treated cohort ($P < 0.001$) [59]. Volanesorsen treatment lowered the TG levels by 77% in comparison to the baseline at 3 months, and as opposed to an 18% increment in the placebo group [59]. The most common adverse events included injection site reactions in 17% of all the injections. Five early terminations in the study were observed because of a decline in the platelet count, which was restored following the treatment with corticosteroids [60]. In spite of the physical and clinical symptoms being well-defined in FCS, the long-term effects on the patient's quality of life (QoL), comprising emotional and cognitive wellness, are poorly studied. To address this concern, a multinational web-based survey known as the Investigation of Findings and Observations Captured in Burden of Illness Survey in FCS Patients (IN-FOCUS) was conducted that evaluated the QoL in FCS patients [11]. Around 22-35% of the patients experienced severe abdominal pain, fatigue, and indigestion, arising from acute pancreatitis. This, in turn, led to the deterioration in the ability of the patients to work efficiently. Half of the patients enrolled in IN-FOCUS reported symptoms in spite of their compliance with the dietary restrictions and lifestyle prescriptions [11].

In order to determine whether volanesorsen-mediated TG reduction had a positive impact on the QoL as observed by the IN-FOCUS study, the Retrospective Findings and Observations Captured in Burden of Illness Survey in FCS Patients (ReFOCUS) was conducted to evaluate the QoL of patients before and during the volanesorsen-therapy [11]. A questionnaire adapted from the IN-FOCUS study was used for an eligible population from the APPROACH OLE study (NCT02658175) to determine whether volanesorsen reduced the clinical and psychological burdens that negatively interfered with the patient's well-being [59]. The number of patients who experienced more than 10 symptoms decreased from 41% to 14% after the volanesorsen treatment, in addition to a significant reduction in the numbers reported for pancreatic pain (8 respondents to 1; $P < 0.05$) [11]. A 60% decrease in the acute pancreatitis symptoms was observed in patients following 3 months of volanesorsen treatment. Additionally, among emotional symptoms, the anxiety related to acute pancreatitis was reduced from 10 to 2 reports ($P < 0.05$) [11]. Since the sample size was small, the average severity of pancreatic pain still needs to be determined.

Overall, the phase III APPROACH study showed that treatment with volanesorsen significantly decreased the TG levels in FCS patients when compared to the placebo [60, 61]. The ReFOCUS study demonstrated that the treatment also resulted in an overall improvement in the QoL of patients with a significant reduction in the number of FCS-associated symptoms when administered volanesorsen along with dietary management [59]. Thus lowering TG levels with the MOE gapmer may be associated with reduced incidences of abdominal pain and acute pancreatitis.

On May 7, 2019, volanesorsen (brand name Waylivra) received conditional marketing authorization from the European Commission (EC) as an adjunct to diet in adults with genetically confirmed FCS and for those who did not respond adequately to diet well to triglyceride-lowering therapy [62]. This approval followed the successful phase III APPROACH study and the ongoing APPROACH OLE study which is supported by the Phase III COMPASS study [62]. In the USA, volanesorsen holds the status of an orphan drug and is under regulatory review for the treatment of FCS [63].

5 Familial Partial Lipodystrophy

5.1 Pathology

Partial lipodystrophy (PL) is a heterogeneous group of disorders characterized by a scarcity of adipose tissue due to the deficiency of leptin, resulting in abnormal metabolism [64]. Lipodystrophy can be classified based on the pattern of adipose loss (generalized or partial) as well as on the basis of whether it is genetic or acquired [65, 66]. The 4 major subtypes of lipodystrophy include:

(a) Congenital generalized lipodystrophy (CGL).

(b) Acquired generalized lipodystrophy (AGL).

(c) Familial partial lipodystrophy (FLPD).

(d) Acquired partial lipodystrophy (APL).

FLPD is a rare autosomal dominant disorder affecting 1 in 100,000 which may be acquired or inherited due to missense mutations in the carboxy-terminal immunoglobulin domain of the *LMNA* (lamin A/C) gene [67, 68]. Six different subtypes of FLPD have been identified, based on the mutations. FLPD Type 2 (Dunnigan Variety) is the most common form of FLPD. The commonly associated characteristics include insulin resistance, diabetes, HTG, and nonalcoholic fatty liver disease (NAFLD) [69]. FLPD patients have fat deposits on their neck and abdominal region, similar to Cushing's syndrome [65]. The phenotype exhibited by FLPD patients is very similar to the phenotype seen in metabolic syndromes or type 2 diabetes mellitus, and therefore careful physical examination is necessary. Since HTG is commonly seen in FLPD patients, acute pancreatitis is a prevalent symptom of FLPD.

5.2 Standard Treatment

Since FLPD is a multidimensional disease comprising an array of symptoms, overlapping with several other diseases, the treatment for FLPD is focused on the specific symptoms seen in each patient [70]. Patients with FLPD are advised with a high carbohydrate, low-fat diet, which can avert the risk of acute pancreatitis. FLPD patients have an increased risk of coronary heart disease and hence are advised to lower the intake of saturated and trans- unsaturated fats [70]. The risk of diabetes can be lowered with regular exercise and maintenance of a healthy weight. Unfortunately, the characteristic feature of FLPD-adipose tissue loss cannot be reversed. Subsequently, cosmetic surgeries and procedures like liposuction may improve the aesthetic appearance and remove unwanted fat accumulation respectively [70].

An analog of leptin known as metreleptin was approved by the FDA for the treatment of metabolic impediments in patients with generalized lipodystrophies but has not been approved for FLPD. Preliminary studies have shown that metreleptin was beneficial in improving some but not all the symptoms of FLPD [70]. In yet another study, they found that patients with lipodystrophy have elevated ApoC-III levels, and the leptin-replacement therapy did not significantly decrease the levels of ApoC-III [71]. Future research is required to investigate whether leptin directly affects ApoC-III.

5.3 Clinical Trials

Patients with FLPD often have moderate to severe HTG [37]. Hence, a randomized, double-blind, placebo-controlled with an open-labeled phase 2/3 study called BROADEN (NCT02527343) was initiated in October 2015 to examine the effects of volanesorsen (IONIS-APOCIIIRx; ISIS 304801), the 20-mer MOE gapmer currently in clinical trials for HTG and FCS. Sixty patients were enrolled in the study, where the primary outcome of the study was to measure the efficacy of volanesorsen as the percent change in fasting TGs from baseline when administered 300 mg SC weekly for 52 weeks [37, 64]. The secondary outcome measured:

(a) Change in hepatic steatosis from baseline (time frame 52 weeks).
(b) Change in hemoglobin A1c (HbA1c) (time frame 52 weeks).
(c) A composite endpoint at the end of 6 months for the percent of patients who have achieved ≥40% reduction in fasting TGs and ≥30% reduction in hepatic fat fraction (time frame 52 weeks).
(d) Change in the patient-reported outcomes (time frame 52 weeks).

After the end of this study, patients may enroll in an OLE study (NCT02639286) or a 13-week post-treatment evaluation period. The study is still on-going and is expected to be completed by September 2021.

6 Challenges and Future Perspectives

The field of AO-based therapy has been growing since the discovery of gapmers in the 80s. Considerable improvements and modifications have been made to make the therapeutic disease-specific. In spite of the approval of 3 gapmers: Mipomersen (brand name Kynamro) for the treatment of HoFH, inotersen (brand name Tegsedi) for the treatment of hereditary transthyretin amyloidosis-associated polyneuropathy, and volanesorsen (brand name Waylivra) for FCS treatment, several gapmer AOs are held back because of low specificity and toxicity [28, 72]. The main issue lies with the low target delivery of the gapmer AO into the site of action in the nucleus or the cytosol within the tissues [5]. The specificity and directed uptake of the gapmer AO can be increased by conjugating the gapmer AO to cell-penetrating peptides (CPPs) that have shown to optimize the delivery of AOs in SMA and DMD by penetrating the blood-brain barrier (BBB) and the vascular endothelial barrier [73–76]. In addition to CPPs, lipid nanoparticles are now being currently tested to deliver the AOs because of their small size and persistence in the circulation; thereby increasing the half-life of AOs [77]. Also, polymeric micelles (core-shell nanoparticles) can be used for AO delivery. For example, poly lactic-co-glycolic acid (PLGA) can be used to form solid nanoparticles through the process of oil-water emulsion techniques [78]. It is possible that these novel delivery approaches can enhance the biodistribution of the AO gapmers and also positively impact the pharmacokinetics and pharmacodynamics of the gapmer moieties. Although these approaches are also associated with some technical challenges.

Another major issue faced by the AOs is toxicity. The majority of the AOs, typically those with MOE modification, accumulate in the liver and kidneys leading to toxicity [5]. This, in turn, activates the inflammatory responses leading to tissue damage and necrosis. Recent studies have reported that a modification made to the backbone of the gapmer, with a single nucleotide conversion in the central gap from a PS-DNA to a PS-2′OMe residue made constrained ethyl gapmers (cEt) and other gapmers less toxic [79–81].

In summary, numerous creative approaches are being tested to offset the limitations faced by gapmer AOs for the treatment of diseases. The judicious application of the strategies can help to deal with the issues associated with the gapmer technology.

7 Conclusion

The field of antisense therapy is evolving, with clinical developments made for approximately 30 years. In 2013, mipomersen (commercially known as Kynamro), a 20-mer MOE gapmer targeting the coding region of ApoB in HoFH patients, was approved in the USA. In addition to lipid disorders, AOs are promising for neurodegenerative and neuromuscular disorders, as evidenced by the approval of nusinersen and eteplirsen for the treatment of SMA and DMD, respectively [2, 3]. Numerous AOs with modified chemistries, including those of gapmers, are now available or are being tested in clinical trials for the treatment of many diseases, beyond the scope of lipid disorders and neuromuscular disorders.

An MOE gapmer known as volanesorsen has made considerable progress in the treatment of several lipid-mediated disorders such as HTG, FCH, and FLPD. Volanesorsen has shown to be effective in lowering the TG levels and reduced the possibility of acute pancreatitis in the phase III trials for HTG and FLPD. Following successful phase III results for FCS, volanesorsen (WAYLIVRA) received market authorization by the EC on May 7, 2019, to treat adults with genetically confirmed FCS for whom the response was inadequate to triglyceride-lowering therapy and diet [62]. Further results are awaited for FPLD and HTG .

In summary, approval of antisense therapies, including gapmers (mipomersen, inotersen, and volanesorsen) and splice-switching oligos (nusinersen and eteplirsen), are considered to be a milestone. In spite of demonstrating good safety and tolerability profiles, drawbacks still remain. The high cost of the treatments poses questions about the accessibility of the drug to patients who live in underinsured parts of the world. Global strategies are yet to be developed that entail physicians, health insurers, and pharmaceutical companies [1]. With the success stories demonstrated by the approval of AO gapmers and much progress being made in understanding, designing, developing, and refining the use of gapmers, this research field may well revolutionize the field of therapeutics in genetic diseases.

References

1. Wurster CD, Ludolph AC (2018) Antisense oligonucleotides in neurological disorders. Ther Adv Neurol Disord 11:1756286418776932. https://doi.org/10.1177/1756286418776932
2. Shimizu-Motohashi Y, Komaki H, Motohashi N et al (2019) Restoring dystrophin expression in duchenne muscular dystrophy: current status of therapeutic approaches. J Pers Med 9(1): E1. https://doi.org/10.3390/jpm9010001
3. Goodkey K, Aslesh T, Maruyama R et al (2018) Nusinersen in the treatment of spinal muscular atrophy. Methods Mol Biol 1828:69–76. https://doi.org/10.1007/978-1-4939-8651-4_4
4. Schoch KM, Miller TM (2017) Antisense oligonucleotides: translation from mouse models to human neurodegenerative diseases. Neuron 94(6):1056–1070. https://doi.org/10.1016/j.neuron.2017.04.010

5. Juliano RL (2016) The delivery of therapeutic oligonucleotides. Nucleic Acids Res 44 (14):6518–6548. https://doi.org/10.1093/nar/gkw236
6. Bennett CF, Swayze EE (2010) RNA targeting therapeutics: molecular mechanisms of antisense oligonucleotides as a therapeutic platform. Annu Rev Pharmacol Toxicol 50:259–293. https://doi.org/10.1146/annurev.pharmtox.010909.105654
7. Geary RS, Norris D, Yu R et al (2015) Pharmacokinetics, biodistribution and cell uptake of antisense oligonucleotides. Adv Drug Deliv Rev 87:46–51. https://doi.org/10.1016/j.addr.2015.01.008
8. Crooke ST (2019) Antisense technology past, Present and Future Ionis Pharmaceuticals. https://ir.ionispharma.com/static-files/. Accessed June 5 2019
9. Cuchel M, Bruckert E, Ginsberg HN et al (2014) Homozygous familial hypercholesterolaemia: new insights and guidance for clinicians to improve detection and clinical management. A position paper from the consensus panel on familial Hypercholesterolaemia of the European atherosclerosis society. Eur Heart J 35(32):2146–2157
10. Raal FJ, Santos RD (2012) Homozygous familial hypercholesterolemia: current perspectives on diagnosis and treatment. Atherosclerosis 223(2):262–268
11. Michael Davidson AH, Ahmad Z, van Lennep JR, Stevenson M (2018) Examining the high disease burden and impact on quality of life in familial Chylomicronemia syndrome. Atherosclerosis 32:1. https://doi.org/10.1016/j.atherosclerosissup.2018.04.199
12. Mahley R (2001) Biochemistry and physiology of lipid and lipoprotein metabolism. In: Becker KL (ed) Principles and practice of endocrinology and metabolism, 3rd edn. Lippincott Williams and Wilkins, Philadelphia
13. Stein CA, Castanotto D (2017) FDA-approved oligonucleotide therapies in 2017. Mol Ther 25(5):1069–1075. https://doi.org/10.1016/j.ymthe.2017.03.023
14. Raal FJ, Santos RD, Blom DJ et al (2010) Mipomersen, an apolipoprotein B synthesis inhibitor, for lowering of LDL cholesterol concentrations in patients with homozygous familial hypercholesterolaemia: a randomised, double-blind, placebo-controlled trial. Lancet 375(9719):998–1006. https://doi.org/10.1016/S0140-6736(10)60284-X
15. Gelsinger C, Steinhagen-Thiessen E, Kassner U (2012) Therapeutic potential of mipomersen in the management of familial hypercholesterolaemia. Drugs 72 (11):1445–1455. https://doi.org/10.2165/11635060-000000000-00000
16. Versmissen J, Oosterveer DM, Yazdanpanah M et al (2008) Efficacy of statins in familial hypercholesterolaemia: a long term cohort study. BMJ 337:a2423. https://doi.org/10.1136/bmj.a2423
17. Baigent C, Keech A, Kearney PM et al (2005) Efficacy and safety of cholesterol-lowering treatment: prospective meta-analysis of data from 90,056 participants in 14 randomised trials of statins. Lancet 366 (9493):1267–1278. https://doi.org/10.1016/S0140-6736(05)67394-1
18. Elis A, Zhou R, Stein EA (2011) Effect of lipid-lowering treatment on natural history of heterozygous familial hypercholesterolemia in past three decades. Am J Cardiol 108(2):223–226. https://doi.org/10.1016/j.amjcard.2011.03.027
19. Pedersen TR, Cater NB, Faergeman O et al (2010) Comparison of atorvastatin 80 mg/day versus simvastatin 20 to 40 mg/day on frequency of cardiovascular events late (five years) after acute myocardial infarction (from the incremental decrease in end points through aggressive lipid lowering [IDEAL] trial). Am J Cardiol 106 (3):354–359. https://doi.org/10.1016/j.amjcard.2010.03.033
20. Pedersen TR, Faergeman O, Kastelein JJ et al (2005) High-dose atorvastatin vs usual-dose simvastatin for secondary prevention after myocardial infarction: the IDEAL study: a randomized controlled trial. JAMA 294 (19):2437–2445. https://doi.org/10.1001/jama.294.19.2437
21. Catapano AL, Reiner Z, De Backer G et al (2011) ESC/EAS guidelines for the management of dyslipidaemias the task force for the management of dyslipidaemias of the European Society of Cardiology (ESC) and the European atherosclerosis society (EAS). Atherosclerosis 217(1):3–46
22. Gebhard C, Huard G, Kritikou EA et al (2013) Apolipoprotein B antisense inhibition-update on mipomersen. Curr Pharm Des 19 (17):3132–3142
23. Yu RZ, Lemonidis KM, Graham MJ et al (2009) Cross-species comparison of in vivo PK/PD relationships for second-generation antisense oligonucleotides targeting apolipoprotein B-100. Biochem Pharmacol 77 (5):910–919. https://doi.org/10.1016/j.bcp.2008.11.005
24. Parham JS, Goldberg AC (2019) Mipomersen and its use in familial hypercholesterolemia.

Expert Opin Pharmacother 20(2):127–131. https://doi.org/10.1080/14656566.2018.1550071

25. Crooke RM, Graham MJ, Lemonidis KM et al (2005) An apolipoprotein B antisense oligonucleotide lowers LDL cholesterol in hyperlipidemic mice without causing hepatic steatosis. J Lipid Res 46(5):872–884. https://doi.org/10.1194/jlr.M400492-JLR200
26. Kastelein JJ, Wedel MK, Baker BF et al (2006) Potent reduction of apolipoprotein B and low-density lipoprotein cholesterol by short-term administration of an antisense inhibitor of apolipoprotein B. Circulation 114(16):1729–1735. https://doi.org/10.1161/CIRCULATIONAHA.105.606442
27. Akdim F, Tribble DL, Flaim JD et al (2011) Efficacy of apolipoprotein B synthesis inhibition in subjects with mild-to-moderate hyperlipidaemia. Eur Heart J 32(21):2650–2659. https://doi.org/10.1093/eurheartj/ehr148
28. Genyzme Corporation. KYNAMRO® (mipomersen sodium) injection prescribing information. (2013). http://www.kynamro.com/~/media/Kynamro/Files/KYNAMRO-PI.pdf. Accessed June 5 2019
29. Lewis GF, Xiao C, Hegele RA (2015) Hypertriglyceridemia in the genomic era: a new paradigm. Endocr Rev 36(1):131–147. https://doi.org/10.1210/er.2014-1062
30. Brahm A, Hegele RA (2013) Hypertriglyceridemia. Nutrients 5(3):981–1001. https://doi.org/10.3390/nu5030981
31. Hegele RA, Ginsberg HN, Chapman MJ et al (2014) The polygenic nature of hypertriglyceridaemia: implications for definition, diagnosis, and management. Lancet Diabetes Endocrinol 2(8):655–666. https://doi.org/10.1016/S2213-8587(13)70191-8
32. Johansen CT, Kathiresan S, Hegele RA (2011) Genetic determinants of plasma triglycerides. J Lipid Res 52(2):189–206. https://doi.org/10.1194/jlr.R009720
33. Johansen CT, Hegele RA (2011) Genetic bases of hypertriglyceridemic phenotypes. Curr Opin Lipidol 22(4):247–253. https://doi.org/10.1097/MOL.0b013e3283471972
34. Johansen CT, Hegele RA (2012) Allelic and phenotypic spectrum of plasma triglycerides. Biochim Biophys Acta 1821(5):833–842. https://doi.org/10.1016/j.bbalip.2011.10.007
35. Gaudet D, Alexander VJ, Baker BF et al (2015) Antisense inhibition of Apolipoprotein C-III in patients with hypertriglyceridemia. N Engl J Med 373(5):438–447. https://doi.org/10.1056/NEJMoa1400283
36. Chan DC, Chen MM, Ooi EM et al (2008) An ABC of apolipoprotein C-III: a clinically useful new cardiovascular risk factor? Int J Clin Pract 62(5):799–809. https://doi.org/10.1111/j.1742-1241.2007.01678.x
37. Gouni-Berthold I (2017) The role of antisense oligonucleotide therapy against apolipoprotein-CIII in hypertriglyceridemia. Atheroscler Suppl 30:19–27. https://doi.org/10.1016/j.atherosclerosissup.2017.05.003
38. Mendivil CO, Zheng C, Furtado J et al (2010) Metabolism of very-low-density lipoprotein and low-density lipoprotein containing apolipoprotein C-III and not other small apolipoproteins. Arterioscler Thromb Vasc Biol 30(2):239–245. https://doi.org/10.1161/ATVBAHA.109.197830
39. Pollin TI, Damcott CM, Shen H et al (2008) A null mutation in human APOC3 confers a favorable plasma lipid profile and apparent cardioprotection. Science 322(5908):1702–1705. https://doi.org/10.1126/science.1161524
40. Tg, Hdl Working Group of the Exome Sequencing Project NHL, Blood I et al (2014) Loss-of-function mutations in APOC3, triglycerides, and coronary disease. N Engl J Med 371(1):22–31. https://doi.org/10.1056/NEJMoa1307095
41. Schirle NTS-GJ, MacRae IJ (2014) Structural basis for microRNA targeting. Science 346(6209):6. https://doi.org/10.1126/science.1258040
42. Dallinga-Thie GM, Berk P II, Bootsma AH et al (2004) Atorvastatin decreases apolipoprotein C-III in apolipoprotein B-containing lipoprotein and HDL in type 2 diabetes: a potential mechanism to lower plasma triglycerides. Diabetes Care 27(6):1358–1364. https://doi.org/10.2337/diacare.27.6.1358
43. Ooi EMWG, Chan DC, Chen MM, Nestel PJ, Sviridov D et al (2008) Dose-dependent effect of rosuvastatin on VLDL-apolipoprotein C-III kinetics in the metabolic syndrome. Diabetes Care 8:6. https://doi.org/10.2337/dc08-0358
44. Swahn E, von Schenck H, Olsson AG (1998) Omega-3 ethyl Ester concentrate decreases Total Apolipoprotein CIII and increases Antithrombin III in Postmyocardial infarction patients. Clin Drug Investig 15(6):473–482. https://doi.org/10.2165/00044011-199815060-00003
45. Milonas D, Tziomalos K (2019) Experimental therapies targeting apolipoprotein C-III for the treatment of hyperlipidemia - spotlight on volanesorsen. Expert Opin Investig Drugs 28

(4):389–394. https://doi.org/10.1080/13543784.2019.1582028
46. Graham MJ, Lee RG, Bell TA 3rd et al (2013) Antisense oligonucleotide inhibition of apolipoprotein C-III reduces plasma triglycerides in rodents, nonhuman primates, and humans. Circ Res 112(11):1479–1490. https://doi.org/10.1161/CIRCRESAHA.111.300367
47. Berglund LBJ, Goldberg AC, Goldberg IJ, Sacks F, Murad MH, Stalenhoef AF, Endocrine society (2012) Evaluation and treatment of hypertriglyceridemia: an Endocrine Society clinical practice guideline. J Clin Endocrinol Metab 97(9):2969–2989
48. Digenio A, Dunbar RL, Alexander VJ et al (2016) Antisense-mediated lowering of plasma Apolipoprotein C-III by Volanesorsen improves Dyslipidemia and insulin sensitivity in type 2 diabetes. Diabetes Care 39 (8):1408–1415. https://doi.org/10.2337/dc16-0126
49. Wu L, Parhofer KG (2014) Diabetic dyslipidemia. Metabolism 63(12):1469–1479. https://doi.org/10.1016/j.metabol.2014.08.010
50. Capell WH, DeSouza CA, Poirier P et al (2003) Short-term triglyceride lowering with fenofibrate improves vasodilator function in subjects with hypertriglyceridemia. Arterioscler Thromb Vasc Biol 23(2):307–313
51. Chyzhyk V, Brown AS (2019) Familial chylomicronemia syndrome: a rare but devastating autosomal recessive disorder characterized by refractory hypertriglyceridemia and recurrent pancreatitis. Trends Cardiovasc Med 30 (2):80–85. https://doi.org/10.1016/j.tcm.2019.03.001
52. Stroes E, Moulin P, Parhofer KG, Rebours V, Löhr JM, Averna M (2017) Diagnostic algorithm for familial chylomicronemia syndrome. Atheroscler Suppl 23:7. https://doi.org/10.1016/j.atherosclerosissup.2016.10.002
53. Sakurai T, Sakurai A, Vaisman BL, Amar MJ, Liu C et al (2016) Creation of Apolipoprotein C-II (ApoC-II) mutant mice and correction of their hypertriglyceridemia with an ApoC-II mimetic peptide. J Pharmacol Exp Ther 356 (2):13. https://doi.org/10.1124/jpet.115.229740
54. Brahm AJ, Hegele RA (2015) Chylomicronaemia--current diagnosis and future therapies. Nat Rev Endocrinol 11(6):352–362. https://doi.org/10.1038/nrendo.2015.26
55. Gotoda T, Shirai K, Ohta T et al (2012) Diagnosis and management of type I and type V hyperlipoproteinemia. J Atheroscler Thromb 19(1):1–12
56. Tenenbaum A, Fisman EZ (2012) Fibrates are an essential part of modern anti-dyslipidemic arsenal: spotlight on atherogenic dyslipidemia and residual risk reduction. Cardiovasc Diabetol 11:125. https://doi.org/10.1186/1475-2840-11-125
57. Blom DJ, O'Dea L, Digenio A et al (2018) Characterizing familial chylomicronemia syndrome: baseline data of the APPROACH study. J Clin Lipidol 12(5):1234–1243. e1235. https://doi.org/10.1016/j.jacl.2018.05.013
58. Gaudet D, Brisson D, Tremblay K et al (2014) Targeting APOC3 in the familial chylomicronemia syndrome. N Engl J Med 371 (23):2200–2206. https://doi.org/10.1056/NEJMoa1400284
59. Arca M, Hsieh A, Soran H et al (2018) The effect of volanesorsen treatment on the burden associated with familial chylomicronemia syndrome: the results of the ReFOCUS study. Expert Rev Cardiovasc Ther 16(7):537–546. https://doi.org/10.1080/14779072.2018.1487290
60. DA Gaudet D, Alexander V et al (2018) The APPROACH study: a randomized, double-blind, placebo-controlled, phase 3 study of volanesorsen administered subcutaneously to patients with familial chylomicronemia syndrome (FCS). Atherosclerosis 32:14–15. https://doi.org/10.1016/j.atherosclerosissup.2018.04.042
61. Hegele RA, Tsimikas S (2019) Lipid-lowering agents. Circ Res 124(3):17. https://doi.org/10.1161/CIRCRESAHA.118.313171
62. Akcea and Ionis Announce Approval of WAYLIVRA® (volanesorsen) in the European Union. (7 May 2019). https://ir.akceatx.com/news-releases/news-release-details/akcea-and-ionis-announce-approval-waylivrar-volanesorsen. Accessed October 30 2019
63. Paik J, Duggan S (2019) Volanesorsen: first global approval. Drugs 79(12):1349–1354. https://doi.org/10.1007/s40265-019-01168-z
64. Joseph J, Shamburek RD, Cochran EK et al (2014) Lipid regulation in lipodystrophy versus the obesity-associated metabolic syndrome: the dissociation of HDL-C and triglycerides. J Clin Endocrinol Metab 99(9):E1676–E1680. https://doi.org/10.1210/jc.2014-1878
65. Handelsman Y, Oral EA, Bloomgarden ZT et al (2013) The clinical approach to the detection of lipodystrophy - an AACE consensus statement. Endocr Pract 19(1):107–116
66. Chan JL, Oral EA (2010) Clinical classification and treatment of congenital and acquired

lipodystrophy. Endocr Pract 16(2):310–323. https://doi.org/10.4158/EP09154.RA

67. Subramanyam L, Simha V, Garg A (2010) Overlapping syndrome with familial partial lipodystrophy, Dunnigan variety and cardiomyopathy due to amino-terminal heterozygous missense Lamin a/C mutations. Clin Genet 78(1):66–73. https://doi.org/10.1111/j.1399-0004.2009.01350.x
68. Garg A (2004) Acquired and inherited lipodystrophies. N Engl J Med 350(12):1220–1234. https://doi.org/10.1056/NEJMra025261
69. Ajluni N, Meral R, Neidert AH et al (2017) Spectrum of disease associated with partial lipodystrophy: lessons from a trial cohort. Clin Endocrinol 86(5):698–707. https://doi.org/10.1111/cen.13311
70. Familial Partial Lipodystrophy. (2015) Rare Disease Database. https://rarediseases.org/rare-diseases/familial-partial-lipodystrophy/. Accessed June 5 2019
71. Kassai A, Muniyappa R, Levenson AE et al (2016) Effect of Leptin administration on circulating Apolipoprotein CIII levels in patients with Lipodystrophy. J Clin Endocrinol Metab 101(4):1790–1797. https://doi.org/10.1210/jc.2015-3891
72. Keam SJ (2018) Inotersen: first global approval. Drugs 78(13):1371–1376. https://doi.org/10.1007/s40265-018-0968-5
73. Guidotti G, Brambilla L, Rossi D (2017) Cell-penetrating peptides: from basic research to clinics. Trends Pharmacol Sci 38(4):406–424. https://doi.org/10.1016/j.tips.2017.01.003
74. Shabanpoor F, Hammond SM, Abendroth F et al (2017) Identification of a peptide for systemic brain delivery of a Morpholino oligonucleotide in mouse models of spinal muscular atrophy. Nucleic Acid Ther 27(3):130–143. https://doi.org/10.1089/nat.2016.0652
75. Yin H, Moulton HM, Seow Y et al (2008) Cell-penetrating peptide-conjugated antisense oligonucleotides restore systemic muscle and cardiac dystrophin expression and function. Hum Mol Genet 17(24):3909–3918. https://doi.org/10.1093/hmg/ddn293
76. Echigoya Y, Nakamura A, Nagata T et al (2017) Effects of systemic multiexon skipping with peptide-conjugated morpholinos in the heart of a dog model of Duchenne muscular dystrophy. Proc Natl Acad Sci U S A 114(16):4213–4218. https://doi.org/10.1073/pnas.1613203114
77. Huang L, Liu Y (2011) In vivo delivery of RNAi with lipid-based nanoparticles. Annu Rev Biomed Eng 13:507–530. https://doi.org/10.1146/annurev-bioeng-071910-124709
78. Babu A, Wang Q, Muralidharan R et al (2014) Chitosan coated polylactic acid nanoparticle-mediated combinatorial delivery of cisplatin and siRNA/plasmid DNA chemosensitizes cisplatin-resistant human ovarian cancer cells. Mol Pharm 11(8):2720–2733. https://doi.org/10.1021/mp500259e
79. Seth PP, Siwkowski A, Allerson CR et al (2008) Design, synthesis and evaluation of constrained methoxyethyl (cMOE) and constrained ethyl (cEt) nucleoside analogs. Nucleic Acids Symp Ser (Oxf) 52:553–554. https://doi.org/10.1093/nass/nrn280
80. Shen W, De Hoyos CL, Migawa MT et al (2019) Chemical modification of PS-ASO therapeutics reduces cellular protein-binding and improves the therapeutic index. Nature biotechnology:1. Nat Biotechnol 37(6):640–650
81. Rinaldi C, Wood MJA (2018) Antisense oligonucleotides: the next frontier for treatment of neurological disorders. Nat Rev Neurol 14(1):9–21. https://doi.org/10.1038/nrneurol.2017.148

Chapter 6

Inotersen for the Treatment of Hereditary Transthyretin Amyloidosis

Maria Mahfouz, Rika Maruyama, and Toshifumi Yokota

Abstract

Hereditary transthyretin amyloidosis (hATTR) is a rare autosomal dominant condition in which mutations in the transthyretin gene cause amyloid fibrils to develop and deposit into tissues, affecting primarily the nerves and heart causing polyneuropathy and cardiomyopathy respectively. Standard treatment has been liver transplants to try and eliminate the mutated transthyretin products as the liver is the main source of transthyretin production. A new drug named inotersen (brand name Tagsedi), also known as IONIS-TTR_{RX}, has been approved by the United States Food and Drug Agency, Health Canada, and European Commission in 2018, and introduced to the market for patients in stage 1 and stage 2 hATTR polyneuropathy. Inotersen is a second-generation antisense oligonucleotide with 2′-*O*-methoxyethyl modification designed to bind to the 3′ untranslated region of the transthyretin mRNA in the nucleus of the liver cells. By doing so, it prevents the production of the mutant and wild-type forms of transthyretin, impeding the progression of the disease. In this article, the mechanism of action and safety profile of inotersen will be discussed along with some future directions following its approval.

Key words Hereditary transthyretin amyloidosis (hATTR), Familial amyloidosis neuropathy (FAP), Inotersen (Tagsedi), Antisense oligonucleotides (ASOs), 2′-*O*-methoxyethyl (2′-MOE), Ribonuclease H1 (RNase H1), Food and Drug Agency (FDA), Health Canada, European Commission, Patisiran

1 Introduction

Amyloidosis is a cluster of diseases defined by amyloid proteins, which are aggregated and misfolded autologous proteins that form amyloid fibrils and are deposited in tissue [1]. There are more than 30 known amyloid fibrils in humans as amyloidosis is classified according to the precursor protein [2]. Hereditary transthyretin amyloidosis (hATTR) is an autosomal dominant multisystem and progressive disease due to the misfolded and extracellular deposition of the transthyretin (TTR) protein [3]. The liver is the main source of TTR; however, a small percentage (<5%) is synthesized by the brain's choroid plexus and the epithelium of retinal pigment [4]. TTR is a highly conserved protein present in human serum that is primarily responsible for retinol-binding protein (RBP4)

Toshifumi Yokota and Rika Maruyama (eds.), *Gapmers: Methods and Protocols*, Methods in Molecular Biology, vol. 2176, https://doi.org/10.1007/978-1-0716-0771-8_6,

transportation and is also involved in the central nervous system regeneration [4, 5]. Single point mutations in the *TTR* gene, which is located on chromosome 18q, can lead to an amino acid substitution that makes the tertiary structure more susceptible to misfolding and forming β-pleated sheets that are insoluble and deposit as amyloid fibrils [3, 4]. This occurs as TTR naturally exists as tetramers; however, tetramers containing the mutated TTR are kinetically unstable leading to their dissociation and releasing of monomers [6, 7]. As those monomers are no longer confined to the tetramer structure, they misfold and aggregate forming toxic oligomers and eventually deposits as amyloid fibrils [6, 7]. A stepwise diagrammatic representation of the amyloid fibril formation is displayed in Fig. 1 [6]. In hATTR, the amyloid fibrils deposit systemically, mainly in the peripheral nerves and the heart leading to polyneuropathy and cardiomyopathy respectively [8]. Two of the most common gene mutations are Val30Met and Val122Ile, where Val30Met mainly correlates with neuropathic phenotypes and Val122Ile is mainly cardiac phenotypic [9]. hAART is an irreversible and progressive disease where symptoms usually present in the third decade and a survival rate of 2–15 years for neuropathy and 2–5 years for cardiomyopathy after diagnosis [5, 8].

Since hATTR is inherited in an autosomal dominant pattern, both the wild-type and mutant forms of TTR will be produced; thus a mixture of both forms can be deposited in tissue and organs [9]. Therefore, hATTR is dependent on the concentration of TTR synthesized and eventually deposited, allowing for a therapeutic target and strategy [7]. The main established disease-modifying treatment for hATTR is liver transplant to remove the main source of mutated TTR [5]. However, many restrictions apply to liver transplants due to the scarce sources, the donor's age, blood type, size, and health; it is only reserved for relatively healthy patients due to side effects from immunosuppressive drugs [5, 10]. Furthermore, wild-type TTR continues to be produced by the transplanted liver allowing for some amyloid deposition; thus liver transplant can only slow the progression of hATTR but cannot eliminate it [5].

One of the newer and most promising therapeutic approaches for hATTR is gene silencing using antisense oligonucleotides (ASO) [5]. They are short synthetic nucleotides complementary to specific regions on mRNA designed to interfere with its metabolism by binding to the complementary sequence. Inotersen (IONIS-TTR_{RX}) is an ASO designed to bind to the 3′ untranslated region (3′ UTR) of human TTR mRNA in the liver that is a conserved part of the mRNA with no known mutation to date. Thus, it can bind to both mutated and wild-type mRNA. Inotersen degrades both forms of TTR mRNA by binding and hybridizing with the mRNA creating a DNA–RNA complex that activates an RNase H1-mediated breakdown of the mRNA. Therefore, inotersen can reduce the total amount of TTR produced (both mutated

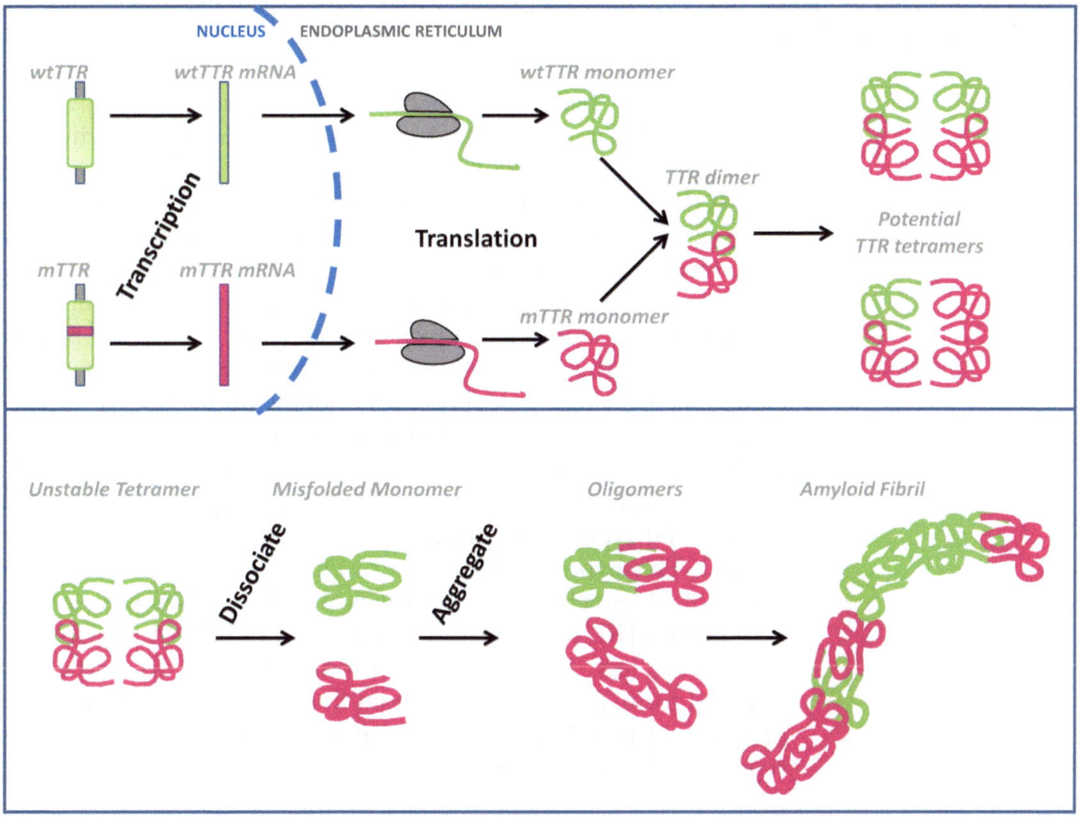

Fig. 1 The mechanism for amyloid fibril formation from mutated transthyretin (mTTR) and wild-type transthyretin (wtTTR). The main source of TTR synthesis is in the liver cells. Transcription occurs in the nucleus, while translation and protein folding occur in the cytoplasm. A single point mutation results in a mutant form of TTR (mTTR) and when combined with the wild-type TTR (wtTTR) it forms an unstable tetramer. After that tetramer is transported outside of the liver into the extracellular space, the monomer that made up the tetramer dissociate and form misfolded monomers. Those misfolded monomers then aggregate together to form toxic insoluble oligomers and if aggregate together once more they form amyloid fibrils that deposit into the tissues. With this repeated cycle, amyloid fibrils composed of both mTTR and wtTTR continue to be deposited into the tissue (particularly nerve and heart) which cause the symptoms of polyneuropathy and restrictive cardiomyopathy respectively

and wild-type) in comparison to liver transplants. Consequently, inotersen can slow down the progression of amyloid deposition.

In 2018, inotersen (brand name Tagsedi) was approved by the European Commission in July and by both the Food and Drug Agency (FDA) and Health Canada in October for the treatment of adults in stages one and two of polyneuropathy hATTR [11]. Stage one includes patients who are ambulatory, while stage two includes patients who are ambulatory but need assistance [12]. Recommended usage is subcutaneous injections once weekly of 284–300 mg dosage that can be self-administered [9, 12]. In this review, the design of the drug, clinical pharmacology, safety, and adverse events will be discussed.

2 Chemistry and Design

Inotersen has been developed by Ionis Pharmaceuticals as a phosphorothioate-linked ASO that is single-stranded and chimeric in origin [13, 14]. Inotersen contains five 2′-*O*-methoxyethyl (2′-MOE) sugar modifications on the 3′ and 5′ ends of the molecule (i.e., a 5-10-5 gapmer); this allows inotersen to be more stable and prevents against nuclease degradation, thus improving its pharmacokinetics [5, 13]. Furthermore, all the pyrimidines in the ribonucleotide sequence (i.e., uracil and cytosine bases) are further methylated in the 5′ position to allow for increased affinity to the complementary base pairs [13]. An example of the first two nucleotide bases of inotersen is represented in Fig. 2a [6, 14]. The 20 base pairs of inotersen (5′ UCUUGGTTACATGAAAUCCC 3′) are designed to bind to the 3′ UTR of hepatic TTR mRNA (both wild-type and mutated), making it highly effective [5, 13]. This hybridized complex of DNA–RNA leads to the activation of RNase H1, which in turn cleaves the mRNA leading to its degradation [5]. By decreasing the amount of mRNA available for translation, less TTR protein is produced (both wild-type and mutated forms); thus fewer amyloid fibrils are formed, halting the progression of hATTR [5]. The mode of action of inotersen is represented in Fig. 2b [6].

3 Preclinical Studies

Inotersen was examined in cell culture using human hepatocellular carcinoma (HCC) cell line HepG2, as well as transgenic mice and cynomolgus monkeys preclinically [15]. In cell culture, HCC HepG2 cell line that expresses wild-type *TTR* gene were treated with inotersen via electroporation [15]. There were lower levels of TTR mRNA observed in the cells treated with inotersen vs non-treated cells [15]. Those results were further confirmed by testing inotersen on hepatic cells isolated from the transgenic mice and cynomolgus monkeys [15].

hATTR patients with Ile84Ser mutations were selected to isolate their genomic DNA from peripheral blood leukocytes [16]. Those DNA samples were cloned and purified to obtain the 15-kb fragment containing the entire Ile84Ser that was subsequently microinjected into mouse embryos to derive the human TTR-Ile84Ser transgenic mouse line [16]. The transgenic mice were injected subcutaneously with either inotersen or with saline [15]. The inotersen doses had a range of 10–100 mg/kg/week that lasted for 4 weeks and the dose level would increase twice/week [15]. At the highest dose group, more than 90% reduction in liver TTR mRNA and serum TTR protein levels were seen compared to

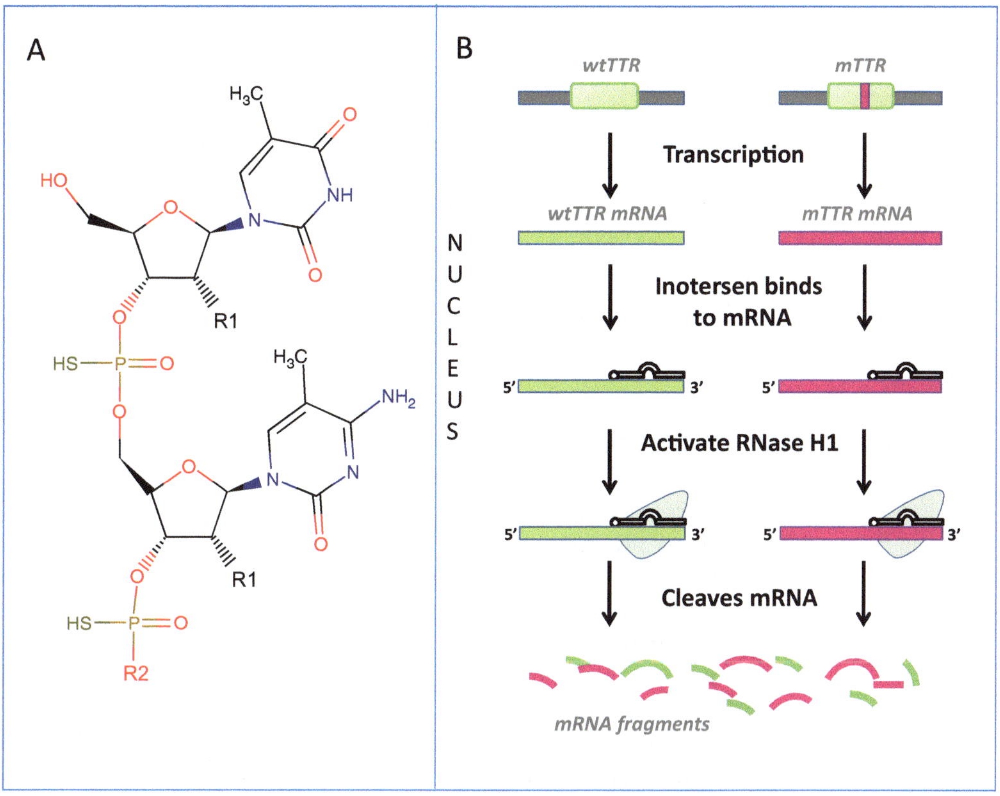

Fig. 2 The structure and inotersen's mechanism for transthyretin mRNA degradation. (**a**) The first two nucleotide bases (ribonucleotide) in the structure of inotersen, a modified 2′-*O*-2 methoxyethyl (MOE) antisense oligonucleotide where R1 = $O(CH_2)_2OCH_3$. Inotersen is composed of 20 nucleotide bases in total (5′UCUUGGTTACATGAAAUCCC3′) where the first and last 5 bases are composed of ribonucleotide bases and the middle 10 are deoxyribonucleotide bases. In comparison to normal nucleotide bases, Inotersen has a phosphorothioate backbone instead of a ribose sugar backbone which allows for easier uptake by the cells and better distribution. Furthermore, all the pyrimidines in the ribonucleotide sequence are further methylated at the 3′ position to allow for more stability and improving the drug's pharmacokinetics. (**b**) The mechanism of action of inotersen for transthyretin (TTR) mRNA degradation via ribonuclease H1 (RNase H1). The 20 base pairs of inotersen are designed to be complementary and bind to the 3′ untranslated region (3'UTR) of the TTR mRNA. As inotersen binds to the mRNA after being transcribed in the nucleus, RNase H1 recognizes this hybridized complex of inotersen and the mRNA as a DNA–RNA complex. In turn, RNase H1 is activated and binds to the complex, which has a targeted mRNA cleavage sequence leading to the mRNA degradation. As seen in the diagram inotersen binds to both mutant transthyretin (mTTR) and wild-type transthyretin (wtTTR) leading to the degradation of both forms of TTR

the control group [15]. Furthermore, there were no detectable reductions in TTR mRNA levels (using RT-PCR) in the brain or the eye (as they are a minor source of TTR production) [15]. This is expected as previous ASO therapies have been proven unable to cross the blood–brain barrier and the blood–retinal barrier [15].

Cynomolgus monkeys with an mRNA sequence 100% complementary to inotersen were also treated with inotersen for 12 weeks [15]. The monkeys were injected subcutaneously three times weekly for the first week with a dose of 75 mg/kg/week followed by twice-weekly subcutaneous injections for the remainder of the 12 weeks with a dose of 50 mg/kg/week [15]. Compared to the control group that was injected with saline with unchanged serum TTR levels, the inotersen treatment group experienced a decrease in serum TTR protein (a mean of 80%) in a time-dependent and progressive pattern where reductions were first seen at 1 week after treatment initiation ($p < 0.05$) [15]. Liver and kidney functions were monitored using blood samples and organ histopathology which showed no adverse advents indicating that the treatment was well tolerated in the monkeys [15].

4 Pharmacodynamics of Inotersen

Inotersen is an ASO with a sequence complementary to the 3′UTR of the TTR mRNA that leads to the mRNA degradation through activating RNase H1, thus reducing the amount of TTR protein produced [5]. In a phase 1 double-blind randomized controlled trial that had been approved by Health Canada, the effects of the drug were tested using single and multiple subcutaneous injections of inotersen into healthy participants for a 4 week treatment period [15]. Throughout the 4 weeks, fasting serum TTR and RBP4 were measured and then were measured again at 10 weeks posttreatment [15]. In the group with multiple doses administered, a reduction of 75% of serum TTR was observed in a dose-dependent manner [5]. Statistically significant reductions were seen in TTR in doses above 50 mg ($p \leq 0.001$) and RBP4 reductions were statistically significant at all dose levels ($p < 0.05$).

Although not yet approved for cardiomyopathy, a phase 2 open-label study investigated the effect of inotersen in patients with proven hATTR via biopsy and cardiomyopathy [9, 17]. Patients had inotersen injections of 300 mg once weekly up to 3 years [9, 17]. In most patients, functional cardiac parameters were stabilized (such as 6 min walk time, left ventricular ejection fraction) [9]. Furthermore, the left ventricle mass was assessed annually using an MRI and no progression was seen during treatment [9]. On the contrary, it had decreased in most patients (8/10 after 2 years) [9].

Internationally, a phase 3 double-blind randomized control trial (NEURO-TTR), was conducted with stage 1 and 2 hATTR polyneuropathy patients for 15 months [18]. The study included a 2:1 ratio of inotersen to placebo where 112 patients received at least one dose of inotersen subcutaneous injections (300 mg once a week) while 60 patients received a placebo. A steady and sustained reduction in serum TTR was observed in the inotersen group by week 13. The decrease of baseline serum TTR was a mean nadir of 74% and a median nadir of 79% from weeks 13 to 65 in the inotersen group. Furthermore, a reduction (indicating improvement) in the baseline scores from the modified Neuropathy Impairment Score+7 and from the Norfolk Quality of Life-Diabetic Neuropathy questionnaire was seen in the inotersen group. Those improvements were seen independently of the mutation type (even though 50% of participants had the Val30Met mutations), disease stage, or cardiomyopathy. Thus, this trial shows the benefits of inotersen in improving the quality of life while also altering the progression of the disease.

5 Pharmacokinetics of Inotersen

In the same phase 1 clinical trial, it was determined that independently of dose concentration, inotersen absorption into the systemic circulation after subcutaneous injection was 1.5–4 h to reach maximum plasma concentration (T_{max}) [15]. However, maximum plasma concentration (C_{max}) was dose-dependent and decline in a biphasic pattern where plasma concentrations decrease more than 90% of C_{max} 24 h after injections [15]. The decline follows a pattern of rapid distribution first then a slower elimination [15]. The same pharmacokinetics patterns were seen in the single and multiple-dose participants [15].

Inotersen is mostly bound to plasma proteins (more than 94%) and is distributed rapidly into the liver and kidney (first phase in plasma concentration decline that prevails the plasma clearance) [15, 19]. After distribution, inotersen is metabolized into short oligonucleotides that are inactive first by endonucleases then by exonucleases [19]. Thus, inotersen is eliminated from the body through the urine as short oligonucleotides that are rendered inactive [19]. Inotersen is slowly metabolized and eliminated from the tissues with a mean elimination plasma half-life of 2–3 weeks [15]. This was expected as the same elimination pattern was noted in other ASO treatments [15, 20, 21].

6 Adverse Effects

In the phase 3 international clinical trial (NEURO-TTR), there were no clinical signs of vitamin A deficiency [18]. However, glomerulonephritis and thrombocytopenia were the main concern [18]. Three patients in the inotersen group (out of 112) experienced glomerulonephritis, and all three had the Val30Met mutation [18]. Renal biopsies were taken and pathology showed a background of amyloidosis and interstitial fibrosis with crescentic glomerulonephritis [18]. One of those three patients was diagnosed at a later stage, leading to a worse prognosis, and the patient was initiated on permanent hemodialysis [18]. Thus a close and frequent renal monitoring program (every 2–3 weeks) and rapid treatment with glucocorticoids to prevent irreversible damage were implemented [18]. A decrease in the postbaseline of platelet count (less than $140,000/m^3$) was seen more frequently in the inotersen group (60/112; 54%) in comparison to the placebo group (8/60; 13%) [18]. Thrombocytopenia (platelet count $<25,000/m^3$) was seen in three patients in the inotersen group [18]. One of these patients presented with a later stage where an intracranial hemorrhage and subsequently died before treatment initiation with glucocorticoids [18]. Thus a close platelet count monitoring program of once-weekly was initiated and no cases with platelet count $<50,000/m^3$ were seen afterward [18]. Furthermore, 2 other patients had a decrease in platelet counts and their inotersen doses were reduced to 150 mg weekly instead of 300 mg weekly [18]. Other side effects that were at least double in the inotersen group compared to the placebo group were nausea, vomiting, pyrexia, chills, and anemia; however, some of these symptoms are often associated with the amyloid disease (i.e., anemia, vomiting) [18]. A total of five deaths occurred during the trial and all five were in the inotersen group [18]. One of these deaths was attributed to the thrombocytopenia adverse event (as discussed above), and the other four were due to hATTR disease progression (cachexia, intestinal perforation, and congestive heart failure) [18].

7 Immunogenicity

In the parent inotersen study (CS2), 30% of patients in the inotersen group had detectable antidrug antibodies (ADA) [14]. Patients were tested for ADA titers on different days throughout the study (from day 1 until day 449); however, the median onset of ADA after 202 days, with a median of another 100 days after onset to reach peak ADA titers [14]. Most titers were relatively low with a median peak of 1:300; however, three patients' titers were above 1:100,000 and one of those patients had to discontinue the study due to a hypersensitivity reaction [14]. A potential cause for the hypersensitivity is thought to be due to complement activation; however, there are limited data regarding C3 activation, and it is unable to provide a definitive correlation between the immunological reaction and C3 activation [14].

In the phase 1 trial, there were no reported injection site reactions with the single-dose participants; on the other hand, injection-site reactions occurred in 4% of the total injections in the multiple-dose group [15]. In the NEURO-TTR study, from the 112 patients in the inotersen group, a mean of 1.1% from all injections had an injection-site reaction [18]. In both studies, the injection-site reactions were mild in nature that spontaneously resolved [15, 18].

8 Pricing

Currently, there is no available information regarding the pricing for inotersen (brand name Tagsedi), however, it is believed to be approximately the same as patisiran (brand name Onpattro) (a siRNA drug that has been approved for the treatment of hATTR polyneuropathy) [13]. The cost of treatment for patisiran is estimated to be $450,000 US dollars and one of the reasons for its high costs is the high investment needed for its development (around 2 billion US dollars) [13]. With such high costs, it is not clear that there will be a return of investment in either patisiran or inotersen especially when they are both competing in the same market and targeting the same disease [13].

9 Current Trials

Most of the ongoing trials of inotersen are conducted to determine the effects of inotersen on hATTR polyneuropathy. One of them is an extension of the global NEURO-TTR study (NCT02175004) is being conducted by Ionis Pharmaceuticals as an open-label extension study to assess the long-term effects of inotersen in hATTR polyneuropathy [9, 22]. The study recruited 135 participants and is studying the effects of inotersen over 5 years (the expected finish date is September 2022) [11]. Some of the primary endpoints of the study include a change in the baseline number of medications used and baseline changes of ophthalmic exams (possible to measure the long-term effects of inotersen on vitamin A levels); while the study's secondary points include changes in baseline TTR and RBP4 and changes in the baseline score numbers of the same questionnaires used in the NEURO-TTR study [22]. Another study is focusing on patients who have discontinued treatment in the NEURO-TTR study (EudraCT2017-005115-13) [9]. It is a phase 3 trial that is noninterventional in nature and focuses on monitoring the progress and survival of patients who have discontinued the inotersen treatment in the NEURO-TTR study or its extension [23].

On the other hand, a phase 2 clinical trial is currently undergoing to study the effects of inotersen in amyloid cardiomyopathy patients (NCT03702829) [11]. It is being conducted as an open-label study for 2 years by recruiting 50 participants with proven amyloid cardiomyopathy via biopsy (due to either wild-type TTR or mutant TTR) [9, 11]. The degree of inhibition by inotersen will be measured via serum TTR and vitamin A levels [11]. Furthermore, cardiac monitoring will be performed to measure disease progression by using the 6-min walk tests, stress tests, and cardiac

MRI [11]. Due to the known adverse effects of glomerulonephritis and thrombocytopenia, a biweekly blood sample will be collected to measure renal function and platelet count [11].

10 Conclusions and Future Directions

With the advances in genetically targeted treatment, inotersen provides new hope for patients with hATTR polyneuropathy. Liver transplantation has been the standard of care historically; however, the disease continues to progress as wild-type TTR continues to be produced and new amyloid fibrils continue to be deposited into tissue and worsen symptoms [18]. On the other hand, inotersen can halt the progression of the amyloid fibrils deposition by targeting the destruction of both mutant and wild-type TTR products, while also improving the quality of life of patients [18]. Following inotersen's (Tagsedi) approval by the FDA, Health Canada, and the European Commission in 2018 for patients in stage 1 and 2 hATTR polyneuropathy, the long-term efficacy and safety of the drug are being tested by multiple preclinical trials [11].

With the promising results from phase 2 trials testing inotersen for patients with hATTR cardiomyopathy, it is hoped that inotersen would be approved for cardiomyopathy as well [9]. In all of the preclinical trials for patients with hATTR polyneuropathy, inotersen shows clear therapeutic benefits for patients on the population level; however, there are still several hurdles. First, it is hard to predict if all patients would respond to the same degree [6]. In all the trials, changes in baseline serum transthyretin levels have been used to compare the differences between the treated and the non-treated groups; however, each patient may not have as strong of a change. Thus the development of a new noninvasive biomarker in the future would be of high clinical interest. Second, delivery of gapmers is a key challenge. Development of tissue-specific delivery methods will greatly advance the field of antisense therapeutics in general. For example, cell-penetrating peptides/moieties, polymers, antibodies, and/or nanoparticles are promising methods to improve delivery [24–28]. Third, antisense chemistry used in gapmers likely has the most influence on toxicity. An overall evaluation of the chemical modifications would be helpful to reduce toxicity. For example, inclusion of 2'-O-methyl oligonucleotides are reported to make gapmers less toxic [29]. With the development of other targeted treatments such as patisiran, the plausible synergistic effects of combined therapeutics of inotersen and patisiran is a new area of interest for research, as they target distinct points in the cascade for the TTR amyloid fibril formation [11].

References

1. Palladini G, Merlini G (2016) What is new in diagnosis and management of light chain amyloidosis? Blood 128(2):159–168. https://doi.org/10.1182/blood-2016-01-629790
2. Sekijima Y (2015) Transthyretin (ATTR) amyloidosis: clinical spectrum, molecular pathogenesis and disease-modifying treatments. J Neurol Neurosurg Psychiatry 86 (9):1036–1043. https://doi.org/10.1136/jnnp-2014-308724
3. Gertz MA (2017) Hereditary ATTR amyloidosis: burden of illness and diagnostic challenges. Am J Manag Care 23(7 Suppl): S107–S112
4. Gertz MA, Benson MD, Dyck PJ et al (2015) Diagnosis, prognosis, and therapy of transthyretin amyloidosis. J Am Coll Cardiol 66 (21):2451–2466. https://doi.org/10.1016/j.jacc.2015.09.075
5. Mathew V, Wang AK (2019) Inotersen: new promise for the treatment of hereditary transthyretin amyloidosis. Drug Des Devel Ther 13:1515–1525. https://doi.org/10.2147/DDDT.S162913
6. Buxbaum JN (2018) Oligonucleotide drugs for transthyretin amyloidosis. N Engl J Med 379(21):2086. https://doi.org/10.1056/NEJMc1810994
7. Johnson SM, Connelly S, Fearns C et al (2012) The transthyretin amyloidoses: from delineating the molecular mechanism of aggregation linked to pathology to a regulatory-agency-approved drug. J Mol Biol 421 (2–3):185–203. https://doi.org/10.1016/j.jmb.2011.12.060
8. Adams D, Gonzalez-Duarte A, O'Riordan WD et al (2018) Patisiran, an RNAi therapeutic, for hereditary transthyretin amyloidosis. N Engl J Med 379(1):11–21. https://doi.org/10.1056/NEJMoa1716153
9. Keam SJ (2018) Inotersen: first global approval. Drugs 78(13):1371–1376. https://doi.org/10.1007/s40265-018-0968-5
10. Mickle K, Lasser KE, Hoch JS et al (2019) The effectiveness and value of patisiran and inotersen for hereditary transthyretin amyloidosis. J Manag Care Spec Pharm 25(1):10–15. https://doi.org/10.18553/jmcp.2019.25.1.010
11. Gales L (2019) Tegsedi (inotersen): an antisense oligonucleotide approved for the treatment of adult patients with hereditary transthyretin amyloidosis. Pharmaceuticals (Basel) 12(2). https://doi.org/10.3390/ph12020078
12. Adams D, Suhr OB, Dyck PJ et al (2017) Trial design and rationale for APOLLO, a phase 3, placebo-controlled study of patisiran in patients with hereditary ATTR amyloidosis with polyneuropathy. BMC Neurol 17 (1):181. https://doi.org/10.1186/s12883-017-0948-5
13. Al Shaer D, Al Musaimi O, Albericio F et al (2019) 2018 FDA tides harvest. Pharmaceuticals (Basel) 12(2). https://doi.org/10.3390/ph12020052
14. Summers JS, Shaw BR (2001) Boranophosphates as mimics of natural phosphodiesters in DNA. Curr Med Chem 8(10):1147–1155
15. Ackermann EJ, Guo S, Benson MD et al (2016) Suppressing transthyretin production in mice, monkeys and humans using 2nd-generation antisense oligonucleotides. Amyloid 23(3):148–157. https://doi.org/10.1080/13506129.2016.1191458
16. Benson MD, Kluve-Beckerman B, Zeldenrust SR et al (2006) Targeted suppression of an amyloidogenic transthyretin with antisense oligonucleotides. Muscle Nerve 33(5):609–618. https://doi.org/10.1002/mus.20503
17. Benson MD, Dasgupta NR, Rissing SM et al (2017) Safety and efficacy of a TTR specific antisense oligonucleotide in patients with transthyretin amyloid cardiomyopathy. Amyloid 24(4):219–225. https://doi.org/10.1080/13506129.2017.1374946
18. Benson MD, Waddington-Cruz M, Berk JL et al (2018) Inotersen treatment for patients with hereditary transthyretin amyloidosis. N Engl J Med 379(1):22–31. https://doi.org/10.1056/NEJMoa1716793
19. Lu QL, Liang HD, Partridge T, Blomley MJ (2003) Microbubble ultrasound improves the efficiency of gene transduction in skeletal muscle in vivo with reduced tissue damage. Gene Ther 10(5):396–405
20. Yu RZ, Grundy JS, Geary RS (2013) Clinical pharmacokinetics of second generation antisense oligonucleotides. Expert Opin Drug Metab Toxicol 9(2):169–182. https://doi.org/10.1517/17425255.2013.737320
21. Geary RS (2009) Antisense oligonucleotide pharmacokinetics and metabolism. Expert Opin Drug Metab Toxicol 5(4):381–391. https://doi.org/10.1517/17425250902877680
22. NIH US National Library of Medicine : Open-Label Extension Assessing Long Term Safety and Efficacy of IONIS-TTR Rx in Familial Amyloid Polyneuropathy (FAP). https://

clinicaltrials.gov/ct2/show/NCT02175004. Accessed 11 July 2019
23. Adams AM, Harding PL, Iversen PL et al (2007) Antisense oligonucleotide induced exon skipping and the dystrophin gene transcript: cocktails and chemistries. BMC Mol Biol 8:57
24. Echigoya Y, Nakamura A, Nagata T et al (2017) Effects of systemic multiexon skipping with peptide-conjugated morpholinos in the heart of a dog model of Duchenne muscular dystrophy. Proc Natl Acad Sci U S A 114 (16):4213–4218. https://doi.org/10.1073/pnas.1613203114
25. Lu QL, Rabinowitz A, Chen YC et al (2005) Systemic delivery of antisense oligoribonucleotide restores dystrophin expression in body-wide skeletal muscles. Proc Natl Acad Sci U S A 102(1):198–203. https://doi.org/10.1073/pnas.0406700102
26. Ezzat K, Aoki Y, Koo T et al (2015) Self-assembly into nanoparticles is essential for receptor mediated uptake of therapeutic antisense oligonucleotides. Nano Lett 15 (7):4364–4373. https://doi.org/10.1021/acs.nanolett.5b00490
27. Walker I, Irwin WJ, Akhtar S (1995) Improved cellular delivery of antisense oligonucleotides using transferrin receptor antibody-oligonucleotide conjugates. Pharm Res 12 (10):1548–1553
28. Echigoya Y, Aoki Y, Miskew B et al (2015) Long-term efficacy of systemic multiexon skipping targeting dystrophin exons 45-55 with a cocktail of vivo-morpholinos in mdx52 mice. Mol Ther Nucleic Acids 4:e225. https://doi.org/10.1038/mtna.2014.76
29. Wen Shen, Cheryl L. De Hoyos, Michael T. Migawa, Timothy A. Vickers, Hong Sun, Audrey Low, Thomas A. Bell, Meghdad Rahdar, Swagatam Mukhopadhyay, Christopher E. Hart, Melanie Bell, Stan Riney, Susan F. Murray, Sarah Greenlee, Rosanne M. Crooke, Xue-hai Liang, Punit P. Seth, Stanley T. Crooke, (2019) Chemical modification of PS-ASO therapeutics reduces cellular protein-binding and improves the therapeutic index. Nature Biotechnology 37(6):640–650

Chapter 7

Degradation of Toxic RNA in Myotonic Dystrophy Using Gapmer Antisense Oligonucleotides

Quynh Nguyen and Toshifumi Yokota

Abstract

Myotonic dystrophy (DM) types 1 (DM1) and 2 (DM2) are caused by autosomal dominant gain-of-function RNA which are, in turn, created by the expansion of repeat sequences in the *DMPK* and *ZNF9* genes, respectively. The expansions are highly unstable and biased for further expansion in somatic cells and across generations. Despite the different genes involved, DM1 and DM2 share several clinical features due to having the similar underlying mechanism of repetitive RNA-mediated toxicity. Both disorders manifest as multisystemic conditions with features including myotonia, cataract development, and abnormalities in cardiac conduction. At present, there is no cure for DM and treatments mostly aim at symptom management. Among the therapeutics being developed, antisense therapy using gapmers is one of the most promising. Compared to other antisense oligonucleotides, gapmers maintain the ability to induce RNase H cleavage while having enhanced target binding affinity and nuclease resistance. This chapter will consolidate the different strategies studied thus far to develop a treatment for DM1 through the targeting of toxic repetitive RNA using gapmers.

Key words ZNF9, Insulin receptor (IR), Cardiac troponin T (TNNT3), CUG-binding protein 1 (CUGBP1), Toxic RNA, IONIS (ISIS)-DMPKRx (IONIS-DMPK-2.5Rx), ClC-1, DMPK, Muscleblind-like (MBNL) family, Locked nucleic acid (LNA)

1 Myotonic Dystrophy

Myotonic dystrophy (dystrophia myotonic, DM) types 1 (DM1) and 2 (DM2) are progressive multisystemic autosomal dominant disorders caused by repetitive RNA-mediated toxicity. The global prevalence of DM varies across sources [1]. DM1 prevalence in Europe was estimated to range from 1 in 8300 to 10,700 individuals [2, 3]. Based on available epidemiological studies, the estimated DM1 gene frequency is 1 in 7400 [1]. Due to the founder effect, DM1 is more prevalent in certain regions such as Quebec, where the incidence is as high as 1 in 500 [4]. There are fewer epidemiological studies on DM2. A study on Finnish blood donors found DM2 to have a similar prevalence as DM1 [5]. In the USA, DM2 is estimated to be five times less common than DM1

Toshifumi Yokota and Rika Maruyama (eds.), *Gapmers: Methods and Protocols*, Methods in Molecular Biology, vol. 2176, https://doi.org/10.1007/978-1-0716-0771-8_7, © Springer Science+Business Media, LLC, part of Springer Nature 2020

[1]. DM2 mutations have been identified predominantly in European descendants and occur on a specific haplotype, suggesting a predisposing mutation from a founding ancestor [6].

DM is caused by RNA toxicity due to a tandem repeat expansion [7]. DM1 individuals have been reported with at least 50 and in some cases up to 3000 CTG repeats in the 3′ UTR of the *DMPK* gene, while healthy people have from 5 to 37 repeats [7]. CTG repeats are highly unstable and biased for further expansion when transmitted from one generation to the next—a phenomenon termed anticipation [8, 9]. Expansion is also unstable in somatic cells, particularly in nondividing cells such as skeletal myocytes, cardiomyocytes, and neurons [10, 11]. Similarly, individuals with DM2 have anywhere from 75 to 1000 of CCTG repeats within the first intron of the *ZNF9* gene, compared to 10–33 repeats in healthy people [12–14]. Anticipation and intergenerational expansion are less pronounced in DM2 [14, 15]. Despite their genetic differences, both DM subtypes share common clinical features due to having the similar underlying mechanism of RNA-caused toxicity [1, 16–18]. Since the majority of preclinical studies were done on DM1 models, the rest of this book chapter will focus only on DM1.

The spectrum of clinical presentation in DM1 patients extends from congenital manifestation to late onset or asymptomatic depending on the repeat expansion size. The majority of patients become symptomatic in adulthood [1, 17]. Myotonia and progressive muscle wasting and weakness are the most common cardinal findings. Neuromuscular manifestations in DM1 are most pronounced in the cranial, trunk, and distal limb muscles [1]. Other systemic features include cardiac conduction defects, cataracts, and cognitive changes, among others [1].

RNA-mediated toxicity is the underlying mechanism of DM pathogenesis [1, 19] (Fig. 1). The expanded repeat in the *DMPK* gene results in the production of mutant mRNA containing numerous CUG repeats. These transcripts are retained within the nucleus and trap CUG-binding proteins such as members of the muscleblind-like (MBNL) family which regulate splicing, transport, and decay of several hundred transcripts [20–24]. Splicing regulation is further compromised by activation of signaling pathways that results in the phosphorylation and stabilization of another splicing factor, CUG-binding protein 1 (CUGBP1) [16, 25, 26]. Together these lead to expression of incorrectly spliced mRNA products. For example, splicing defects of the muscle chloride channel (*ClC-1*), the insulin receptor (*IR*) and cardiac troponin T (*TNNT3*) transcripts lead to myotonia, insulin resistance and cardiac abnormalities, respectively [26–28]. Furthermore, transcripts from the mutant *DMPK* allele alter the expression of various transcription factors and miRNAs, which lead to changes in gene expression [16, 21]. Repetitive RNAs

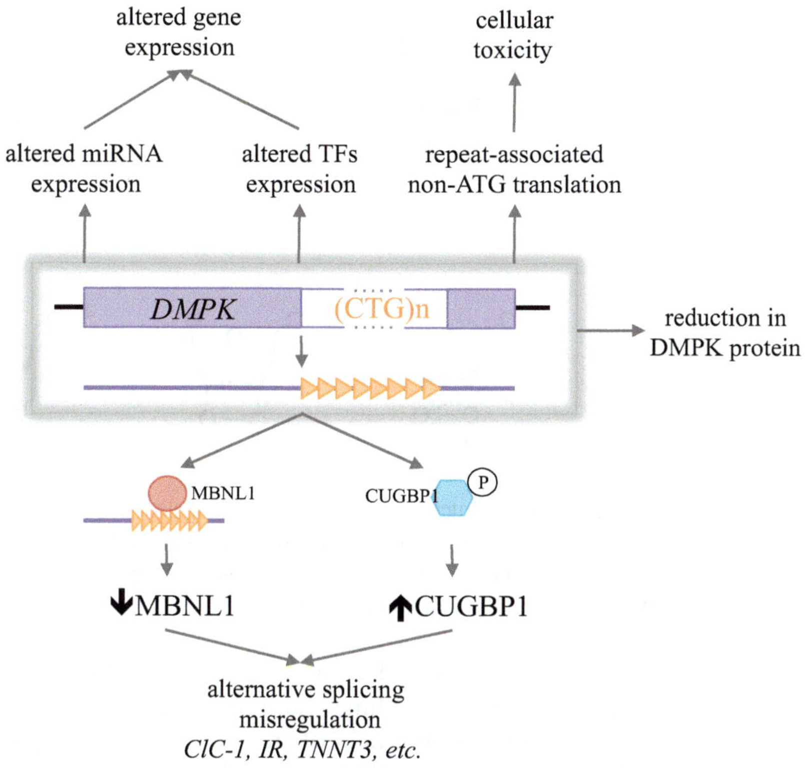

Fig. 1 Pathophysiology of myotonic dystrophy type I. The expanded CTG repeat in the *DMPK* gene results in the production of mutant mRNA containing numerous repeats, which mediate numerous deleterious intracellular events. *TFs* transcription factors, *MBNL1* muscleblind-like protein 1, *CUGBP1* CUG-binding protein 1, *ClC-1* chloride channel 1, *IR* insulin receptor, *TNNT3* troponin T3

may also have unusual interactions with the translation machinery, producing proteins that are potentially toxic for the cell [29]. Reduction in DMPK protein levels due to retained transcripts in the nucleus may also contribute to the overall pathogenesis [30].

Currently, there is no cure for DM. Medical therapy mostly aims at preserving function and preventing disease progression [1, 16–18]. A lot of preclinical studies have been looking at alternative approaches to lessen the severity of DM1. In this book chapter, we will discuss the use of gapmer antisense oligonucleotides (ASOs) in the treatment of DM1.

2 Antisense Oligonucleotides

ASOs are short, synthetic nucleic acids that are capable of altering gene expression. Here, we only consider ASOs that act by promoting the degradation of complementary transcripts via an RNase H-dependent mechanism [31–33]. Currently, such ASOs are mostly used as gapmers, which consist of a central DNA sequence

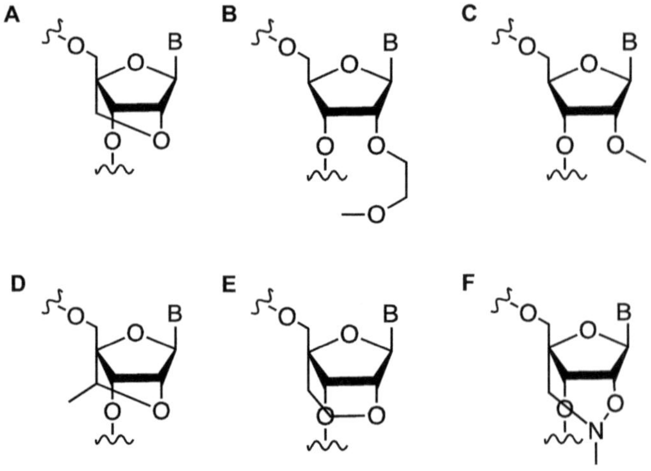

Fig. 2 Chemical structures of antisense oligonucleotides mentioned in this book chapter. (**a**) locked nucleic acid, (**b**) 2′-*O*-methoxyethyl, (**c**) 2′-*O*-methyl, (**d**) constrained ethyl, (**e**) ethylene-bridged nucleic acid, (**f**) 2′-*O*,4′-C-aminomethylene bridged nucleic acid

that induces RNase H activity, and chemically modified ends to increase ASO binding affinity and nuclease resistance [34]. Advances in the field of antisense therapy are partly attributable to progress in ASO chemical modification. These include alternative backbones, such as phosphorothioate (PS) bridges; 2′ sugar modifications, such as 2′-*O*-methyl (2′-OMe) and 2′-*O*-methoxyethyl (2′-MOE); and bridged sugar rings, such as locked nucleic acid (LNA), ethylene-bridged nucleic acid (ENA), constrained ethyl (cEt), and 2′-*O*,4′-C-aminomethylene bridged nucleic acid (BNA^{NC}) (Fig. 2).

The PS backbone is among the most important modifications for recent works on antisense therapy [35]. The negatively charged oxygen is replaced by a sulfur atom, making the PS backbone less susceptible to nuclease recognition. The PS backbone also has a higher protein binding affinity, therefore decreasing renal clearance and increasing ASO half-life. Notably, the PS modification retains the capacity to induce RNase H activity and is the most commonly used backbone for making gapmers [35].

Alterations at the 2′ position of the ribose sugar have yielded a class of ASOs with improved safety and efficacy profiles. The most common 2′-modified chemistries are the 2′-OMe and 2′-MOE [34, 36]. These modified chemistries do not support RNase H activity; instead, they are often used at the ends of gapmers to increase target binding affinity and ASO nuclease resistance. Alternatively, the LNA, ENA, cEt and BNA^{NC} chemistries all involve bridging of the sugar ring into conformations that have better binding affinity and nuclease resistance [37, 38]. Similar to

2′ modifications, these chemistries do not possess an RNase H-inducing ability and often flank both sides of the central DNA sequence in gapmers.

3 Gapmers for the Treatment of Myotonic Dystrophy

This section will discuss antisense approaches using gapmers to target toxic RNA in DM1. ASOs would be advantageous for the treatment of monogenic disorders caused by gain-of-function mutations like DM1. Strategies are organized according to the ASO chemistry used. Not all experiments are mentioned in this chapter, but a selected few will be discussed for each approach.

3.1 LNA Gapmers

The first LNA gapmers used for DM1 treatment were fully phosphorothioated and contained a central stretch of ten DNA nucleotides flanked by four LNA residues on the 3′ and 5′ ends [39]. These ASOs were designed to bind the CUG expansion (CUG^{exp}) in mutant *DMPK* transcripts. To test these gapmers, the authors generated a variant of the HT1080 human fibrosarcoma cell line that had 800 DMPK 3′UTR CTG repeats stably integrated into its genome. Treatment of this line with LNA gapmers significantly reduced the percentage of cells with CUG^{exp} nuclear foci to 6% compared to 49% in the non-treated group. Addition of gapmers to the culture media reduced the levels of *DMPK* 3′ UTR and CUG^{exp} transcripts in *DMPK*-transfected cells to levels similar to the nontransfected control group. The repeat instability in gapmer-treated cells was significantly reduced, and the average CUG repeat expansion was smaller than in untreated cells.

In another study, by Lee and colleagues, treatment with LNA gapmers resulted in a 70% reduction in DMPK-containing plasmids as well as a significant reduction of RNA foci in the nucleus of transfected COSM6 cells [40]. This is an interesting finding since the target was RNA, and yet plasmid levels were also affected. The data also suggested that LNA gapmers displayed selective knockdown, preferentially targeting transcripts containing longer repeats. However, treatment specificity in the context of the presence of other non-*DMPK* transcripts remains to be seen.

Despite encouraging in vitro results, in vivo treatment only showed a modest effect [39]. Gapmer-treated tissues had a lower frequency of unstable alleles and reduced somatic instability compared to their non-treated counterparts. The low efficacy may reflect a limitation in the experimental system, as the baseline repeat instability could not be reliably quantified prior to ASO administration in the treated animals.

3.2 2′-MOE Gapmers

Among all the gapmers assessed thus far, those with the 2′-MOE chemistry have advanced the most in therapeutic development for DM1. 2′-MOE gapmers were first tested as a therapy for DM1 by Lee and colleagues in 2012 [40]. In vitro treatment with CAG-targeting 2′-MOE gapmers resulted in an 80% reduction in *DMPK*-containing plasmid compared to controls. Intramuscular injection of these gapmers led to a 50% decrease in mutant transcript levels and a 40% decrease in the average number of RNA foci per nuclei.

The application of 2′-MOE gapmers for the in vivo correction of DM1 was most extensively evaluated by Wheeler and colleagues [41]. A mouse model of DM1 was injected subcutaneously with 2′-MOE gapmers targeting the CUG repeats at 25 mg/kg twice weekly and euthanized 4 weeks postinjection. CUG^{exp} levels were reduced by more than 80% in the hind limb muscles. Reduction of CUG^{exp} RNA was shown to correct downstream splicing abnormalities caused by sequestered MBNL protein, as demonstrated by the normalized alternative splicing of MBNL-dependent genes. Microarray data showed an overall improvement in the muscle transcriptome, with more than 85% of transcripts having normalized or partially corrected expression. The therapeutic effects of the administered 2′-MOE gapmers persisted for up to 1 year following the discontinuation of ASO treatment. Long-term reduction of CUG^{exp} levels was demonstrated to be muscle-protective, as evidenced by a reduced frequency of central nuclei and improved muscle fiber diameter in treated mice. Since RNase H cleavage is mainly a nuclear process, mutant *DMPK* transcripts are advantageously more sensitive to RNase H-mediated treatments such as gapmers due to their high residence time in the nucleus. A study by Hu and colleagues further found that conjugation of a C16 lipophilic group at the 5′ end of 2′-MOE gapmers further improved their efficacy [42].

One 2′-MOE gapmer candidate from Ionis Pharmaceuticals, named IONIS-DMPKRx or IONIS-DMPK-2.5Rx, was eventually tested in a clinical trial (NCT02312011) [10, 40–42]. However, further development was discontinued owing to limited uptake into muscles and poor efficacy [43].

3.3 2′-OMe Gapmers

In an effort to identify the specific properties of ASOs required for therapeutic efficacy in DM1, Gonzalez-Barriga and colleagues investigated different ASO lengths, nucleotide chemistries and sequence designs for their ability to knockdown *DMPK* transcripts containing expanded CUG repeats in DM1 cell models [44]. In terms of gapmer conformation, two chemistries were tested, 2′-OMe and ENA. ENA gapmers appeared to be ineffective, whereas 2′-OMe gapmers were able to knock down 80% of CUG^{exp} transcripts in vitro. However, the 2′-OMe gapmers only showed modest selectivity between expanded and normal-sized *DMPK* transcripts. This is contrary to previous studies which demonstrated that LNA gapmers preferentially targeted transcripts containing longer CUG repeats [39, 40]. The discrepancy was speculated to be due to differences in oligonucleotide chemistry and experimental condition; for instance, the poor internalization of ENA gapmers in the nuclei makes it difficult for them to target toxic RNAs that had high nuclear residence times [44].

3.4 cEt Gapmers

Another chemistry that was extensively evaluated in vivo is the cEt gapmer. The cEt ASOs have improved in vivo efficacy compared to 2′-MOE ASOs, as well as a favorable safety profile [45, 46]. Pandey and colleagues evaluated cEt gapmers for targeting mutant *DMPK* transcripts in both murine and primate models of DM1 [47]. Subcutaneous injection at 12.5 or 25 mg/kg body weight twice a week for 6 weeks in wild type mice and rats showed a dose-dependent reduction of up to 90% in mutant *DMPK* transcripts in the liver and skeletal muscle. Treatment was well tolerated, although a significant increase in spleen weight was observed in treated rats. In a mouse model of DM1, systemic administration of the cEt gapmer at 12.5, 25, or 50 mg/kg body weight twice weekly for 6 weeks resulted in a dose-dependent reduction in *DMPK* mRNA levels of up to 60% in the heart and 70% in the diaphragm. Gapmer activity against mutant transcripts showed long-term efficacy following treatment discontinuation. Mutant *DMPK* expression remained strongly suppressed at 15 weeks posttreatment and below baseline at 31 weeks posttreatment. From a safety perspective, treatment was generally well tolerated with a transient elevation of plasma creatinine observed.

Cynomolgus monkeys were treated systemically with 40 mg/kg at day 1, 3, 5, and 7, and then weekly for a total of 13 weeks [47]. Transcript knockdown was observed as early as week 7 of treatment with up to 80% reduction of *DMPK* mRNA. Quantitative PCR results showed 50–70% reduction of mutant mRNAs in several muscles including the heart. However, treatment had no effect on the cardiac conduction phenotype in treated animals. At 13 weeks posttreatment, some muscles had mutant mRNA levels that were 60% of those found in controls, whereas in other muscles these amounts returned to pretreatment levels. Treatment was generally well tolerated with no major adverse effects noted. Liver weights were significantly increased, but the values were still within the normal range.

3.5 BNA^{NC} Gapmers

The BNA^{NC} chemistry is a member of the bridged nucleic acid family. BNA^{NC} and LNA are structurally similar; however, BNA^{NC} has been shown to be better tolerated in animal models [38]. In a study by Manning and colleagues, gapmers were designed to target either the CUG repeat (CAG gapmer) or a nonrepetitive region of the *DMPK* gene (DMPK gapmer) [48]. DMPK gapmers have the potential to reduce off-target effects, given the relatively higher ubiquity of the CAG repeat in the genome. In vitro results indicated that BNA^{NC} gapmers are effective at knocking down CUG^{exp} RNA. DMPK gapmers were more effective than CAG gapmers at a higher dose, but CAG gapmers displayed more potent knockdown than DMPK gapmers at the lower dose. DMPK gapmers preferentially targeted CUGexp transcripts without affecting expression of the nonexpanded allele, explained as potentially due to the retention of expanded RNAs in the nucleus. This selective knockdown was not observed with CAG gapmers. In terms of functional rescue, treatment with BNA^{NC} gapmers reduced the number of nuclear RNA foci. Despite only modestly improving MBNL activity, splicing abnormalities were, nevertheless, corrected as demonstrated by significant reverted splicing of MBNL-dependent transcripts.

4 Conclusion

In summary, antisense therapy using gapmers is a promising approach to treat monogenic genetic disorders caused by gain-of-function mutations such as DM1. The gapmer conformation is favorable as it has enhanced binding affinity and nuclease resistance while maintaining the ability to support RNase H activity. Since RNase H functions predominantly in the nucleus, gapmers are ideal for DM1 treatment because repetitive RNAs are trapped within the nucleus. Gapmer-mediated treatment has the potential for selective knockdown, as it preferentially targets mutant transcripts with longer nucleus residency time while sparing cytoplasmic transcripts.

With the exception of the BNA^{NC} DMPK gapmer, other sequences were designed to target the CTG repeat. More thorough studies on the issue of off-target effects are needed to evaluate the feasibility of translating these preclinical approaches to the clinic. In addition, the majority of studies reported on DM1 treatment focused on skeletal features. DM1 is a multisystemic disorder with other signs such as cataract and abnormal cardiac conduction phenotypes. It would be interesting to see if the reduction in repeat expansion and repeat instability is adequate for reversing or preventing the progression of such other manifestations in DM1 animal models. Preclinical studies on larger animal models such as cynomolgus monkeys are also valuable since these models more closely resemble the human body in terms of size and physiology.

Finally, delivery to target tissues has always been a major issue in antisense therapy. As a case in point, the first clinical trial for DM1 using 2′-MOE gapmers was recently completed but yielded unsatisfactory results due to inefficient delivery of the drug into muscles. Recent advances in the antisense field such as the development of peptides and nanoparticles for ASO delivery may improve uptake in tissues such as the myocardium [32, 49]. It would be interesting to see how these new methods of delivery will affect the efficacy, safety, tolerability, as well as pharmacokinetics of antisense drugs in humans, particularly for DM. Continued work on these efforts should help lead to a cure or intervention that will significantly reduce the symptom burden experienced by DM patients in the near future.

References

1. Thornton CA (2014) Myotonic dystrophy. Neurol Clin 32(3):705–719. https://doi.org/10.1016/j.ncl.2014.04.011
2. Siciliano G, Manca M, Gennarelli M et al (2001) Epidemiology of myotonic dystrophy in Italy: re-apprisal after genetic diagnosis. Clin Genet 59(5):344–349
3. Magee A, Nevin NC (1999) The epidemiology of myotonic dystrophy in Northern Ireland. Community Genet 2(4):179–183. https://doi.org/10.1159/000016209
4. Yotova V, Labuda D, Zietkiewicz E et al (2005) Anatomy of a founder effect: myotonic dystrophy in Northeastern Quebec. Hum Genet 117 (2-3):177–187. https://doi.org/10.1007/s00439-005-1298-8
5. Suominen T, Bachinski LL, Auvinen S et al (2011) Population frequency of myotonic dystrophy: higher than expected frequency of myotonic dystrophy type 2 (DM2) mutation in Finland. Eur J Hum Genet 19(7):776–782. https://doi.org/10.1038/ejhg.2011.23
6. Bachinski LL, Udd B, Meola G et al (2003) Confirmation of the type 2 myotonic dystrophy $(CCTG)_n$ expansion mutation in patients with proximal myotonic myopathy/proximal myotonic dystrophy of different European origins: a single shared haplotype indicates an ancestral founder effect. Am J Hum Genet 73 (4):835–848. https://doi.org/10.1086/378566
7. Brook JD, McCurrach ME, Harley HG et al (1992) Molecular basis of myotonic dystrophy: expansion of a trinucleotide (CTG) repeat at the 3′ end of a transcript encoding a protein kinase family member. Cell 68(4):799–808
8. Redman JB, Fenwick RG Jr, Fu YH et al (1993) Relationship between parental trinucleotide GCT repeat length and severity of myotonic dystrophy in offspring. JAMA 269 (15):1960–1965
9. De Temmerman N, Sermon K, Seneca S et al (2004) Intergenerational instability of the expanded CTG repeat in the DMPK gene: studies in human gametes and preimplantation embryos. Am J Hum Genet 75(2):325–329. https://doi.org/10.1086/422762
10. Ashizawa T, Dubel JR, Harati Y (1993) Somatic instability of CTG repeat in myotonic dystrophy. Neurology 43(12):2674–2678
11. Thornton CA, Johnson K, Moxley RT 3rd (1994) Myotonic dystrophy patients have larger CTG expansions in skeletal muscle than in leukocytes. Ann Neurol 35(1):104–107. https://doi.org/10.1002/ana.410350116
12. Liquori CL, Ricker K, Moseley ML et al (2001) Myotonic dystrophy type 2 caused by a CCTG expansion in intron 1 of ZNF9. Science 293 (5531):864–867. https://doi.org/10.1126/science.1062125
13. Bachinski LL, Czernuszewicz T, Ramagli LS et al (2009) Premutation allele pool in myotonic dystrophy type 2. Neurology 72 (6):490–497. https://doi.org/10.1212/01.wnl.0000333665.01888.33
14. Day JW, Ricker K, Jacobsen JF et al (2003) Myotonic dystrophy type 2: molecular, diagnostic and clinical spectrum. Neurology 60 (4):657–664
15. Schneider C, Ziegler A, Ricker K et al (2000) Proximal myotonic myopathy: evidence for anticipation in families with linkage to chromosome 3q. Neurology 55(3):383–388
16. Meola G, Cardani R (2015) Myotonic dystrophies: an update on clinical aspects, genetic, pathology, and molecular pathomechanisms. Biochim Biophys Acta 1852(4):594–606. https://doi.org/10.1016/j.bbadis.2014.05.019
17. Turner C, Hilton-Jones D (2014) Myotonic dystrophy: diagnosis, management and new therapies. Curr Opin Neurol 27(5):599–606. https://doi.org/10.1097/WCO.0000000000000128
18. Udd B, Krahe R (2012) The myotonic dystrophies: molecular, clinical, and therapeutic challenges. Lancet Neurol 11(10):891–905. https://doi.org/10.1016/S1474-4422(12)70204-1
19. Osborne RJ, Thornton CA (2006) RNA-dominant diseases. Hum Mol Genet 15

(2):R162–R169. https://doi.org/10.1093/hmg/ddl181
20. Ho TH, Charlet BN, Poulos MG et al (2004) Muscleblind proteins regulate alternative splicing. EMBO J 23(15):3103–3112. https://doi.org/10.1038/sj.emboj.7600300
21. Lee JE, Cooper TA (2009) Pathogenic mechanisms of myotonic dystrophy. Biochem Soc Trans 37(Pt 6):1281–1286. https://doi.org/10.1042/BST0371281
22. Miller JW, Urbinati CR, Teng-Umnuay P et al (2000) Recruitment of human muscleblind proteins to (CUG)(n) expansions associated with myotonic dystrophy. EMBO J 19 (17):4439–4448. https://doi.org/10.1093/emboj/19.17.4439
23. Mulders SA, van den Broek WJ, Wheeler TM et al (2009) Triplet-repeat oligonucleotide-mediated reversal of RNA toxicity in myotonic dystrophy. Proc Natl Acad Sci U S A 106 (33):13915–13920. https://doi.org/10.1073/pnas.0905780106
24. Wang ET, Cody NA, Jog S et al (2012) Transcriptome-wide regulation of pre-mRNA splicing and mRNA localization by muscleblind proteins. Cell 150(4):710–724. https://doi.org/10.1016/j.cell.2012.06.041
25. Kuyumcu-Martinez NM, Wang GS, Cooper TA (2007) Increased steady-state levels of CUGBP1 in myotonic dystrophy 1 are due to PKC-mediated hyperphosphorylation. Mol Cell 28(1):68–78. https://doi.org/10.1016/j.molcel.2007.07.027
26. Mankodi A, Takahashi MP, Jiang H et al (2002) Expanded CUG repeats trigger aberrant splicing of ClC-1 chloride channel pre-mRNA and hyperexcitability of skeletal muscle in myotonic dystrophy. Mol Cell 10 (1):35–44
27. Lueck JD, Mankodi A, Swanson MS et al (2007) Muscle chloride channel dysfunction in two mouse models of myotonic dystrophy. J Gen Physiol 129(1):79–94. https://doi.org/10.1085/jgp.200609635
28. Savkur RS, Philips AV, Cooper TA (2001) Aberrant regulation of insulin receptor alternative splicing is associated with insulin resistance in myotonic dystrophy. Nat Genet 29 (1):40–47. https://doi.org/10.1038/ng704
29. Zu T, Gibbens B, Doty NS et al (2011) Non-ATG-initiated translation directed by microsatellite expansions. Proc Natl Acad Sci U S A 108 (1):260–265. https://doi.org/10.1073/pnas.1013343108
30. Maeda M, Taft CS, Bush EW et al (1995) Identification, tissue-specific expression, and subcellular localization of the 80- and 71-kDa forms of myotonic dystrophy kinase protein. J Biol Chem 270(35):20246–20249
31. Nguyen Q, Yokota T (2017) Immortalized muscle cell model to test the exon skipping efficacy for Duchenne muscular dystrophy. J Pers Med 7:E13. https://doi.org/10.3390/jpm7040013
32. Nguyen Q, Yokota T (2019) Antisense oligonucleotides for the treatment of cardiomyopathy in Duchenne muscular dystrophy. Am J Transl Res 11(3):1202–1218
33. Lee J, Yokota T (2013) Antisense therapy in neurology. J Pers Med 3:144–176. https://doi.org/10.3390/jpm3030144
34. Juliano RL (2016) The delivery of therapeutic oligonucleotides. Nucleic Acids Res 44:6518–6548. https://doi.org/10.1093/nar/gkw236
35. Eckstein F (2014) Phosphorothioates, essential components of therapeutic oligonucleotides. Nucleic Acid Ther 24(6):374–387. https://doi.org/10.1089/nat.2014.0506
36. Juliano RL, Carver K (2015) Cellular uptake and intracellular trafficking of oligonucleotides. Adv Drug Deliv Rev 87:35–45. https://doi.org/10.1016/j.addr.2015.04.005
37. Campbell MA, Wengel J (2011) Locked vs. unlocked nucleic acids (LNA vs. UNA): contrasting structures work towards common therapeutic goals. Chem Soc Rev 40(12):5680–5689. https://doi.org/10.1039/c1cs15048k
38. Rahman SM, Seki S, Obika S et al (2008) Design, synthesis, and properties of 2′,4′-BNA(NC): a bridged nucleic acid analogue. J Am Chem Soc 130(14):4886–4896. https://doi.org/10.1021/ja710342q
39. Nakamori M, Gourdon G, Thornton CA (2011) Stabilization of expanded (CTG)*(CAG) repeats by antisense oligonucleotides. Mol Ther 19(12):2222–2227. https://doi.org/10.1038/mt.2011.191
40. Lee JE, Bennett CF, Cooper TA (2012) RNase H-mediated degradation of toxic RNA in myotonic dystrophy type 1. Proc Natl Acad Sci U S A 109(11):4221–4226. https://doi.org/10.1073/pnas.1117019109
41. Wheeler TM, Leger AJ, Pandey SK et al (2012) Targeting nuclear RNA for in vivo correction of myotonic dystrophy. Nature 488 (7409):111–115. https://doi.org/10.1038/nature11362
42. Hu N, Antoury L, Baran TM et al (2018) Non-invasive monitoring of alternative splicing outcomes to identify candidate therapies for myotonic dystrophy type 1. Nat Commun 9

(1):5227. https://doi.org/10.1038/s41467-018-07517-y

43. Ionis Pharmaceuticals (2017) Ionis Pharmaceuticals Reports on DMPKRx Phase 1/2 Clinical Trial. https://us8.campaign-archive.com/?u=8f5969cac3271759ce78c8354&id=8cc67ae9b8&e=cd1f4d18fe. Accessed 30 September 2019
44. Gonzalez-Barriga A, Mulders SA, van de Giessen J et al (2013) Design and analysis of effects of triplet repeat oligonucleotides in cell models for myotonic dystrophy. Mol Ther Nucleic Acids 2:e81. https://doi.org/10.1038/mtna.2013.9
45. Seth PP, Siwkowski A, Allerson CR et al (2009) Short antisense oligonucleotides with novel 2′-4′ conformationaly restricted nucleoside analogues show improved potency without increased toxicity in animals. J Med Chem 52 (1):10–13. https://doi.org/10.1021/jm801294h
46. Burel SA, Han SR, Lee HS et al (2013) Preclinical evaluation of the toxicological effects of a novel constrained ethyl modified antisense compound targeting signal transducer and activator of transcription 3 in mice and cynomolgus monkeys. Nucleic Acid Ther 23 (3):213–227. https://doi.org/10.1089/nat.2013.0422
47. Pandey SK, Wheeler TM, Justice SL et al (2015) Identification and characterization of modified antisense oligonucleotides targeting DMPK in mice and nonhuman primates for the treatment of myotonic dystrophy type 1. J Pharmacol Exp Ther 355(2):329–340. https://doi.org/10.1124/jpet.115.226969
48. Manning KS, Rao AN, Castro M et al (2017) BNA(NC) Gapmers revert splicing and reduce RNA foci with low toxicity in Myotonic dystrophy cells. ACS Chem Biol 12(10):2503–2509. https://doi.org/10.1021/acschembio.7b00416
49. Echigoya Y, Nakamura A, Nagata T et al (2017) Effects of systemic multiexon skipping with peptide-conjugated morpholinos in the heart of a dog model of Duchenne muscular dystrophy. Proc Natl Acad Sci U S A 114 (16):4213–4218. https://doi.org/10.1073/pnas.1613203114

Part II

Design of Gapmers and Strategies

Chapter 8

DNA–RNA Heteroduplex Oligonucleotide for Highly Efficient Gene Silencing

Rintaro Iwata Hara, Kotaro Yoshioka, and Takanori Yokota

Abstract

Heteroduplex oligonucleotides (HDOs) were a novel type of nucleic acid drugs based on an antisense oligonucleotide (ASO) strand and its complementary RNA (cRNA) strand. HDOs were originally designed to improve the properties of RNase H-dependent ASOs and we reported in our first paper that HDOs conjugated with an α-tocopherol ligand (Toc-HDO) based on a gapmer ASO showed 20 times higher silencing effect to liver apolipoprotein B (apoB) mRNA in vivo than the parent ASO. Thereafter the HDO strategy was found to be also effective for improving the properties of ASOs modulating blood–brain barrier function and ASO antimiRs which are RNase H-independent ASOs. Therefore, the HDO strategy has been shown to be versatile technology platform to develop effective nucleic acid drugs.

Key words Nucleic acid drug, Heteroduplex oligonucleotide (HDO), Antisense oligonucleotide (ASO), Drug delivery system (DDS)

1 Introduction

1.1 Introduction of Current Nucleic Acid Therapeutics

Among all types of nucleic acid therapeutics, antisense oligonucleotides (ASOs) are the most successful ones, and five ASOs have been approved by the FDA in the USA by May, 2019. ASOs are composed of a single-stranded oligonucleotide and are mainly categorized into two groups according to their action mechanisms: RNase H-dependent ASOs and RNase H-independent ASOs [1]. Gapmer ASOs are a typical example of RNase H-dependent ASOs, [2] which contain a stretch of DNA region flanked by 2′-modified RNA derivatives such as 2′-*O*-methyl (2′-OMe) RNA, 2′-*O*-methoxyethyl (2′-*O*-MOE) RNA, and locked nucleic acid (LNA). In addition, generally they are modified with phosphorothioate (PS) at each internucleotide linkage. Gapmer ASOs can bind to target RNAs and subsequently induce their catalytic cleavage by RNase H.

The development of a drug delivery system (DDS) is necessary to maximize the potency of nucleic acid drugs, especially for systemic administration. One of the standard approaches for

Toshifumi Yokota and Rika Maruyama (eds.), *Gapmers: Methods and Protocols*, Methods in Molecular Biology, vol. 2176, https://doi.org/10.1007/978-1-0716-0771-8_8,

delivering nucleic acid drugs including ASOs to a target tissue is utilization of delivery ligand molecules. Normally, a ligand molecule is covalently conjugated to a terminus of nucleic acid drugs via a linker of adequate length. This covalent conjugation strategy has been especially successful using lipid ligands including cholesterol and α-tocopherol (vitamin E), [3–8] and *N*-acetylgalactosamine (GalNAc) ligand, which can enhance cellular uptake mediated by asialoglycoprotein receptor (ASGR). [9–11] On the other hand, delivery ligand molecules themselves are intrinsically needless once nucleic acid drugs enter into inside target cells, and the covalently attached ligand molecules may inhibit the activity of nucleic acid drugs in some cases by steric hindrance or other reasons. Therefore, the ligand molecules on nucleic acid drugs should ideally be released from the drug after they finished their roles.

1.2 Heteroduplex Oligonucleotides (HDOs) Composed of a Gapmer–RNA Hybrid

We invented a heteroduplex oligonucleotide (HDO) in order to both take advantage of ligand-directed delivery and induce dissociation of the ligand molecules after internalization of nucleic acid drugs into target cells. As shown in Fig. 1, a typical HDO is based on the heteroduplex of a gapmer ASO and its complementary RNA strand (cRNA). At a terminus of cRNA, a ligand molecule was covalently conjugated. A stretch of unmodified RNA in the cRNA strand can be cleaved by RNase H, which can recognize a stretch of DNA–RNA hybrid duplex structure. Therefore, once an HDO is internalized in a cell, the cRNA strand can be cleaved to release the ligand-free ASO strand. The released ASO binds to its target RNAs and induce the cleavage of them by RNase H. That is, RNase H can work twice in the HDO mechanism to cleave cRNA and target RNA (Fig. 2).

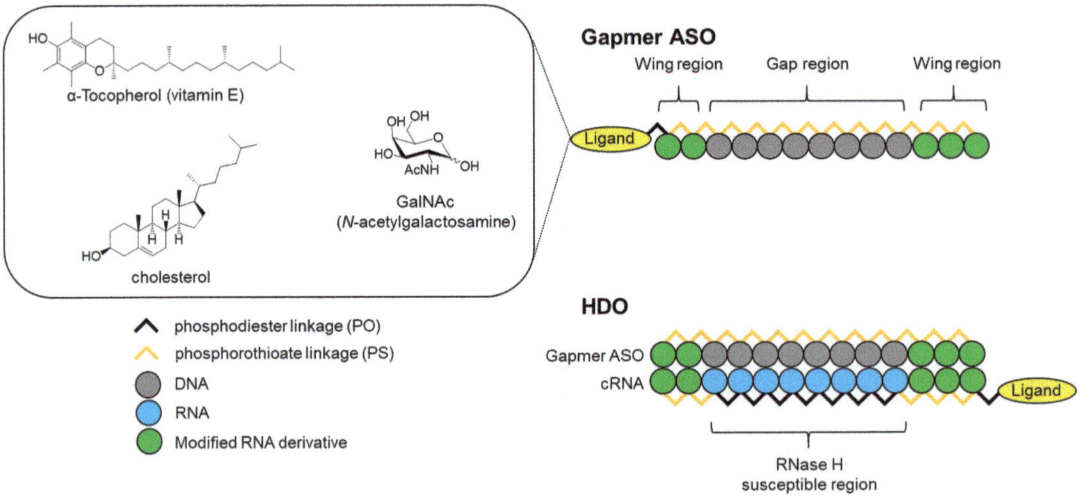

Fig. 1 Typical structures of a gapmer ASO and an HDO conjugated with ligand molecules

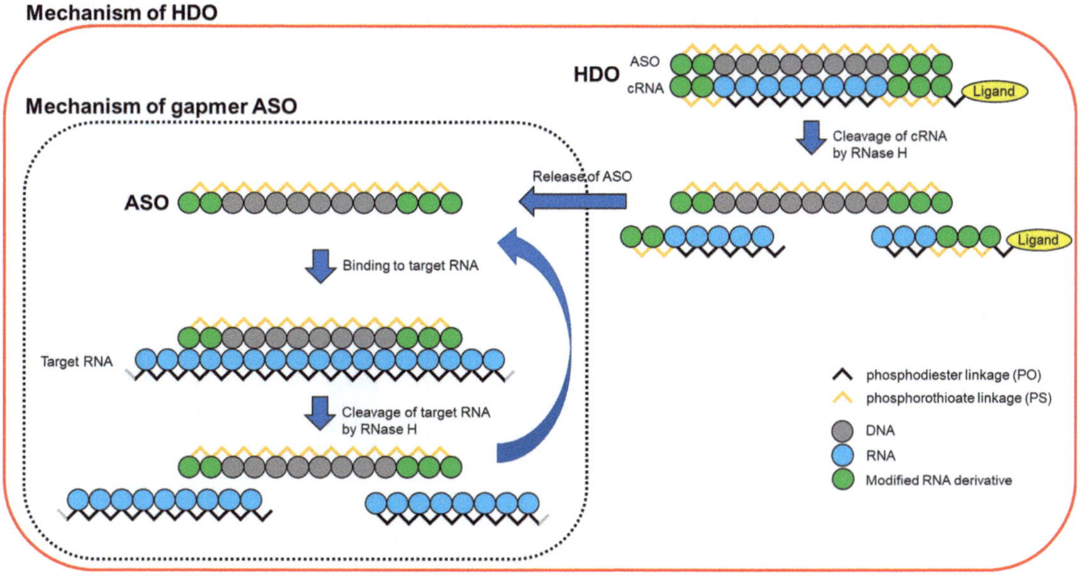

Fig. 2 Action mechanism of gapmer ASOs and HDOs

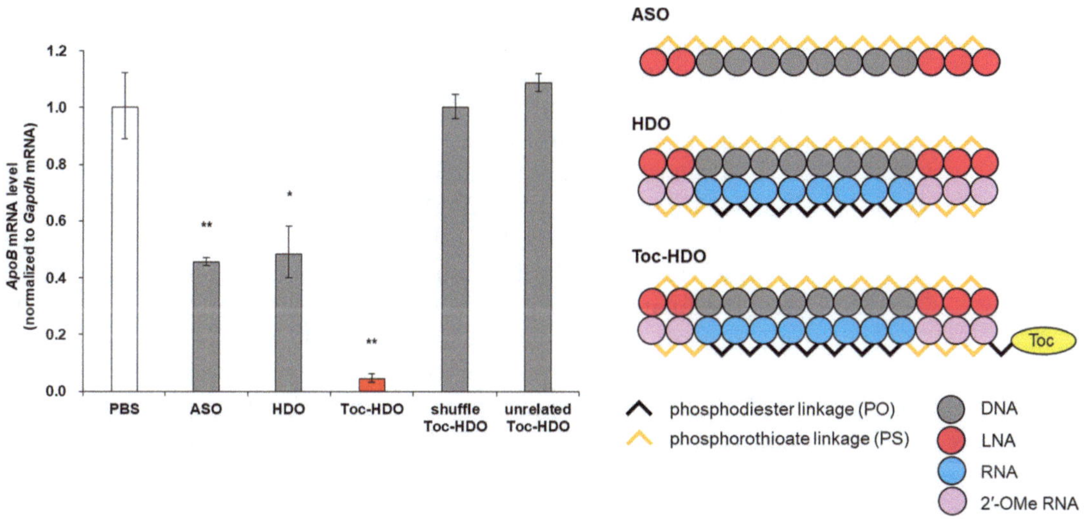

Fig. 3 Stronger knockdown of hepatic apoB mRNA by a 13mer Toc-HDO than the parent ASO in vivo

We first demonstrated that HDOs conjugated with an α-tocopherol ligand (Toc-HDO), whose ASO strands were LNA–DNA gapmers, were significantly effective at reducing apoB mRNA levels in mouse liver tissues in vivo compared to their parent ASOs (Fig. 3) [12]. Comparing their effective dose 50 values (ED_{50}), a 13mer Toc-HDO was demonstrated to show higher silencing effect to liver apoB mRNA in vivo than the parent ASO. It is noteworthy that the degree of efficacy improvement by HDO

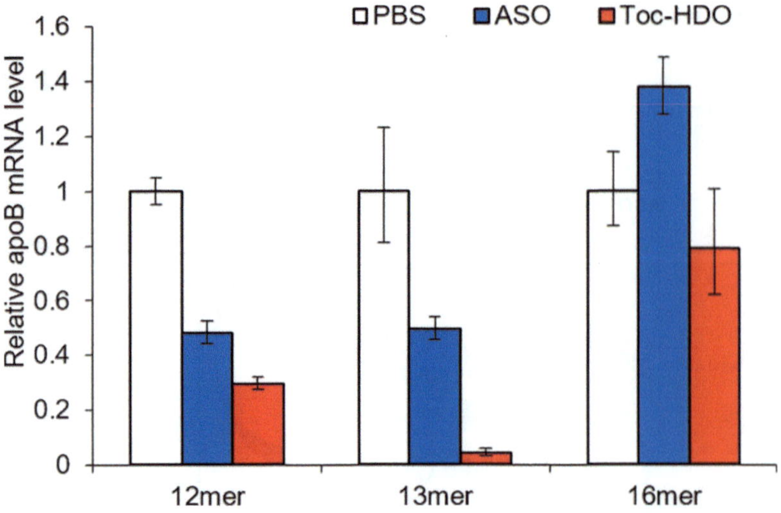

Fig. 4 Knockdown of hepatic apoB mRNA by different length of ASOs and Toc-HDOs in vivo

formation was variable dependent on the length of ASOs. The best ASO length of Toc-HDO to apoB is 13mer (Fig. 4) but may be longer in those to other sequence or targets. The length must affect not only the silencing potential of the parent ASO strands but also the serum stability and off-target effects of HDOs. In vivo efficacy screening of HDOs to find their best length is required in each kind of ASO from the early stage.

While to achieve liver delivery of HDOs had we used the Toc ligand, thereafter a Toc-HDO was found to be also effective for targeting other tissues. We found Toc-HDOs composed of a gapmer–RNA hybrid duplex could reach blood–brain barrier (BBB) and modulate its function. In the study, a 13mer Toc-HDO targeting organic anion transporter 3 (OAT3) mRNA were shown to downregulate the expression of OAT3 gene in brain microvascular endothelial cells and decrease the ion efflux transport function at the BBB in vivo [13]. Therefore, the HDO strategy could be a powerful approach to improve the pharmacokinetics of ASOs regardless of their target tissues.

1.3 HDOs for RNase H-Independent-Type ASOs

While the HDOs were designed to release their cRNA strands by RNase H cleavage, we reported the HDO strategy can also improve the properties of RNase H-independent-type ASOs in our more recent report. Anti-microRNA drugs (antimiRs) are oligonucleotides which inhibit microRNA (miRNA) function by RNase H-independent mechanism. Their function is generally steric blocking, [14] direct inhibition of targeted miRNA by antimiR binding [15]. We designed HDO-antimiR based on LNA–DNA ASO antimiRs (Fig. 5) [16]. Considering their chemical structure, the cRNA strand is not expected to be cleaved by RNase H.

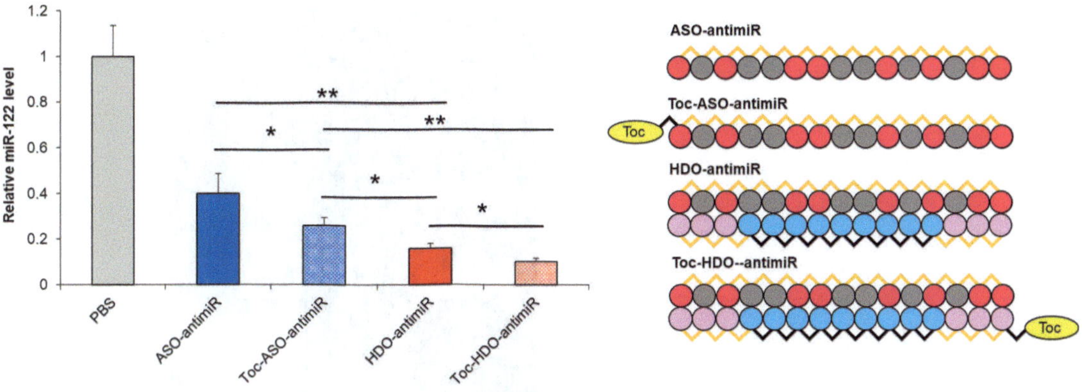

Fig. 5 Efficient knockdown of hepatic miR-122 by 13mer HDOs in vivo

However, this type of HDO showed ten times higher efficacy in liver tissues in downregulating miR-122 level than the parent ASO antimiR in vivo, and these results suggested that the release of the cRNA strand from an HDO occurred at least partly. In this case, it was surprising that an HDO even without Toc ligands showed higher in vivo efficacy in liver tissues than single-stranded ASO antimiR with a Toc ligand. These results clearly indicate the formation of HDO could improve the properties of anitmiRs. Therefore, the HDO strategy was shown to be effective to improve the properties of not only RNase H-dependent ASOs but also RNase H-independent ASO antimiRs in the study.

2 Materials (A Typical Example of a Toc-HDO)

1. ASO solution (*see* **Note 1**).
2. Toc-cRNA solution (*see* **Notes 2** and **3**).
3. 6× nondenaturing gel loading buffer.
4. 10× TBE.
5. 15% nondenaturing acrylamide gel.
6. Female ICR mice aged 4 weeks.

3 Methods (A Typical Procedure for Silencing apoB mRNA by a Toc-HDO)

3.1 Preparation of HDO

1. Mix equimolar amounts of the ASO solution and Toc-cRNA solution.
2. Heat the solution to 95 °C and maintain it at the temperature for 5 min.
3. Cool the solution to 37 °C over 60 min and then store it at 4 °C.

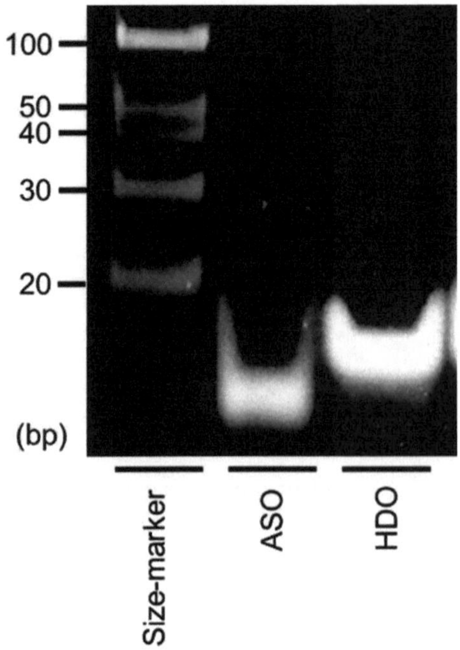

Fig. 6 Acrylamide gel analysis of the HDO annealing. 100 pmol of the single-stranded ASO and HDO solution in 10 μL PBS were mixed individually with 2 μL of 6× nondenaturing gel loading buffer. These solutions were subjected to electrophoresis (15% nondenaturing acrylamide gel), run at 100 V in 1× TBE running buffer for 1 h. As shown in this figure, the double-stranded HDO showed a higher band compared with the single-stranded ASO

3.2 Check of HDO Annealing

1. Put 10 μL of the ASO and HDO solution (1–10 μM) in each nuclease-free tube. Add 2 μL of 6× nondenaturing gel loading buffer and mix well, respectively.
2. Use 1× TBE as the gel running buffer (*see* **Note 4**).
3. Run gels at ~100 V constant voltage until the bromophenol blue (*see* **Note 5**).
4. Stain the nucleic acids in the gel by soaking for 5–10 min in 0.5–1 μg/mL ethidium bromide in water or 1× TBE. Visualize the gel on a UV transilluminator.
5. Typical results are shown in Fig. 6.

3.3 Administration of HDO

1. Dilute the HDO solution prepared at Subheading 3.1. according to the doses of HDO.
2. Administrate the HDO solution by bolus intravenous or subcutaneous injection to each mouse.
3. Sacrifice each mouse 3 days after the IV injection.

4 Notes

1. Typically 100–1800 μM solution in PBS.
2. Typically 100–1800 μM solution in PBS.
3. The sequence of RNA is complementary to that of the ASO.
4. Follow the manufacturer's instructions for the details of attaching gels to the running apparatus.
5. The faster migrating dye has moved about 3/4 of the length of the gel (for example 1 h).

References

1. Kurreck J (2003) Antisense technologies. Eur J Biochem 270:1628–1644
2. Seth PP, Swayze EE (2019) The Medicinal Chemistry of RNase H-activating Antisense Oligonucleotides In: Agrawal S, Gait MJ (eds) Advances in nucleic acid therapeutics. RSC publishing, United Kingdom
3. Wada S, Yasuhara H, Wada F et al (2016) Evaluation of the effects of chemically different linkers on hepatic accumulations, cell tropism and gene silencing ability of cholesterol-conjugated antisense oligonucleotides. J Control Release 226:57–65
4. Godard G, Boutorine AS, Saison-Behmoaras E, Hélène C (1995) Antisense effects of cholesterol-Oligodeoxynucleotide conjugates associated with poly(alkylcyanoacrylate) nanoparticles. Eur J Biochem 232:404–410
5. Soutschek J, Akinc A, Bramlage B et al (2004) Therapeutic silencing of an endogenous gene by systemic administration of modified siRNAs. Nature 432:173
6. Lorenz C, Hadwiger P, John M et al (2004) Steroid and lipid conjugates of siRNAs to enhance cellular uptake and gene silencing in liver cells. Bioorg Med Chem Lett 14:4975–4977
7. Nishina T, Numata J, Nishina K et al (2015) Chimeric antisense oligonucleotide conjugated to α-Tocopherol. Mol Ther Nucleic Acids 4: e220
8. Nishina K, Unno T, Uno Y et al (2008) Efficient in vivo delivery of siRNA to the liver by conjugation of α-Tocopherol. Mol Ther 16:734–740
9. Nair JK, Willoughby JL, Chan A et al (2014) Multivalent N-Acetylgalactosamine-conjugated siRNA localizes in hepatocytes and elicits robust RNAi-mediated gene silencing. J Am Chem Soc 136:16958–16961
10. Prakash TP, Graham MJ, Yu J et al (2014) Targeted delivery of antisense oligonucleotides to hepatocytes using triantennary N-acetyl galactosamine improves potency 10-fold in mice. Nucleic Acids Res 42:8796–8807
11. Yu RZ, Graham MJ, Post N et al (2016) Disposition and pharmacology of a GalNAc3-conjugated ASO targeting human lipoprotein (a) in mice. Mol Ther Nucleic Acids 5:e317
12. Nishina K, Piao W, Yoshida-Tanaka K et al (2015) DNA/RNA heteroduplex oligonucleotide for highly efficient gene silencing. Nat Commun 6:7969
13. Kuwahara H, Song J, Shimoura T et al (2018) Modulation of blood-brain barrier function by a heteroduplex oligonucleotide in vivo. Sci Rep 8:4377
14. Hogan DJ, Vincent TM, Fish S et al (2014) Anti-miRs competitively inhibit microRNAs in Argonaute complexes. PLoS One 9:e100951
15. Li Z, Rana TM (2014) Therapeutic targeting of microRNAs: current status and future challenges. Nat Rev Drug Discov 13:622–638
16. Yoshioka K, Kunieda T, Asami Y et al (2019) Highly efficient silencing of microRNA by heteroduplex oligonucleotides. Nucleic Acids Res 47(14):7321–7332. https://doi.org/10.1093/nar/gkz492

Chapter 9

Tips for Successful lncRNA Knockdown Using Gapmers

Kim A. Lennox and Mark A. Behlke

Abstract

Long noncoding RNAs (lncRNAs) are a recently discovered class of RNA that have diverse intracellular regulatory and structural roles. Because of their wide assortment of functions, lncRNAs can have varied distributions in the nucleus and/or cytoplasm of a cell. However, even though tens of thousands of human lncRNAs have been identified, currently less than 3% have empirically validated functions. RNA knockdown is now a relatively commonplace laboratory technique used to functionally characterize an RNA. These techniques (most commonly antisense therapy and RNA interference) can even have therapeutic benefit to treat a wide variety of genetic or infectious diseases as evidenced by the several RNA knockdown reagents currently in clinical trials. This protocol describes the use of validated gapmer antisense oligonucleotides (ASOs) to knockdown human MALAT1, a nuclear-retained lncRNA that is upregulated in multiple cancer cells. Methods used include cationic lipid transfection into HeLa cells, RNA isolation, and RT-qPCR analysis of the RNA knockdown levels.

Key words Long noncoding RNAs (lncRNAs), Oligonucleotides, Antisense, ASOs, Gapmer, Chemical modifications, Knockdown

1 Introduction

Over the years, improved technologies such as next-generation sequencing have enabled a more accurate "big picture" examination of how the genome and transcriptome align. With this, it has become fully appreciated that the genome of many diverse species (including human) is pervasively transcribed, yet less than 2% of the human genome actually encodes proteins [1]. These noncoding RNAs that represent the majority of transcription products come in a variety of lengths and structures, ranging from small RNAs (e.g., miRNAs, snoRNAs, piRNAs) and circularized RNAs (circRNAs) to the now commonly studied long noncoding RNAs (lncRNAs). LncRNAs are characterized by having a length >200 nt and can be transcribed from intergenic or intronic regions, or from either the sense or antisense strand that overlaps protein coding genes [2]. Historically, it was assumed that noncoding transcripts were byproducts of transcriptional noise and played no valuable role in

Toshifumi Yokota and Rika Maruyama (eds.), *Gapmers: Methods and Protocols*, Methods in Molecular Biology, vol. 2176, https://doi.org/10.1007/978-1-0716-0771-8_9, © Springer Science+Business Media, LLC, part of Springer Nature 2020

the inner workings of a cell. Now, it is widely accepted that lncRNAs can play critical intracellular roles in transcriptional or posttranscriptional regulation, can be involved in chromatin organization, can function as scaffolds or as molecular decoys, and may even encode small peptides [3–8]. Because lncRNAs play such diverse roles in cellular functions, they can shuttle to a wide variety of subcellular locations in both the nucleus and cytoplasm [4, 9, 10]. Dysregulation of lncRNA expression is implicated in many disease states such as cancer, neurodegeneration, and autoimmunity [11–18]. In fact, lncRNAs exhibit variable levels of expression in several diseased states enabling them to be excellent biomarkers for a variety of pathologies, tumor types and infectious diseases [19–23]. Currently, there are just over 55,000 human lncRNA transcripts annotated in LNCipedia 5, yet only 1555 of these transcripts have documented functional information associated with them [24]. Being able to effectively reduce lncRNA levels will assist with defining functional relevance and potentially have therapeutic benefit when targeting disease-associated lncRNAs.

Currently, the two most widely employed RNA knockdown techniques are antisense technology and RNA interference (RNAi). These two knockdown strategies are optimal in different intracellular locations, with antisense technology having higher activity in the nucleus and RNAi functioning predominantly in the cytoplasm [25]. Since lncRNAs can populate a wide range of cellular locations, selecting the optimal knockdown tool for the targeted lncRNA can be critical to the success of the knockdown experiment. All transcripts originate in the nucleus, so antisense technology has the potential to degrade both nuclear-retained RNAs as well as cytoplasmic-bound RNAs prior to nuclear export. Low levels of RNase H1 (the active enzyme employed in antisense technology, described below) have also been reported to be present and active in the cytoplasm, enabling modest antisense degradation of some cytoplasmic RNAs [26–29]. While RNAi can sometimes degrade nuclear-retained RNAs, this technique is generally more limited to cytoplasmic targets [25, 30]. Since antisense technology can access all RNAs at their point of origin in the nucleus and additionally has some cytoplasmic activity, it is the preferred option for lncRNA knockdown. This is especially true if the lncRNA localization is unknown [25].

RNase H1 is an endogenous enzyme that recognizes a DNA–RNA heteroduplex and cleaves the RNA strand. Antisense oligonucleotides (ASOs) are single-stranded DNA oligonucleotides that bind to a complementary RNA target to intentionally form a DNA–RNA heteroduplex and trigger RNase H1-mediated degradation of the RNA. However, it is well documented that unmodified DNA makes a poor ASO as it is rapidly degraded in serum and the intracellular environment, so it is critical to use chemical modifications to help protect the ASO from endo- and exonuclease degradation [31–33]. Chemical modification to DNA can be made on the ribose sugar, nucleotide base, and/or the phosphate backbone to provide various levels of T_m alterations and nuclease protection (reviewed in [34, 35]. Chemical modifications must be strategically incorporated in a manner that enhances ASO binding affinity and nuclease resistance, yet is still compatible with RNase H1 recruitment. The most common design strategy employed is a chimeric ASO, termed a "gapmer," that has T_m-enhancing modifications positioned in the flanking ends while retaining a central DNA core. Modifying the phosphate backbone with phosphorothioate (PS) linkages (substituting a nonbridging oxygen with a sulfur) imparts nuclease resistance and still maintains RNase H1 recognition. Importantly, incorporation of PS linkages causes the ASO to bind nonspecifically to proteins which increases the in vivo bioavailability of the ASO and promotes nuclear uptake [36–39]. PS linkages also lower the binding affinity toward its RNA complement and as a result can negatively impact the potency of an ASO. This can be offset with a gapmer ASO design, which incorporates flanking 2′-ribose modifications that alter the furanose ring conformation to favor the C3′-endo sugar pucker and thus increase the ASO binding affinity to the target RNA [40–42]. The commercially available modifications at the 2′ ribose position most commonly employed in gapmer ASOs are 2′-*O*-Methyl RNA (2′OMe), 2′-*O*-Methoxyethyl RNA (2′MOE), and locked nucleic acid (LNA®). 2′OMe is a T_m-enhancing natural modification that is endonuclease resistant; however, 2′OMe residues are still susceptible to exonuclease attack so including PS linkages in a 2'OMe oligonucleotide is often recommended to increase the stability [32]. 2′MOE residues have slightly higher binding affinity than 2′OMe residues [43] and are nuclease resistant with the exception that the 3′ terminal 2′MOE in a phosphodiester oligonucleotide can be nibbled off in serum or liver cell lysates (Lennox, unpublished data). We have consistently observed that 2′MOE ASOs demonstrate increased potency when compared to 2′OMe ASOs with the same sequence and chemical modification placement (Lennox, unpublished observation for both gapmer and steric blocking ASOs). There are several other trade names in the oligonucleotide industry for the same Locked Nucleic Acid, commonly referred to by its trademarked acronym, LNA®. For example, Eurofins markets a product called LocNA™, while IDT markets a product called Affinity Plus™. Each are identical chemically to LNA®. LNA® residues impart the highest T_m-enhancement (4–8 °C per insert) by locking the sugar into the C3′-endo conformation with a methylene bridge between the 2′- and 4′-positions [44, 45]. When optimally designed, LNA® gapmers are the most potent commercially available ASO compounds to date; however, LNA® ASOs can also be toxic due to problematic sequence motifs or by undesired knockdown of off-target RNAs [25, 32, 46–52]. Due to the high performance of the 2′MOE PS and LNA® PS gapmer ASOs, this protocol will focus on these two chemical modification strategies. However, we find that the same transfection protocol can be used even if other chemical modifications are employed.

Gapmer ASOs are typically 15–20 nucleotides (nts) long, and the desired length selected can be influenced by GC content as well as by the number and/or type of chemical modifications incorporated (e.g., high affinity LNA® gapmer ASOs are typically ~15–16-mers, while 2′OMe and 2′MOE gapmer ASOs are typically ~18–20-mers). Generally speaking, LNA® gapmer ASOs are a 3–10–3 design, while 2′OMe and 2′MOE gapmer ASOs are a 5–10–5 design, where the central DNA core is 10 nts long with 3 or 5 modifications on each flanking end, respectively. Design parameters for effective gapmer ASO site selection take into consideration GC content (binding affinity), T_m-balancing to ensure an adequate ASO dissociation rate from the cleaved target [53], RNA target accessibility [54], mitigating inherent ASO secondary structure or self-dimerization potential, avoiding protein binding motifs [55], and eliminating off-target knockdown. Even with these design guidelines, it is oftentimes difficult to identify a highly potent gapmer ASO site. We recommend screening 5–10 sites predicted to have high performance from the design parameters and empirically identifying the most active site from this subset. Support vector machine (SVM) design algorithms have been developed to help predict active gapmer ASO sites, and one validated algorithm is available for 20mer 2′OMe or 2′MOE gapmer ASOs through Integrated DNA Technologies customer care services. Additionally, assistance in designing 16mer LNA® gapmer ASOs can be provided by Integrated DNA Technologies. ASO LNA® GapmeRs can also be designed by Exiqon (Qiagen) on their website or by contacting their support team.

This protocol describes a method to degrade lncRNAs in cell culture with cationic lipid transfection. To verify that the gapmer ASO transfection protocols and reagents are working properly, a positive control example is provided using a 2′MOE or LNA® gapmer targeting human MALAT1 (metastasis associated in lung adenocarcinoma transcript 1). MALAT1 is localized in the nucleus and is highly expressed in many cell types (e.g., HeLa, HCT116, and A549 cells) [25]; http://v15.proteinatlas.org/ENSG00000279576-MALAT1/cell). A comparable chemically modified negative control sequence is also provided to ensure that knockdown effects are specific to the on-target gapmer ASO and to assess potential toxicity that sometimes occurs during knockdown experiments. Knockdown results are measured using reverse transcription quantitative PCR (RT-qPCR) and comparing MALAT1 RNA levels between the on-target gapmer ASO, the negative control gapmer ASO and a mock control containing only the cationic lipid transfection reagent. It is recommended that two 5′ nuclease qPCR assays distantly located on the transcript are compared for the targeted RNA quantification and that the knockdown data are concordant between both assays. This helps to control against inaccurate results by ruling out mis-quantification from (1) the absence or presence of unannotated isoforms overlapping the qPCR assay, (2) artifactually retained RNA fragments that didn't degrade after RNA cleavage or (3) assay enzyme interference (e.g., blocking either the reverse transcriptase or the DNA polymerase with the ASO), giving false-negative or false-positive results. Additionally, data is normalized against two internal reference controls: HPRT (NM_000194) and SFRS9 (NM_003769).

2 Materials

All reagents and tools should be sterile and nuclease-free.

2.1 Cell Culture

1. Cell line expressing the target lncRNA. This specific protocol uses HeLa cells.
2. Culture media and supplements. HeLa cells require Dulbecco's Modified Essential Medium (DMEM) supplemented with 10% fetal bovine serum (FBS).
3. Phosphate-buffered saline (1× PBS, pH 7.4).
4. Trypsin–EDTA Solution.

2.2 Cell Transfections

1. Opti-Minimal Essential Medium (Opti-MEM®; ThermoFisher Scientific).
2. Optimal cell line specific transfection reagent, such as Lipofectamine® 2000 or Lipofectamine® RNAiMAX (ThermoFisher Scientific). For gapmer ASO experiments, we find that Lipofectamine® 2000 slightly outperforms Lipofectamine® RNAiMAX.
3. 1.5 mL polypropylene microcentrifuge tubes.
4. 96-well tissue culture plates.
5. Predesigned target-specific and negative control (NC) 2′MOE and/or LNA® gapmer ASOs. The example protocol in this chapter uses gapmer ASOs targeting human MALAT1. The sequences here are presented using Integrated DNA Technologies ordering format with the appropriate modification codes, where uppercase is DNA, /2MOErX/ is 2′MOE, +X is LNA® and the asterisks represent phosphorothioate (PS) linkages. Please *see* Table 1 for an outlay of the sequences in a format that is easier to visualize.
   (a) MALAT1 site-5042 2′MOE PS ASO:
   /52MOErG/*/i2MOErG/*/i2MOErC/*/i2MOErA/*/i2MOErT/

Table 1
Sequences of targeting and control gapmer ASOs

Sequence Name	Oligonucleotide Sequence (5' → 3')
MALAT1 site-5042 2'MOE PS ASO	G*G*C*A*T*A*T*G*C*A*G*A*T*A*A* T*G*T*T*C
NC 2'MOE PS ASO	G*C*G*A*C*T*A*T*A*C*G*C*G*C*A* A*T*A*T*G
MALAT1 site-1800 LNA® PS ASO	C*G*T*T*A*A*C*T*A*G*G*C*T* T*T*A
NC LNA® PS ASO	G*A*C*T*A*T*A*C*G*C*G*C*A* A*T*A

Black = DNA; Red = 2'MOE; Black underscore = LNA®; "*" = PS linkages

∗A∗T∗G∗C∗A∗G∗A∗T∗A∗A∗/i2MOErT/∗/i2MOErG/∗/i2MOErT/∗/i2MOErT/∗/32MOErC/.

(b) NC 2′MOE PS ASO: /52MOErG/∗/i2MOErC/∗/i2MOErG/∗/i2MOErA/∗/i2MOErC/∗T∗A∗T∗A∗C∗G∗C∗G∗C∗A∗/i2MOErA/∗/i2MOErT/∗/i2MOErA/∗/i2MOErT/∗/32MOErG/

(c) MALAT1 site-1800 LNA® PS ASO: +C∗ + G∗ + T∗T∗A∗A∗C∗T∗A∗G∗G∗C∗T∗ + T∗ + T∗ + A.

(d) NC LNA® PS ASO: +G∗ + A∗ + C∗T∗A∗T∗A∗C∗G∗C∗G∗C∗A∗ + A∗ + T∗ + A.

6. IDTE pH 7.5 buffer (sterile Tris–EDTA (TE); Integrated DNA Technologies).

2.3 RT-qPCR Reagents

1. SV 96 Total RNA Isolation System (Promega) (if fewer samples need to be processed, *see* **Note 1**).
2. Vac-Man® 96 Vacuum Manifold (Promega).
3. Vacuum pump capable of 15–20 in. of Hg and vacuum trap.
4. 95% ethanol, RNase-free.
5. SuperScript™ IV First-Strand Synthesis System (ThermoFisher Scientific).
6. 96-well PCR plates.
7. MicroAmp™ Optical Adhesive Film (ThermoFisher).
8. PrimeTime® Gene Expression Master Mix (Integrated DNA Technologies).
9. 384-well qPCR plates.
10. PrimeTime® qPCR Probe Assays (Integrated DNA Technologies). Sequences can be manually entered as custom PrimeTime® qPCR Probe Assays on the Integrated DNA Technologies website. Selecting the desired 5′dye/3′quencher

Table 2
Sequences of PrimeTime® primers and probes

Assay	Sequence name	Oligonucleotide sequence (5′ → 3′)
MALAT1 2638-2726	MALAT1 For-2638 MALAT1 Rev-2726 MALAT1 Probe-2668 HEX	GTGCTACACAGAAGTGGATTC CCTCAGTCCTAGCTTCATCA /5HEX/AGGAAGACA/ZEN/GCAGCAGACAGGATT/3IABkFQ/
MALAT1 7443-7535	MALAT1 For-7443 MALAT1 Rev-7535 MALAT1 Probe-7481 FAM	CACCGAAGGCTTAAAGTAGGACA GCTGACACTTCTCTTGACCTTAG /56-FAM/CGCTTTGTT/ZEN/GTCTCTCCTGCCACA/3IABkFQ/
HPRT 517-591	HPRT For-517 HPRT Rev-591 HPRT Probe-554 FAM	GACTTTGCTTTCCTTGGTCAG GGCTTATATCCAACACTTCGTGGG /56-FAM/ATGGTCAAG/ZEN/GTCGCAAGCTTGCTGGT/3IABkFQ/
SFRS9 594-690	SFRS9 For-594 SFRS9 Rev-690 SFRS9 Probe-625 HEX	GTCGAGTATCTCAGAAAAGAAGACA CTCGGATGTAGGAAGTTTCACC /5HEX/ATGCCCTGC/ZEN/GTAAACTGGATGACA/3IABkFQ/

combination when ordering will automatically input the fluorophore and quencher modification codes for the end user. The positioning of the fluorophore and quenchers can be found in the sequences in Table 2.

(a) MALAT1 2638-2726 (MALAT1 For-2638 GTGCTACACAGAAGTGGATTC, MALAT1 Rev-2726 CCTCA GTCCTAGCTTCATCA, MALAT1 Probe-2668 HEX A GGAAGACAGCAGCAGACAGGATT).

(b) MALAT1 7443-7535 (MALAT1 For-7443 CACCGAA GGCTTAAAGTAGGACA, MALAT1 Rev-7535 GCTG ACACTTCTCTTGACCTTAG, MALAT1 Probe-7481 FAM CGCTTTGTTGTCTCTCCTGCCACA).

(c) HPRT 517-591 (HPRT For-517 GACTTTGCTTTCC TTGGTCAG, HPRT Rev-591 GGCTTATATCCAAC ACTTCGTGGG, HPRT Probe-554 FAM ATGGTC AAGGTCGCAAGCTTGCTGGT).

(d) SFRS9 594-690 (SFRS9 For-594 GTCGAGTATCTCAG AAAAGAAGACA, SFRS9 Rev-690 CTCGGATGTAGG AAGTTTCACC, SFRS9 Probe-625 HEX ATGCCCTG CGTA
AACTGGATGACA).

11. gBlocks® Gene Fragment qPCR amplicons for standard curve analysis (Integrated DNA Technologies).

(a) MALAT1 2638-2726 and MALAT1 7443-7535 amplicons: GTGCTACACAGAAGTGGATTCAGTGAATCT AGGAAGACAGCAGCAGACAGGATTCCAGGAACC AGTGTTTGATGAAGCTAGGACTGAGGTTTTTCAC CGAAGGCTTAAAGTAGGACAACCATGGAGCCTT CCTGTGGCAGGAGAGACAACAAAGCGCTATTAT CCTAAGGTCAAGAGAAGTGTCAGC.

(b) HPRT 517-591 and SFRS9 594-690 amplicons: GACTTTGCTTTCCTTGGTCAGGCAGTATAATCCA AAGATGGTCAAGGTCGCAAGCTTGCTGGTGAAA AGGACCCCACGAAGTGTTGGATATAAGCCTTTTT GTCGAGTATCTCAGAAAAGAAGACATGGAATAT GCCCTGCGTAAACTGGATGACACCAAATTCCGC TCTCATGAGGGTGAAACTTCCTACATCCGAG.

3 Methods

This section details the methodology used to assess lncRNA knockdown with gapmer ASOs. Appropriate controls and normalization recommendations for data analysis are included. The selected lncRNA target for this protocol is human MALAT1, and knockdown is performed in HeLa cells which are known to highly express this lncRNA [25]. Volumes included in this protocol are calculated to perform biological triplicates for each transfection condition and qPCR triplicates of each transfection well (i.e., 9 qPCR data points per transfection condition).

3.1 Reverse Transfection

3.1.1 Nucleic Acids

1. Centrifuge lyophilized gapmer ASOs prior to opening and reconstitute in sterile IDTE pH 7.5 buffer at a stock concentration of 100 μM (e.g., add 50 μL of IDTE to 5 nmoles of gapmer ASO).
2. Quantify the gapmer ASO using a nucleic acid quantification system such as the NanoDrop™.
3. Prepare a 10 μM working dilution in IDTE pH 7.5 buffer. For this 3 nM, 1 nM, 0.3 nM, and 0.1 nM dose response experiment, prepare serial gapmer ASO dilutions of 0.3 μM, 0.1 μM, 30 nM, and 10 nM in IDTE pH 7.5 buffer (*see* **Note 2**). Gapmer ASO stock solutions and dilutions should be stored at −20 °C. If accidental contamination of the gapmer ASO occurs, filter the gapmer ASO stock solution through a 0.2 μm syringe filter (Pall) and quantify prior to use.

3.1.2 Reverse Transfection

Here we outline the protocol for reverse transfection of the MALAT1 2′MOE PS, NC 2′MOE PS, MALAT1 LNA® PS, and NC LNA® PS gapmer ASOs (four gapmer ASOs each with four doses) in a 96-well plate format. If different sized plates are required (e.g., if the lncRNA of interest has low expression in the selected cell type and the amount of RNA needs to be increased), adjust the volumes accordingly by surface area. For example, all reagent volumes can be doubled if using a 48-well plate versus a 96-well plate. This protocol uses HeLa cells; if a different cell line is used, cells should be plated at the specified confluency. Include a "reagent only" control where only the transfection reagent (no gapmer ASO) is applied to the cells as a control to help assess the cause of any toxicity or contamination that might occur. For the MALAT1 ASOs, a dosing range of 3 nM, 1 nM, 0.3 nM and 0.1 nM is recommended to start (*see* **Note 2**).

1. For a single transfection well, dilute 0.5 μL of Lipofectamine® 2000 into 24.5 μL of Opti-MEM® (total volume is 25 μL) (*see* **Note 3**). For triplicate transfections, multiply these values by 3.3 to account for pipetting errors (e.g., combine 1.65 μL of Lipofectamine® 2000 with 80.9 of μL Opti-MEM®), and multiply again by the total number of transfection conditions used. With the number of transfection conditions in this experiment (four gapmer ASOs with four doses each and a "reagent only" control), this Lipofectamine® 2000/Opti-MEM® dilution may be mixed in a 5 mL polypropylene tube to account for the volume.
2. Mix by briefly vortexing or pipetting up and down several times.
3. In separate 1.5 mL microcentrifuge tubes dilute 1.5 μL of the appropriate gapmer ASO dilution (e.g., use the 0.3 μM dilution for 3 nM dose, *see* **Note 1**) into 23.5 μL of Opti-MEM® (total volume is 25 μL) (*see* **Note 4**). For triplicate transfections, multiply these values by 3.2 to account for pipetting errors (e.g., combine 4.8 μL of gapmer ASO with 75.2 μL of Opti-MEM® for a total volume of 80 μL).
4. Add 80 μL of the diluted Lipofectamine® 2000 to the diluted gapmer ASO (total volume is 160 μL).
5. Gently mix the lipid complexes by pipetting up and down several times.
6. Incubate the lipid complexes for 20 min at room temperature.
7. During the 20 min incubation, trypsinize the HeLa cells and calculate the cells/mL using a hemocytometer.
8. Dilute the HeLa cells to 24,000 cells per 100 μL (240,000 cells per 1 mL) in DMEM containing 10% FBS. Each well in the 96-well plate will be seeded with 100 μL of diluted cells to achieve ~90% confluency the next day (*see* **Note 5**). Scale up the total volume of diluted cells required for the number of wells being transfected.

9. Add 50 μL of the gapmer ASO–lipid complex into triplicate wells in a 96-well transfection plate and wait for the duration of the incubation time.
10. After the 20-min incubation, add 100 μL of diluted HeLa cells on top of the lipid complexes for a final volume of 150 μL. Carefully rock the plate periodically for the first 15 min to allow cells to form an evenly confluent monolayer.
11. Incubate the cells at 37 °C and 5% CO_2 for 24 h.
12. After 24 h, visualize the cells to ensure there is no toxicity or contamination (*see* **Note 6**).

3.2 RNA Isolation and RT-qPCR Analysis

Here we outline the protocols for RNA isolation, cDNA synthesis in 96-well plate format and qPCR in 384-well plate format. For more efficient throughput, use a multichannel pipette for liquid transfer. Ensure that proper care is taken when handling RNA (e.g., always wear clean gloves) and all reagents and plasticware are RNase-free (*see* **Note 7**).

3.2.1 RNA Isolation

All buffer solutions for this procedure are provided in the SV 96 Total RNA Isolation System kit.

1. Prepare reagents for the SV 96 Total RNA Isolation System by adding 0.5 mL of β-mercaptoethanol (BME) to 50 mL of RNA Lysis Buffer, 275 μL of nuclease-free water to the lyophilized DNase I (do not vortex), 100 mL of 95% ethanol to the RNA Wash Solution, and 20 mL of 95% ethanol to the DNase Stop Solution.
2. Aspirate the media off the cells and wash cells with 100 μL of sterile 1× PBS.
3. Add 100 μL of RNA Lysis Buffer to each well. If needed, the cell lysates can be stored in the transfection plate at temperatures from −20 °C to −80 °C, taking care that the plate is sealed (e.g., with tape or parafilm) to reduce evaporation potential.
4. Attach the vacuum manifold base to the vacuum via the vacuum port and position an RNA binding plate into the vacuum manifold base.
5. Pipette up and down to mix the cell lysate, and carefully transfer the entire lysate volume to the membranes in the 96-well RNA binding plate. Apply vacuum until lysate is drawn through (*see* **Note 8**).
6. Add 500 μL of RNA Wash Solution to each well. Apply vacuum until the RNA Wash Solution is drawn through.
7. Prepare the DNase I solution by combining (in this order) 20 μL of Yellow Core Buffer, 2.5 μL of 0.09 M $MnCl_2$, and

2.5 μL of DNase I for each well. Multiply these volumes by the number of wells required for the experiment, accounting for pipetting errors. Mix by gentle pipetting.

8. Add 25 μL of DNase I solution to the binding plate membrane for each sample and incubate at room temperature for 10 min. Unused DNase I can be stored at −20 °C.
9. Add 200 μL of DNase Stop Solution to each well. Apply vacuum until the DNase Stop Solution is drawn through.
10. Add 500 μL of RNA Wash Solution to each well. Apply vacuum until the RNA Wash Solution is drawn through and continue to apply the vacuum for 10 additional minutes to dry the membranes.
11. Remove the binding plate and gently blot the bottom tips onto a clean Kimwipes or paper towel.
12. Position a 96-well elution plate into the vacuum manifold bed, place the manifold collar on top of the elution plate and put the binding plate on top of the manifold collar.
13. Add 100 μL of nuclease-free water to each well, ensuring that the water is in contact with the membrane. Incubate at room temperature for 1 min.
14. Apply the vacuum for 1 min to draw the RNA through.
15. Cover the plate with a plate sealant and store RNA at temperatures between −20 °C and −80 °C until ready to use.
16. Accurate RT-qPCR quantification requires the RNA to be intact and of high quality. Quantify RNA samples and assess the sample purity prior to cDNA synthesis. This can be performed using a nucleic acid quantification system such as the NanoDrop™. To assess RNA purity, calculate the A_{260}/A_{280} ratio; a value within the 1.8–2.1 range indicates high sample purity. Additionally, residual chemical contaminants often absorb around 230 nm or can shift the 260 nm peak to a higher wavelength. Visualize the position of the RNA peak making sure there is no significant shift from 260 nm. Calculate the A_{260}/A_{230} ratio and ensure it is in the range of 2.0–2.2; a lower ratio can indicate the presence of a contaminant that may inhibit downstream enzymatic reactions. Some variability in RNA quantification is acceptable for this application, as fluctuations will be accounted for with the two validated reference genes. However, best practice is to normalize the RNA samples to the same approximate concentration. If any of the samples significantly deviate from the average RNA input (e.g., less than twofold concentration), this may indicate toxicity during transfection or inefficient RNA extraction and/or RNA degradation, and the sample should be excluded from analysis.

3.2.2 cDNA Synthesis

All components for this procedure are provided in the SuperScript™ IV First-Strand Synthesis System. A "no reverse transcriptase (RT)" control will be included to assess genomic DNA contamination carryover as this can affect accurate quantification of the RNA.

1. Prepare a master mix of the following components: per reaction, mix 1 μL of 50 μM oligo dT, 1 μL of 50 ng/μL random hexamers, and 1 μL of 10 mM dNTP mix. Multiply each volume by the total number of RNA samples. For example, this protocol is analyzing 51 RNA samples and a no-RT control (52 total), so prepare enough master mix for 54 samples by combining 54 μL of each of the three components and vortexing to mix.
2. Pipette 3 μL of the master mix into 52 wells of a 96-well PCR plate.
3. Add 10 μL of each RNA sample, pipetting briefly to mix. RNA from the "reagent only" transfection control can be used in the no-RT control reaction.
4. Cover the plate with a plate sealant such as MicroAmp™ Optical Adhesive Film. We typically do not use plate sealants at this step, but the sterility of the thermocycler would need to be validated prior to leaving the plate uncovered at this step.
5. Incubate the plate at 65 °C for 5 min, and place on ice for at least 1 min. If condensation has occurred, briefly centrifuge the plate prior to removing the plate sealant.
6. Prepare a master mix of the following components: per reaction, mix 4 μL of 5× SSIV Buffer, 1 μL of 100 mM DTT, 1 μL of Ribonuclease Inhibitor and 1 μL of SuperScript™ IV. Multiply each volume by the total number of RNA samples, accounting for pipetting errors. For example, this protocol is preparing 52 cDNA reactions, so prepare enough for 54 samples by combining 216 μL of 5× SSIV Buffer and 54 μL of each of the other four components, vortexing to mix.
7. Carefully remove the plate sealant from the 96-well PCR plate. Pipette 7 μL of the master mix (containing the enzymes) into each of the 52 wells, pipetting to mix. Cover the plate with a plate sealant and briefly centrifuge the plate.
8. Incubate the plate at 23 °C for 10 min, 50 °C for 10 min, 80 °C for 10 min, and then 4 °C until ready to use.
9. Dilute each cDNA reaction twofold by adding 20 μL of nuclease-free water prior to use.

3.2.3 qPCR Methods

Use PCR template-free reagents, consumables and equipment during this procedure. Keep gBlocks® Gene Fragment qPCR amplicons for standard curve analysis separate from all PrimeTime®

qPCR Probe Assays and components and designate a different set of clean pipettes for the qPCR assay setup. Please refer to the *M*inimum *I*nformation necessary for *Q*uantitative real-time PCR *E*xperiments (MIQE) guidelines for accurate and comprehensive data collection and reporting of results [56, 57].

1. Briefly centrifuge PrimeTime® qPCR Probe Assay tubes and resuspend to 20× stock concentration in IDTE pH 7.5 buffer using template-free pipettes in a clean PCR hood to reduce contamination. For example, if using the PrimeTime® Std qPCR Assay (500 reaction size), add 500 μL of IDTE pH 7.5 buffer and vortex to mix. Store unused PrimeTime® qPCR Probe Assays at −20 °C.
2. Prepare qPCR master mixes for the lncRNA target and the reference (normalizer) genes. Thaw the PrimeTime Gene Expression Master Mix and reference dye on ice and vortex to mix. If using a QuantStudio™ Flex Systems qPCR machine, add 4 μL of reference dye to every 1 mL of PrimeTime® Gene Expression Master Mix (this is a "low" reference dye qPCR system) (*see* **Note 9**).
3. Calculate the amount of master mix required for the number of qPCR reactions to be run. For example, this protocol is analyzing 51 transfection wells, a no-RT control cDNA reaction, 6 dilutions of a standard curve and a no template control (NTC) to assay for template contamination that can affect quantification accuracy. Two multiplexed master mixes will be prepared in 1.5 mL microcentrifuge tubes: (1) lncRNA assays MALAT1 2638–2726 and MALAT1 7443–7535 together in one tube, and (2) normalizer assays HPRT 517-591 and SFRS9 594-690 together in a second tube. All qPCR reactions are performed in triplicate, giving a total of 177 reactions per multiplexed assay mix. Using a 10 μL per reaction volume setup, prepare enough master mix for 200 reactions to account for pipetting errors. For each of the two multiplexed master mixes, combine 1 mL of PrimeTime® Gene Expression Master Mix (with reference dye added) and 100 μL of each of the 20× PrimeTime® qPCR Assays (i.e., both MALAT1 assays in one tube and both normalizer assays in a separate tube), giving a total of 1.2 mL of each multiplexed master mix. Vortex to mix.
4. Prepare the qPCR standard curves in an area designated for template handling. Briefly centrifuge the lyophilized gBlocks® Gene Fragments to ensure product is at the bottom of the tube. With a separate set of pipettes from those used for qPCR setup (to mitigate potential template contamination), hydrate the gBlocks® Gene Fragments in IDTE pH 7.5 buffer to 10 ng/μL and briefly vortex.

5. Incubate the gBlocks® Gene Fragments at 50 °C for 15–20 min to assist with resuspension. Vortex the samples and briefly centrifuge to collect the contents at the bottom of the tube.

6. Using a nucleic acid quantification platform such as the NanoDrop™, quantify the gBlocks® Gene Fragments to verify concentration.

7. Calculate the copy number/μL by using the following formula: $(C)(M)(1 \times 10^{-15}$ mol/fmol)(Avogadro's number) = copy number/μL, where C is the gBlocks® Gene Fragments concentration in ng/μL, M is the molecular weight in fmol/ng (found on the spec sheet provided), and Avogadro's number = 6.022×10^{23}. For example, a 10 ng/μL stock of the MALAT1 gBlocks® Gene Fragment is as follows: (10 ng/μL) (8.66 fmol/ng)(1×10^{-15} mol/fmol) (6.022×10^{23}) = 5.22×10^{10} copies/μL.

8. Prepare tenfold serial dilutions in IDTE pH 7.5 buffer from 2.5×10^6 to 2.5×10^1 copies/μL (*see* **Note 10**). As 4 μL of template will be loaded per well, this will provide a standard curve from 1×10^7 to 1×10^2 copies.

9. Using a multichannel pipette, load 177 wells in a 384-well qPCR plate with 6 μL of the MALAT1 multiplexed master mix.

10. In triplicate wells, dispense 4 μL of each cDNA reaction (including the no-RT control) and each dilution of the MALAT1 standard curve. Dispense 4 μL of nuclease-free water for the NTC.

11. Seal the plate with MicroAmp™ Optical Adhesive Film and briefly centrifuge to ensure the contents are collected at the bottom of each well.

12. Repeat this plate setup (**steps 9–11**) for the normalizer assays in a separate 384-well plate.

13. Run the qPCR plates on a real-time PCR instrument using the following cycling parameters: enzyme activation at 95 °C for 3 min, followed by 40 cycles of two-step PCR (95 °C for 15 s, 60 °C for 1 min). Input sample names and standard curve values.

3.2.4 Data Analysis

1. Analyze your standard curves to assess assay efficiency, amplification quality and pipetting accuracy. Slopes ranging between −3.1 and −3.6 equate to 90–110% efficient (−3.32 is 100% efficient) and are considered acceptable. The R^2 value measures the pipetting accuracy and assay performance and should be greater than 0.99. All test samples must lie within the range of the standard curve.

2. Verify that the no-RT control and the NTC negative control amplification reactions did not amplify any contaminating carryover genomic DNA or template, respectively. Look at the amplification curves generated for the reference genes for all samples and ensure that minimal deviation is occurring. If the spread between all samples combined is greater than 2 Cqs, this may be indicative of cell toxicity, poor sample integrity or that the amplification process was not robust.
3. For each sample, calculate the average copy number values for the reference genes and for each assay of the target gene (MALAT1), and divide the target gene averaged value to its respective reference gene averaged value. Divide the normalized copy number value of each ASO treatment sample by the normalized copy number value of the appropriate negative control (e.g., LNA ASO treatment should be compared to LNA negative control ASOs at the same dose). All treatments should also be compared to the "reagent only" control to verify that the negative control ASO treatment did not largely affect MALAT1 expression levels. Perform the calculations separately for both MALAT1 qPCR assays and compare the results to assess assay concordance. The MALAT1 knockdown results for this protocol can be found in an example transfection shown in Fig. 1.

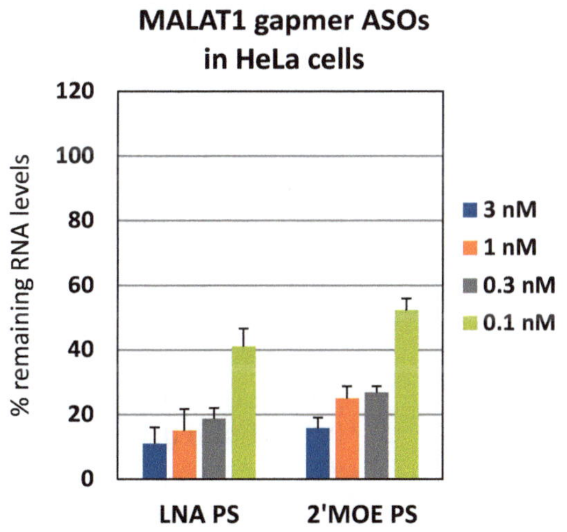

Fig. 1 Knockdown of the nuclear human MALAT1 lncRNA with 2′MOE PS or LNA® PS gapmer ASOs. Gapmer ASOs were transfected in triplicate into HeLa cells with Lipofectamine® 2000. RNA levels were measured by RT-qPCR 24 h after transfection. LncRNA levels were calculated by normalizing to the internal reference genes HPRT and SFRS9 and comparing to negative control gapmer ASOs with similar chemical composition at a similar dose

4 Notes

1. The 96-well plate format for RNA isolation is convenient when processing multiple samples. If fewer samples are being processed, the tube-based SV Total RNA Isolation System is also available.
2. Working dilutions should be prepared so that the same volume of gapmer ASO is added to the transfection mix, regardless of the dose. When testing several gapmer ASO dilutions, we simplify pipetting by preparing our serial dilutions in 96-well deep well plates to enable the use of a multichannel pipette. The ideal dose range can vary due to differences in the transfection efficiency of various cell lines. For example, a higher concentration of gapmer ASO may be needed in cells that are difficult to lipofect to compensate for inefficient delivery. In cells easy to lipofect (such as HeLa cells), a relatively high gapmer ASO dose may result in toxicity. We recommend starting with a 30 nM, 10 nM, and 3 nM dose response study and to subsequently adjust concentrations up or down according to results and toxicity observed. It is advised to use to lowest dose possible that achieves maximum knockdown effect to mitigate toxicity and the potential for OTEs. In the current protocol, the MALAT1 gapmer ASOs are validated to have high potency and demonstrate a nice dose response in the 3 nM to 0.1 nM dose range.
3. Optimal transfection reagents can be cell-line dependent. While we find that Lipofectamine® 2000 efficiently delivers gapmer ASOs to a wide variety of commonly employed cell lines used for gapmer ASO screens, an alternative transfection reagent may be more suitable for different cell types. Typically, a transfection reagent that is optimal in a particular cell line when using a similar small nucleic acid, such as an siRNA, will also demonstrate high transfection efficiency with gapmer ASOs. Holding the volume of the lipid transfection reagent constant while titrating down the gapmer ASO for the dose response gives optimal lipofection results. For example, in this protocol the gapmer ASOs will be transfected at 3 nM, 1 nM, 0.3 nM, and 0.1 nM, all with 0.5 μL of Lipofectamine® 2000/well in a 96-well plate.
4. To assist with pipetting efficiency, the lipoplexes can be prepared in a 96-well plate instead of 1.5 mL microcentrifuge tubes. This enables use of a multichannel pipette, simplifying and speeding up the liquid transfer. In this format, triplicate transfection wells are oriented in rows (e.g., the first biological triplicate would be in well positions A1, A2, and A3).

5. Cell confluency can influence the intracellular gapmer ASO concentration in the transfected cells. We have observed that a lower cell confluency can have higher gapmer ASO activity than a higher cell confluency when using the same transfection conditions. To help control for inter-experimental variability, ensure that the confluency is similar between experiments. Additionally, transfection of chemically modified oligonucleotides can sometimes cause cell toxicity. We have found that having a higher confluency at the time of transfection can reduce the toxicity potential from a wide variety of chemically modified oligonucleotides.
6. Significant bacterial contamination during the transfection step can easily be macroscopically visualized, as the media typically takes on a cloudy appearance and can also change color due to a change in pH. Slight bacterial contamination can be visualized microscopically as tiny moving granules, and every well should be examined to ensure sterility. If contamination occurs, review aseptic techniques making sure all tools, equipment and consumables are sterile. Filter all suspect nucleic acids through a 0.2 μm syringe filter and quantify prior to use. Cell toxicity can also sometimes result from residual chemicals during the gapmer ASO synthesis and/or purification process. Ethanol precipitating the gapmer ASOs prior to use can remove trace toxic substances. If toxicity persists, lowering the gapmer ASO dose, increasing cell confluency (*see* **Note 5**) or redesigning the gapmer ASO sequence can help resolve the issue. The cause of toxicity could also be attributed to the lncRNA knockdown itself, especially if the targeted lncRNA is involved in a cellular pathway important for viability.
7. If RNA degradation has occurred, identifying the source of the RNases can be critical for the success of future experiments. Surfaces and equipment can be treated with Nuclease Decontamination Solution (Integrated DNA Technologies) to inactivate RNases. Reagents can be tested for RNase contamination by using RNaseAlert (Integrated DNA Technologies).
8. If the liquid isn't flowing through the membrane because the vacuum pressure is too low, cover unused wells with a plate sealer such as the MicroAmp™ Optical Adhesive Film. Keep covered throughout the RNA isolation process. The RNA binding plate can be saved (e.g., wrapped in clean paper towel and/or placed in a clean plastic storage bag) and stored at room temperature for later use of any unused wells.
9. Each qPCR system has specifications for the required amount of reference dye. We use a QuantStudio™ Flex Systems (Thermo Fisher Scientific) that requires a "low" concentration of the reference dye (4 μL of reference dye per 1 mL of

PrimeTime® Gene Expression Master Mix). Some qPCR systems require a "high" concentration of the reference dye (40 μL of reference dye per 1 mL of PrimeTime® Gene Expression Master Mix). Please refer to the guide for PrimeTime® Gene Expression Master Mix to ensure the appropriate amount of reference dye is added for your specific qPCR system.

10. Serial dilutions should be prepared fresh, as the DNA can stick to the side of the tubes and cause inaccurate quantification. This is particularly problematic for the lower concentrated dilutions where there are only a small number of molecules present. Adding 100 ng/μL yeast tRNA (ThermoFisher Scientific) helps to stabilize the dilutions by acting as a carrier molecule and adhering to the side of the microcentrifuge tubes, coating the surface and reducing the opportunity for the standard curve DNA to bind.

References

1. Djebali S, Davis CA, Merkel A, Dobin A, Lassmann T, Mortazavi A, Tanzer A, Lagarde J, Lin W, Schlesinger F et al (2012) Landscape of transcription in human cells. Nature 489:101–108
2. Ma L, Bajic VB, Zhang Z (2013) On the classification of long non-coding RNAs. RNA Biol 10:925–933
3. Quinn JJ, Chang HY (2016) Unique features of long non-coding RNA biogenesis and function. Nat Rev Genet 17:47–62
4. Ulitsky I, Bartel DP (2013) lincRNAs: genomics, evolution, and mechanisms. Cell 154:26–46
5. Geisler S, Coller J (2013) RNA in unexpected places: long non-coding RNA functions in diverse cellular contexts. Nat Rev Mol Cell Biol 14:699–712
6. Rinn JL, Chang HY (2012) Genome regulation by long noncoding RNAs. Annu Rev Biochem 81:145–166
7. Dey BK, Mueller AC, Dutta A (2014) Long non-coding RNAs as emerging regulators of differentiation, development, and disease. Transcription 5:e944014
8. Matsumoto A, Pasut A, Matsumoto M, Yamashita R, Fung J, Monteleone E, Saghatelian A, Nakayama KI, Clohessy JG, Pandolfi PP (2017) mTORC1 and muscle regeneration are regulated by the LINC00961-encoded SPAR polypeptide. Nature 541:228–232
9. van Heesch S, van Iterson M, Jacobi J, Boymans S, Essers PB, de Bruijn E, Hao W, MacInnes AW, Cuppen E, Simonis M (2014) Extensive localization of long noncoding RNAs to the cytosol and mono- and polyribosomal complexes. Genome Biol 15:R6
10. Cabili MN, Dunagin MC, McClanahan PD, Biaesch A, Padovan-Merhar O, Regev A, Rinn JL, Raj A (2015) Localization and abundance analysis of human lncRNAs at single-cell and single-molecule resolution. Genome Biol 16:20
11. Nie L, Wu HJ, Hsu JM, Chang SS, Labaff AM, Li CW, Wang Y, Hsu JL, Hung MC (2012) Long non-coding RNAs: versatile master regulators of gene expression and crucial players in cancer. Am J Transl Res 4:127–150
12. Niland CN, Merry CR, Khalil AM (2012) Emerging roles for long non-coding RNAs in cancer and neurological disorders. Front Genet 3:25
13. Wang J, Wei F, Zhou H (2018) Advances of lncRNA in autoimmune diseases. Front Lab Med 2:79–82
14. Yarani R, Mirza AH, Kaur S, Pociot F (2018) The emerging role of lncRNAs in inflammatory bowel disease. Exp Mol Med 50:161
15. Aune TM, Crooke PS 3rd, Patrick AE, Tossberg JT, Olsen NJ, Spurlock CF 3rd. (2017) Expression of long non-coding RNAs in autoimmunity and linkage to enhancer function and autoimmune disease risk genetic variants. J Autoimmun 81:99–109
16. Hu G, Niu F, Humburg BA, Liao K, Bendi S, Callen S, Fox HS, Buch S (2018) Molecular mechanisms of long noncoding RNAs and their role in disease pathogenesis. Oncotarget 9:18648–18663

17. Wei CW, Luo T, Zou SS, Wu AS (2018) The role of long noncoding RNAs in central nervous system and neurodegenerative diseases. Front Behav Neurosci 12:175
18. Sanchez Calle A, Kawamura Y, Yamamoto Y, Takeshita F, Ochiya T (2018) Emerging roles of long non-coding RNA in cancer. Cancer Sci 109:2093–2100
19. Chen ZL, Wei LL, Shi LY, Li M, Jiang TT, Chen J, Liu CM, Yang S, Tu HH, Hu YT et al (2017) Screening and identification of lncRNAs as potential biomarkers for pulmonary tuberculosis. Sci Rep 7:16751
20. Liu W, Ding C (2017) Roles of LncRNAs in viral infections. Front Cell Infect Microbiol 7:205
21. Bolha L, Ravnik-Glavac M, Glavac D (2017) Long noncoding RNAs as biomarkers in cancer. Dis Markers 2017:7243968
22. Jiang X, Lei R, Ning Q (2016) Circulating long noncoding RNAs as novel biomarkers of human diseases. Biomark Med 10:757–769
23. Li L, Wang L, Li H, Han X, Chen S, Yang B, Hu Z, Zhu H, Cai C, Chen J et al (2018) Characterization of LncRNA expression profile and identification of novel LncRNA biomarkers to diagnose coronary artery disease. Atherosclerosis 275:359–367
24. Volders PJ, Anckaert J, Verheggen K, Nuytens J, Martens L, Mestdagh P, Vandesompele J (2019) LNCipedia 5: towards a reference set of human long non-coding RNAs. Nucleic Acids Res 47:D135–D139
25. Lennox KA, Behlke MA (2016) Cellular localization of long non-coding RNAs affects silencing by RNAi more than by antisense oligonucleotides. Nucleic Acids Res 44:863–877
26. Suzuki Y, Holmes JB, Cerritelli SM, Sakhuja K, Minczuk M, Holt IJ, Crouch RJ (2010) An upstream open reading frame and the context of the two AUG codons affect the abundance of mitochondrial and nuclear RNase H1. Mol Cell Biol 30:5123–5134
27. Lima WF, Wu H, Crooke ST (2008) The RNase H Mechanism. In Antisense Drug Technology: Principles, Strategies, and Applications, CRC Press:Boca Raton, vol. 2:47–74
28. Wu H, Sun H, Liang X, Lima WF, Crooke ST (2013) Human RNase H1 is associated with protein P32 and is involved in mitochondrial pre-rRNA processing. PLoS One 8:e71006
29. Liang XH, Sun H, Nichols JG, Crooke ST (2017) RNase H1-dependent antisense oligonucleotides are robustly active in directing RNA cleavage in both the cytoplasm and the nucleus. Mol Ther 25:2075–2092
30. Robb GB, Brown KM, Khurana J, Rana TM (2005) Specific and potent RNAi in the nucleus of human cells. Nat Struct Mol Biol 12:133–137
31. Eder PS, DeVine RJ, Dagle JM, Walder JA (1991) Substrate specificity and kinetics of degradation of antisense oligonucleotides by a 3′ exonuclease in plasma. Antisense Res Dev 1:141–151
32. Lennox KA, Owczarzy R, Thomas DM, Walder JA, Behlke MA (2013) Improved performance of Anti-miRNA oligonucleotides using a novel non-nucleotide modifier. Mol Ther Nucleic Acids 2:e117
33. Kurreck J (2003) Antisense technologies. Improvement through novel chemical modifications. Eur J Biochem 270:1628–1644
34. Shen X, Corey DR (2018) Chemistry, mechanism and clinical status of antisense oligonucleotides and duplex RNAs. Nucleic Acids Res 46:1584–1600
35. Khvorova A, Watts JK (2017) The chemical evolution of oligonucleotide therapies of clinical utility. Nat Biotechnol 35:238–248
36. Dowdy SF (2017) Overcoming cellular barriers for RNA therapeutics. Nat Biotechnol 35:222–229
37. Eckstein F (2014) Phosphorothioates, essential components of therapeutic oligonucleotides. Nucleic Acid Ther 24:374–387
38. Geary RS, Norris D, Yu R, Bennett CF (2015) Pharmacokinetics, biodistribution and cell uptake of antisense oligonucleotides. Adv Drug Deliv Rev 87:46–51
39. Koller E, Vincent TM, Chappell A, De S, Manoharan M, Bennett CF (2011) Mechanisms of single-stranded phosphorothioate modified antisense oligonucleotide accumulation in hepatocytes. Nucleic Acids Res 39:4795–4807
40. Wan WB, Seth PP (2016) The medicinal chemistry of therapeutic oligonucleotides. J Med Chem 59:9645–9667
41. Freier SM, Altmann KH (1997) The ups and downs of nucleic acid duplex stability: structure-stability studies on chemically-modified DNA:RNA duplexes. Nucleic Acids Res 25:4429–4443
42. Lennox KA, Behlke MA (2010) A direct comparison of anti-microRNA oligonucleotide potency. Pharm Res 27:1788–1799
43. Manoharan M (1999) 2′-carbohydrate modifications in antisense oligonucleotide therapy: importance of conformation, configuration and conjugation. Biochim Biophys Acta 1489:117–130

44. Koshkin AA, Nielsen P, Meldgaard M, Rajwanshi VK, Singh SK, Wengel J (1998) LNA (locked nucleic acid): an RNA mimic forming exceedingly stable LNA:LNA duplexes. J Am Chem Soc 120:2
45. Eichert A, Behling K, Betzel C, Erdmann VA, Furste JP, Forster C (2010) The crystal structure of an "All Locked" nucleic acid duplex. Nucleic Acids Res 38:6729–6736
46. Lennox KA, Sabel JL, Johnson MJ, Moreira BG, Fletcher CA, Rose SD, Behlke MA, Laikhter AL, Walder JA, Dagle JM (2006) Characterization of modified antisense oligonucleotides in Xenopus laevis embryos. Oligonucleotides 16:26–42
47. Swayze EE, Siwkowski AM, Wancewicz EV, Migawa MT, Wyrzykiewicz TK, Hung G, Monia BP, Bennett CF (2007) Antisense oligonucleotides containing locked nucleic acid improve potency but cause significant hepatotoxicity in animals. Nucleic Acids Res 35:687–700
48. Hagedorn PH, Yakimov V, Ottosen S, Kammler S, Nielsen NF, Hog AM, Hedtjarn M, Meldgaard M, Moller MR, Orum H et al (2013) Hepatotoxic potential of therapeutic oligonucleotides can be predicted from their sequence and modification pattern. Nucleic Acid Ther 23:302–310
49. Burel SA, Hart CE, Cauntay P, Hsiao J, Machemer T, Katz M, Watt A, Bui HH, Younis H, Sabripour M et al (2016) Hepatotoxicity of high affinity gapmer antisense oligonucleotides is mediated by RNase H1 dependent promiscuous reduction of very long pre-mRNA transcripts. Nucleic Acids Res 44:2093–2109
50. Dieckmann A, Hagedorn PH, Burki Y, Brugmann C, Berrera M, Ebeling M, Singer T, Schuler F (2018) A sensitive in vitro approach to assess the hybridization-dependent toxic potential of high affinity Gapmer oligonucleotides. Mol Ther Nucleic Acids 10:45–54
51. Burdick AD, Sciabola S, Mantena SR, Hollingshead BD, Stanton R, Warneke JA, Zeng M, Martsen E, Medvedev A, Makarov SS et al (2014) Sequence motifs associated with hepatotoxicity of locked nucleic acid--modified antisense oligonucleotides. Nucleic Acids Res 42:4882–4891
52. Kamola PJ, Maratou K, Wilson PA, Rush K, Mullaney T, McKevitt T, Evans P, Ridings J, Chowdhury P, Roulois A et al (2017) Strategies for in vivo screening and mitigation of hepatotoxicity associated with antisense drugs. Mol Ther Nucleic Acids 8:383–394
53. Pedersen L, Hagedorn PH, Lindholm MW, Lindow M (2014) A kinetic model explains why shorter and less affine enzyme-recruiting oligonucleotides can be more potent. Mol Ther Nucleic Acids 3:e149
54. Markham NR, Zuker M (2008) UNAFold: software for nucleic acid folding and hybridization. Methods Mol Biol 453:3–31
55. Paz I, Kosti I, Ares M Jr, Cline M, Mandel-Gutfreund Y (2014) RBPmap: a web server for mapping binding sites of RNA-binding proteins. Nucleic Acids Res 42:W361–W367
56. Johnson G, Nour AA, Nolan T, Huggett J, Bustin S (2014) Minimum information necessary for quantitative real-time PCR experiments. Methods Mol Biol 1160:5–17
57. Bustin SA, Benes V, Garson JA, Hellemans J, Huggett J, Kubista M, Mueller R, Nolan T, Pfaffl MW, Shipley GL et al (2009) The MIQE guidelines: minimum information for publication of quantitative real-time PCR experiments. Clin Chem 55:611–622

Chapter 10

Calcium-Mediated In Vitro Transfection Technique of Oligonucleotides with Broad Chemical Modification Compatibility

Fumito Wada, Shin-ichiro Hori, Satoshi Obika, and Tsuyoshi Yamamoto

Abstract

Oligonucleotide drugs (ODs) have gained increasing attention owing to their promising therapeutic potential. One major obstacle that ODs have been facing is the lack of appropriate in vitro validation systems that can predict in vivo activity and toxicity. We have devised a transfection method called CEM (Ca^{2+}-enrichment method), where the simple enrichment of calcium ion with calcium chloride in culture medium potentiates the activity of various types of naked oligonucleotides including gapmers, siRNA, and phosphorodiamidate morpholino antisense oligonucleotides (PMO) in many cultured cell lines with limited cytotoxicity. We here describe a precise procedure of the method. Besides the benefit of the CEM's predictive power to accurately estimate in vivo activity of ODs of your interest in drug discovery and development settings, this cost-efficient, easy-to-access method can be a robust laboratory technique to modulate gene expressions with ODs with a variety of mechanisms of action.

Key words Oligonucleotide drugs, 2′4′-BNA/LNA gapmers, LNA, siRNA, Phosphorodiamidate morpholino oligonucleotide, Transfection, Nanoparticles, Naked oligonucleotides

1 Introduction

Oligonucleotide drugs (ODs) including antisense oligonucleotides (ASOs), siRNAs, anti-miRNAs and miRNA mimics have gained increasing attention owing to their promising therapeutic potential [1–3]. Chemical modification of oligonucleotides has largely overcome the weakness of ODs by improving their PK/PD to provide robust systemic activity even in the absence of delivery carriers [4–6]. One major obstacle that ODs have been facing is the lack of appropriate in vitro validation systems that can predict in vivo activity and toxicity due mainly to the fact that in vivo fate of naked ODs has not been fully illustrated yet [6, 7]. Unformulated gapmers show an excellent knockdown in the liver of animals including humans through the controversial pathways, however, their relatively narrow therapeutic window limit clinical

Toshifumi Yokota and Rika Maruyama (eds.), *Gapmers: Methods and Protocols*, Methods in Molecular Biology, vol. 2176, https://doi.org/10.1007/978-1-0716-0771-8_10,

applications [8–11]. In fact, using a conventional carrier-dependent in vitro transfection (e.g., lipofection) often yields false-positives mainly because the cellular uptake mechanism is entirely different from one happening in vivo. [12]

In this context, several alternative in vitro assay systems that could reproduce in vivo native uptake have been developed. It is known that primary hepatocytes maintain the ability to take up naked oligonucleotides and elicit a gene silencing effect of ASOs; however, this uptake is lost within 24–36 h after isolation [13]. Koller et al. have established a mouse hepatocellular SV40 large T-antigen carcinoma cell line (MHT) from a specific transgenic mouse to maintain this function of primary hepatocytes [14]. One issue associated with this technique is that MHT is not a human cell line, limiting translational applications of the system for drug development. On the other hand, it was revealed that naked 2′, 4′-BNA/LNA-modified gapmers are taken up by various cell lines, including human cell, after a week-long exposure without the use of any transfection reagents. Interestingly, this in vitro free-uptake system was shown to reflect in vivo activity [12, 15, 16]. However, despite this clear advantage, this system is compatible only with biologically stable ASOs, and the efficiency varies widely depending on cell lines, and also requires long-term exposure to medium.

We have sought to device a new in vitro system and revealed that simple enrichment of calcium ion with calcium chloride (from $CaCl_2$) in culture medium potentiates the activity of various types of naked oligonucleotides including gapmers and siRNA but not that of plasmids, in many cultured cell lines with limited cytotoxicity [17]. We have also shown that our method, called CEM (Ca^{2+} enrichment method), has a higher positive correlation with in vivo activity than does a conventional lipofection. In addition, CEM can be applied to PMO (phosphorodiamidate morpholino) [17] and PNA (peptide nucleic acids) [18], two representative charge neutral ODs, both of which in general require special transfection reagents. Mechanistically, cellular uptake of ODs is greatly enhanced under CEM condition. Highly monodispersed particles with a size of approximately 100 nm, which most likely derive from albumin, were identified in the regular culture medium with an excess of Ca^{2+} (~9 mM). These particles are suspected to be involved in CEM mechanism.

We here describe a precise and comprehensive procedure of our CEM transfection method. This cost-efficient, easy-to-access method has become a robust and standard laboratory technique to modulate gene expression with ODs with a variety of mechanisms of action, [19–25] as well as provides you a more accurate prediction of in vivo activity of ODs of your interest in drug discovery and development settings.

2 Materials

2.1 Oligonucleotides

1. 2′,4′-BNA/LNA-modified antisense oligonucleotides (Table 1, Fig. 1, Gene Design, Inc., Osaka, Japan).
2. Stealth RNAi siRNAs (*see* **Note 1**) targeting ApoB (Table 1) and negative control siRNA with a medium GC (Invitrogen, Carlsbad, CA, USA).
3. The ST6-PMO and NC-PMO (Table 1, Fig. 1, Gene-Tools, LLC, Corvallis, OR, USA) [26].

2.2 Cell Lines

Huh-7, HLE, HeLa, HEK293, A549.

2.3 Cell Culture

1. Dulbecco's Modified Eagle's Medium (DMEM) supplemented with 10% heat-inactivated fetal bovine serum (FBS) and antibiotics (100 units/mL of penicillin, 100 μg/mL of streptomycin, and 0.25 μg/mL of Fungizone®).
2. 10-cm culture dish.
3. 2.5 g/L trypsin, 1 mM EDTA solution.
4. 0.4 w/v% trypan blue solution.
5. Automated cell counter.

2.4 Ca^{2+}-Mediated Transfection

1. Calcium chloride (molecular biology grade) (*see* **Note 2**).
2. 96- or 12-well culture plates.
3. Dulbecco's Modified Eagle's Medium (DMEM) supplemented with 10% heat-inactivated fetal bovine serum (FBS).
4. 100 μM or 1 mM antisense oligonucleotides.

2.5 RT-PCR Assay

1. In vitro: CellAmp Direct RNA Prep Kit for RT-PCR (TaKaRa Bio Inc., Shiga, Japan).
2. In vivo: High Capacity cDNA Reverse Transcription kit (Life Technologies, Gaithersburg, MA, USA).

2.6 qRT-PCR Assay

1. One Step SYBR PrimeScript PLUS RT-PCR Kit (TaKaRa Bio Inc.) or Fast SYBR Green Master Mix (Applied Biosystems, Foster City, CA, USA).
2. Suggested primer sets.
   (a) Human APOB (forward: 5′-GCACCGTCAAGGCTGAGAAC-3′; reverse:5′-TGGTGAAGACGCCAGTGGA-3′);
   (b) Human GAPDH (forward: 5′-GCACCGTCAAGGCTGAGAAC-3′; reverse:5′-TGGTGAAGACGCCAGTGGA-3′);
   (c) Mouse apoB (forward: 5′-TCCTCGGTGAGTTCAATGACTTTC-3′; reverse: 5′-TGGACCTGCTGTAGCTTGTAGGA-3′);

Table 1
Exemplars of oligonucleotide CEM can be applied to [17]

Oligo ID	Sequences
ApoB-521-BNA(13)	5′-A^A^t^g^g^c^c^a^g^c^T^T^G-3′
ApoB-1261-BNA(13)	5′-A^A^g^g^c^t^t^g^t^a^A^A^G-3′
ApoB-2101-BNA(13)	5′-T^C^a^a^a^t^a^t^a^a^G^A^T-3′
ApoB-5141-BNA(13)	5′-A^A^g^c^t^c^t^g^c^a^T^T^C-3′
ApoB-5701-BNA(13)	5′-A^C^a^c^c^c^t^g^a^a^C^C^T-3′
ApoB-8701-BNA(13)	5′-G^T^g^t^g^t^a^a^a^c^T^T^G-3′
ApoB-9621-BNA(13)	5′-A^T^g^a^t^t^g^c^t^t^T^G^T-3′
ApoB-10177-BNA(13) [27]	5′-G^C^a^t^t^g^g^t^a^t^T^C^A-3′
ApoB-10461-BNA(13)	5′-T^G^g^t^a^t^t^t^c^c^A^T^T-3′
ApoB-11001-BNA(13)	5′-G^G^t^t^c^t^t^a^g^t^G^T^T-3′
ApoB-12241-BNA(13)	5′-T^C^a^g^t^t^t^t^g^a^A^T^A-3′
Survivin-BNA(16) or (EZN-3042) [29]	5′-C^T^C^A^a^t^c^c^a^t^g^g^C^A^G^c-3′
NC-BNA(16)	5′-T^A^G^c^t^t^g^t^c^c^c^a^t^C^T^C-3′
ApoB-si1	5′-*CCUGAAGUUUGUAACUCAAGCAGAA*-3′ 3′-*GGACUUCAAACAUUGAGUUCGUCUU*-5′
ApoB-si2	5′-*GGGACCACAGAUGUCUGCUUCAGUU*-3′ 3′-*CCCUGGUGUCUACAGACGAAGUCAA*-5′
ApoB-si3	5′-*CCAAUACAAGUUGAAGGAGACUAUU*-3′ 3′-*GGUUAUGUUCAACUUCCUCUGAUAA*-5′
NC-PMO [26]	GGATGTCTTGCTGGACGTCGTTA
ST6-PMO [26]	CATTTTCTGTTCTAGATCCTGTT

Uppercase: 2′,4′-BNA/LNA, Lowercase: DNA, Italic: RNA,
Underscore: PMO, ^: phosphorothioate linkage
*The structural information of Stealth RNAi™ siRNAs is not open to public

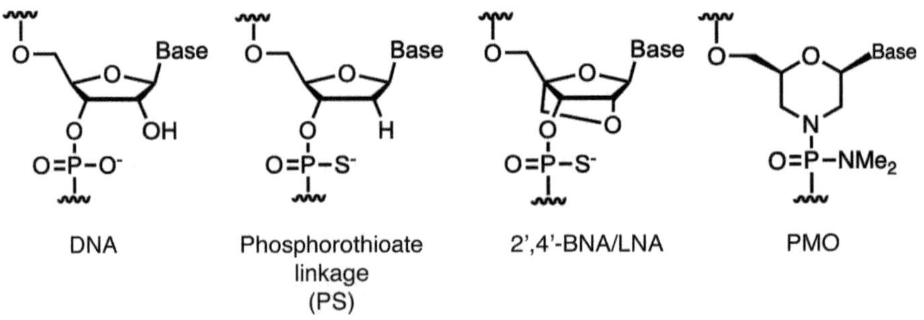

Fig. 1 Structures of chemical modification for oligonucleotide drugs mentioned in this method

(d) Human SURVIVIN (forward: 5′-AGTCTGGCGTAAGA TGATGGATTTG-3′; reverse: 5′-CACAGCAGTGTTTG AAATGACAGG-3′);

(e) Human STAT3(forward: 5′-GGAATCCCGTGGGTTG CTTAC-3′; 5′-TTGAATGCAGTGGCCAGGAC-3′);

(f) Mouse GAPDH (forward: 5′-TGTGTCCGTCGTGGAT CTGA-3′; reverse: 5′-TTGCTGTTGAAGTCGCAGGA G-3′).

3. Real-Time qPCR system.

2.7 Confirming the Skipping of STAT3-Exon 6

1. 10% Tris–Borate–EDTA (TBE) gel.
2. Ethidium bromide.
3. Electrophoresis tank.
4. Gel imaging system.

2.8 Western Blotting Analysis

1. Radioimmunoprecipitation assay (RIPA) buffer.
2. Complete Protease Inhibitor Cocktail.
3. BCA Protein Assay Kit.
4. 5–20% gradient polyacrylamide gel.
5. Polyvinylidene difluoride (PVDF) membranes.
6. Blocking solution (e.g., Blocking One (Nacalai Tesque, Kyoto, Japan)).
7. Tris-buffered saline (TBS) with 0.05%v/v Tween 20 (TBS-T).
8. Bovine serum albumin.
9. Anti-bodies: monoclonal rabbit anti-Stat3 antibody (Cell Signaling Technology, Danvers, MA, USA); monoclonal mouse anti-β-actin (Sigma-Aldrich); HRP-conjugated antibodies (Amersham Biosciences, Buckinghamshire, UK).
10. Chemiluminescent reagent (e.g., ECL Prime Kit (Amersham Biosciences)).
11. Gel imager (e.g., ImageQuant LAS 4000 (FujiFilm, Tokyo, Japan)).

2.9 In vivo Experiment

1. C57BL/6J male mice, 7 weeks old (CLEA Japan, Tokyo, Japan).
2. Syringes with 29-gauge needle (Terumo, Tokyo, Japan).
3. Saline.
4. Isoflurane (FujiFilm).
5. QuickGene RNA tissue kit SII (FujiFilm).
6. QuickGene-810 (FujiFilm).
7. Isoflurane (FujiFilm).
8. Tissue homogenizer (e.g., TissueLyser II (QIAGEN)).

3 Methods

3.1 Generation of Library of 2′4′-BNA/LNA Gapmers

1. Access the NCBI website (https://www.ncbi.nlm.nih.gov/).
2. Select the category of "Nucleotide" to obtain target mRNA sequences (e.g., apoB here) and other related information needed.
3. Pick out as many 13–15 nucleotide-long target regions as possible out of the whole mRNA sequence and convert the sequences into the reverse complementary sequences to obtain a 2′4′-BNA/LNA gapmer sequence library. (Fig. 2) (*see* **Notes 2–4**) [27, 28].
4. Purchase as many members of the 2′4′-BNA/LNA gapmer library as possible from Gene Design, Inc. (Osaka, Japan) in a 0.2 μmol scale (*see* **Note 5**).
5. Dissolve each oligonucleotide with an appropriate amount of RNase-free water to prepare a 0.1–1 mM stock solution.

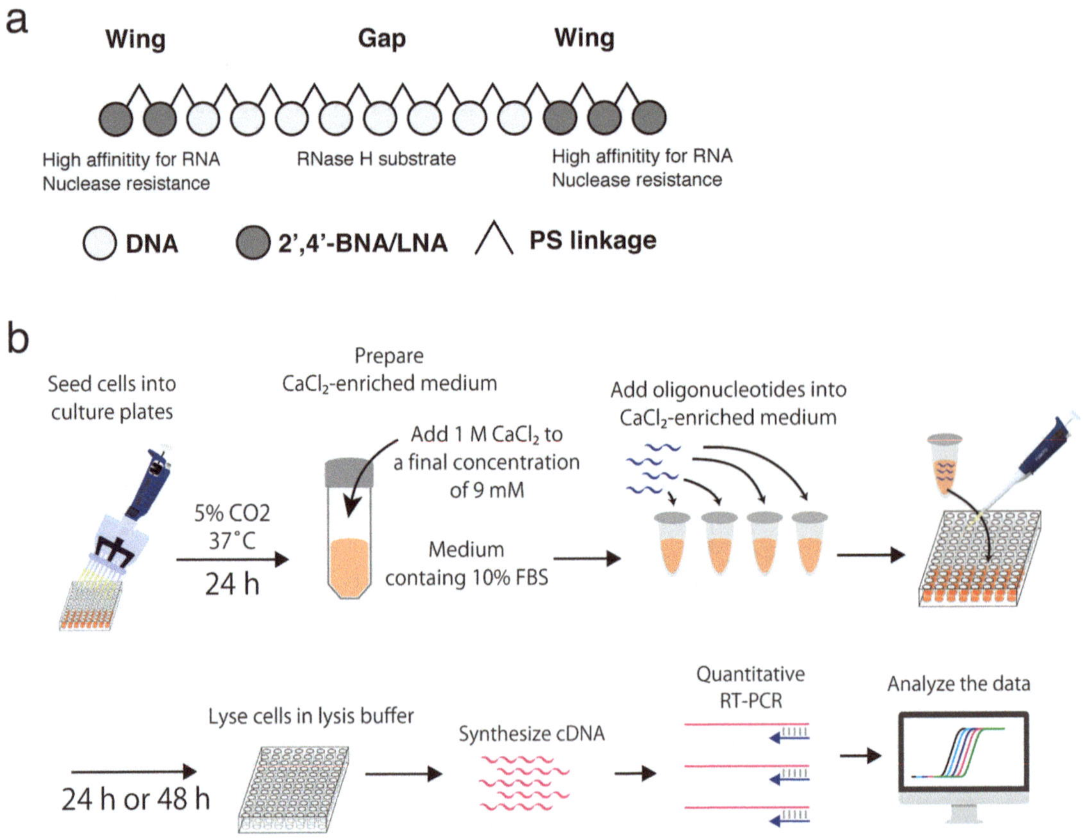

Fig. 2 Graphical illustration of a representative gapmer structure (**a**) and schematic procedure of CEM-mediated transfection of oligonucleotides (**b**)

3.2 Cell Culture and Seeding

1. Grow the cells at 37 °C and 5% CO_2 in the medium in 10 cm-culture dishes.
2. Passage the cells with trypsin-EDTA solutions when they reach to ~80% confluence.
3. Count the suspended cells using a TC20™ Automated Cell Counter and seed them at 8.0×10^3 cells/well in 96-well plates or 9.5×10^4 cells/well in 12-well plates.

3.3 Ca^{2+}-Mediated Oligonucleotide Transfection (See Note 6)

1. Dissolve 1 g of $CaCl_2$ in 9 mL of sterile water to make a 1 M stock solution.
2. Filter the stock solution using a 0.22 μm PES membrane filter.
3. Add 1 M $CaCl_2$ to DMEM containing 10% FBS to make a final Ca^{2+} concentration 9 mM (*see* **Note 7**).
4. Add oligonucleotides to be tested to the above-prepared Ca^{2+}-enriched medium.

   Suggested final concentration ranges

   (a) 2′4′-BNA/LNA gapmers: ~1 μM.

   (b) Stealth RNAi siRNA™: ~1 μM. (*see* **Note 8**).

   (c) PMO: ~20 μM.
5. Wash the cells twice with Ca^{2+}-enriched DMEM+10%FBS without oligonucleotides and apply them to the cultured cells in 96-well or 12-well plates. The optimum amount of the solution to be added is 100 μL/well for the 96-well plates and 1 mL/well for the 12 well-plates.
6. After 24–48 h of incubation at 37 °C and 5% CO_2, wash the wells gently with PBS(−) twice using an 8-channel pipette (*see* **Note 9**).

3.4 Complementary DNA (cDNA) Synthesis

1. Lyse the cells using the lysate using CellAmp Direct RNA Prep Kit.
2. Synthesize complementary DNA (cDNA) of the whole transcripts from the lysates using a PCR instrument in accordance with the instructions of the kit.

3.5 Quantification of Target RNAs

1. Design target-specific primer sets that can amplify any one of the exon–exon boundaries of your target, as well as any housekeeping genes (e.g., GAPDH) using primer design tools (e.g., Primer3Plus and Primer-BLAST) (*see* **Note 10**). Alternatively, use one of the predesigned TaqMan probes at ThermoFisher Scientific, Inc. websites.
2. Prepare real-time qPCR cocktails by following the instructions of the quantitative PCR kits listed above.
3. Perform quantitative PCR in accordance with a protocol of the qPCR kit that you use on a Real-Time qPCR system.

4. Measure PCR threshold cycle numbers (C_t), the intersection between an amplification curve and a threshold line, for the target of your interest and household genes, and normalize the C_t of the target with the corresponding C_t value of the housekeeping gene to estimate relative gene expression levels for gapmers and siRNAs, or exon skipping efficiency for PMOs, where ST-PMO were designed to skip the exon 6 of STAT3 (Fig. 3b).

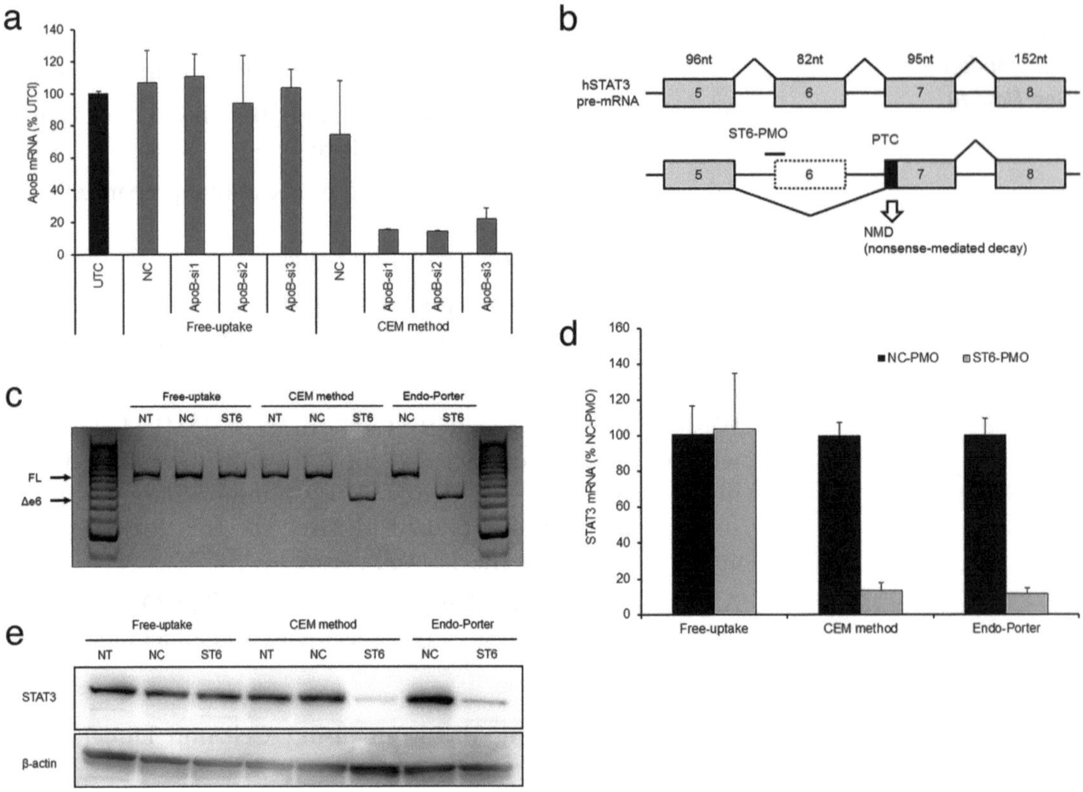

Fig. 3 A few applications to show the effect of the CEM on the activity of siRNA and PMO. (**a**) CEM effect on the activity of ApoB-siRNAs. Huh-7 cells were treated with 1 μM ApoB-siRNAs with or without 9 mM $CaCl_2$ for 24 h. ApoB mRNA levels were analyzed using qRT-PCR. The relative quantification of ApoB mRNA was normalized against expression of the GAPDH gene. Each data point represents the mean + SD of three independent experiments. (**b**) ST6-PMO targets the boundary region of intron 5 and exon 6 in the human STAT3 pre-mRNA provides exon 6 skipping, and induces RNA degradation via nonsense-mediated mRNA decay. PTC: premature termination codon. In this experiment, Huh-7 cells were treated with 20 μM of ST6-PMO in the presence or absence of 9 mM $CaCl_2$. After 48 h, STAT3 pre-mRNA and protein expressions were analyzed. (**c**) STAT3 pre-mRNA splicing was visualized by polyacrylamide gel separation of RT-PCR products amplified with a specific primer set to confirm the exclusion of exon 6. *FL* full-length, *Δe6* exon 6 skipping, *NT* non-treatment, *NC* negative control. (**d** and **e**) qRT-PCR and western blotting analysis were performed to confirm a STAT3 expression change as described in Subheadings 3.5 and 3.6, respectively. For qRT-PCR, the relative quantification of STAT3 mRNA was normalized against expression of the GAPDH gene. Each data point represents the mean + SD of three independent experiments. β-actin was the loading control for western blotting. Endo-Porter was used as the positive control for PMO transfection. (Reproduced from Ref. 17)

3.6 Western Blot Analysis

We here describe a detection method of Stat3 protein after CEM treatment of Stat3-targeting PMOs to exemplify a general western blotting procedure. Many other proteins that you may be interested in will follow suit.

1. After the CEM-mediated introduction of Stat3-targeting PMOs, lyse the cells seeded in 12-well plates with radioimmunoprecipitation assay (RIPA) buffer (*see* **Note 11**).
2. Determine a total protein concentration of each sample using a total protein assay kit, BCA Protein Assay Kit.
3. Mix each sample containing 10 μg of total proteins with DTT (dithiothreitol)-containing loading buffer.
4. Heat the samples at 100 °C for 5 min, then cool them in an ice-cold bath.
5. Spin them down for 10 s with a benchtop centrifuge.
6. Run a gel electrophoresis for these samples on a 5–20% SDS–polyacrylamide gradient at 40 mA, 80 min.
7. Blot the gels onto PVDF membranes by standard semidry for 90 min.
8. Perform blocking of the membranes for 1 h in a blocking solution at room temperature.
9. Wash the membranes with TBS-T for 15 min three times at room temperature.
10. Immunostain the membranes at 4 °C overnight in a presence of a monoclonal rabbit anti-Stat3 antibody (1:1000) in 5% bovine serum albumin-containing TBS-T, or for 1 h at room temperature in a blocking buffer containing monoclonal mouse anti--β-actin (1:1000). Then, wash the membranes in TBS-T for 15 min three times.
11. Immunostain the membranes for 1 h in TBS-T containing 5% Blocking One with an HRP-conjugated antibody (1: 3000).
12. Visualize Stat3 and β-actin bands using ECL Prime Kit and a gel imaging system to estimate the exon skipping efficiency (Fig. 3e).

3.7 In vivo Experiment

1. Purchase 6-week-old male mice from CLEA Japan.
2. Maintain the mice in a 12 h light–dark cycle, being fed ad libitum.
3. Dissolve each oligonucleotide of your interest in saline and fill it up to 100–200 μL for injection: if you give 10 mg/kg of dosage, add 177 μL of saline to 23 μL of a 10 μg/μL 2′4′-BNA/LNA gapmer solution when the bodyweight of mice is around 23 g (*see* **Note 12**).

4. Inject saline or oligonucleotides subcutaneously to mice (*see* **Note 13**).
5. Three days after the administration, anesthetize the mice with isoflurane, and harvest the livers and cut them into pieces (Fig. 4a) (*see* **Note 14**).

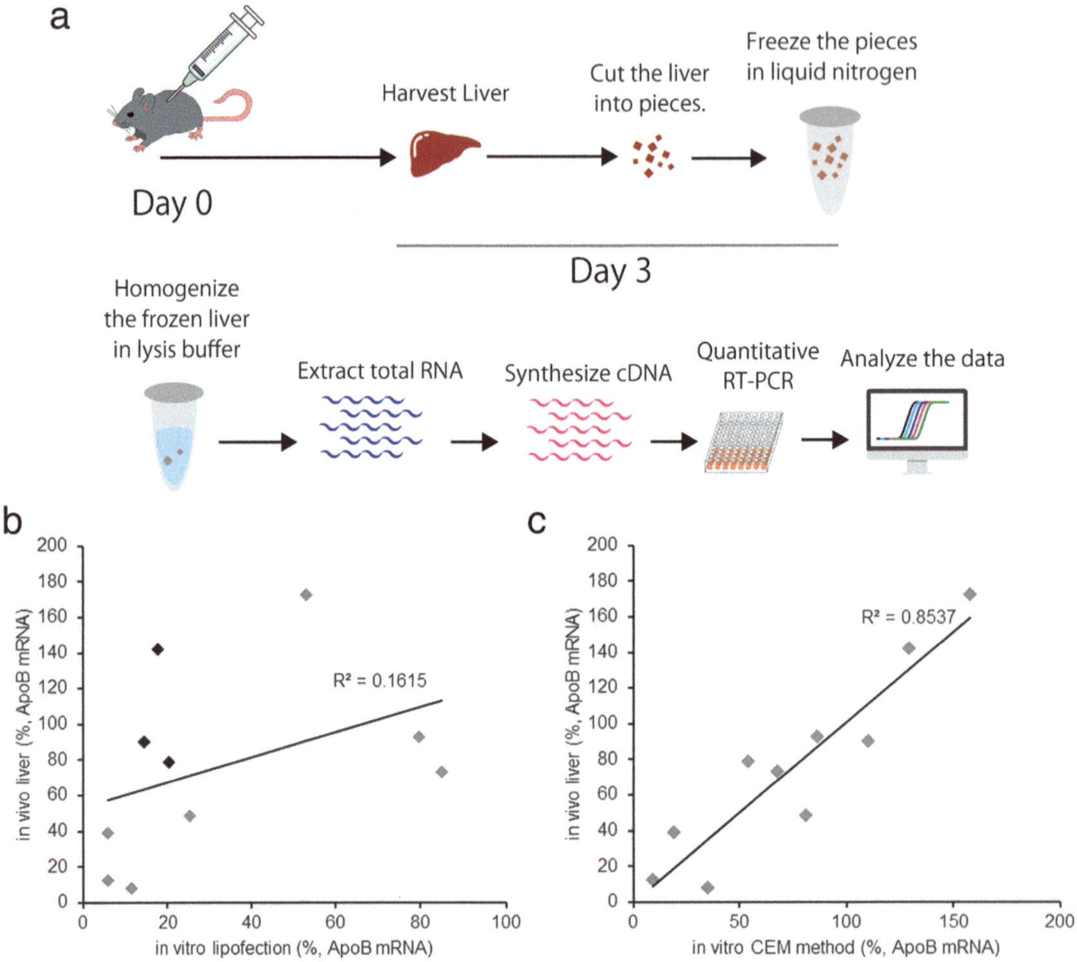

Fig. 4 CEM outcomes better reflect in vivo activity of 2′4′-BNA/LNA gapmers than those of lipofection. (**a**) Schematic overview of the in vivo experiment described in Subheading 3.7. (**b, c**) To compare the in vitro activity of CEM or lipofection versus in vivo activity, Huh-7 cells were treated with ten different ApoB-ASOs shown in Table 1 at 10 nM under CEM condition or using RNAiMAX (Invitrogen). After 24 h, total RNA was extracted and ApoB mRNA was quantitated by qRT-PCR as described in Subheadings 3.4 and 3.5. On the other hand, mice (C57BL/6, $n = 3$/group) were injected subcutaneously at a dose of 10 mg/kg of each 10 ApoB-ASOs. Mice were sacrificed 72 h later and livers were analyzed for reductions in ApoB mRNA levels in accordance with Subheading 3.7. The relative quantification of ApoB mRNA was normalized against the expression of the GAPDH gene. Each data point represents the mean of three independent experiments. The mean value of knockdown activity was plotted in the two-dimensional space to analyze the correlation between transfection methods. (Reproduced from Ref. 17)

6. Homogenize 15–30 mg of the segments at 30 Hz for 5 min in the lysis buffer of QuickGene RNA tissue kit SII using a tissue homogenizer.
7. Isolate total RNA from the liver tissue by following the instructions of QuickGene RNA tissue kit SII.
8. Generate cDNA from 2 μg of total RNA using an RT-PCR reagent, High Capacity cDNA Reverse Transcription kit (Life Technologies).
9. Perform quantitative RT-PCR according to Subheading 3.5 (Fig. 4b, c).

4 Notes

1. Avoid sequences containing a CpG motif and high GC content in designing ODs.
2. Calcium chloride must be nuclease and protease free.
3. The use of 5-methyl cytosine instead of cytosine could be helpful in reducing the CpG-derived toxicity.
4. Gapmers, in general, have wing–gap–wing structures as shown in Fig. 2a. The wings are typically 2- to 3-mer RNA stretches with high-affinity modifications such as 2′,4′-BNA/LNA. The stretch of the gap is as long as 7–10 nucleotide-long DNA. All the linkages of gapmers should be phosphorothioate.
5. As far as we know, it has yet been difficult to accurately predict the efficacy of 2′4′-BNA/LNA gapmers only from the sequence information of mRNA. One benefit of using our CEM technique is to enable efficient in vitro screening of a library of ODs.
6. The procedure of the CEM is illustrated in Fig. 2b. A few representative examples are outlined in Figs. 5a–c and 4c for 2′4′-BNA/LNA gapmers; Fig. 3a for siRNAs; c–e for PMOs.
7. To maximize the effect of the CEM, both $CaCl_2$ and ODs should be added at the same time. Under the condition, we have confirmed the formation of highly monodispersed particles with a size of approximately 100 nm by using both dynamic light scattering analysis and transmission electron microscopy analysis.
8. A negative control siRNA with a medium GC content were obtained from Invitrogen (Carlsbad, CA, USA) as Stealth RNAi siRNAs.
9. Only very limited cytotoxicity has been observed even after 96 h of incubation in the presence of 9 mM Ca^{2+} in a variety of cell lines.

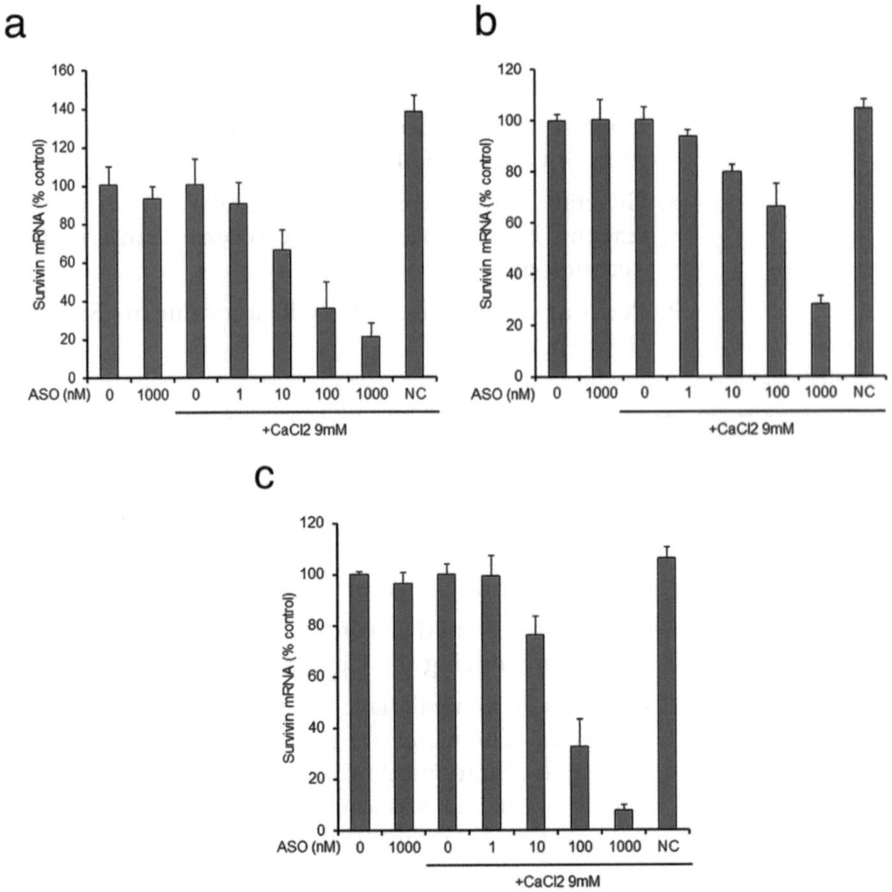

Fig. 5 A few representative examples of the CEM effect on a gene, named Survivin, on different cell lines. HeLa (**a**), HEK293 (**b**), or A549 (**c**) cells were seeded in 96-well plates containing 10%FBS/DMEM. After 24 h, Survivin-targeting 2′4′-BNA/LNA gapmer antisense oligonucleotide (ASO) (Survivin-BNA (16)) was added, and cells were cultured in medium in the presence or absence of the additional Ca^{2+}. After 24 h, total RNA was extracted and Survivin mRNA was quantitated by qRT-PCR as described Subheadings 3.4 and 3.5. The level of target (Survivin) gene expression was normalized to that of GAPDH. Each data point represents the mean + SD of three independent experiments. (Reproduced from Ref. 17)

10. See guidelines and procedures for the designing and preparation of optimal real-time PCR primers elsewhere.
11. Use a larger number of cells (=bigger well plates) for protein detection/quantitation due to limitations in sensitivity of western blotting analysis.
12. Prior to apply ODs in vivo, you may have to begin with estimating proper dose ranges for unformulated siRNAs and PMOs of your interest to evaluate as the information about in vivo systemic effects of naked ODs other than 2′4′-BNA/LNA gapmers is very limited.

13. ODs can be applied from any routes such as intravenously, subcutaneously, and intraperitoneally.
14. Right after being harvested, the livers can be snap-frozen in liquid nitrogen and keep them in −80 °C freezer to store.

Acknowledgments

The method described in this unit is mainly based on our previous work published in Nucleic Acids Research, Oxford University Press in 2015 [17]. This work was supported by JSPS KAKENHI Grant-in-Aid for Scientific Research (B) (Grant No. 19H03656, 19H03393) and JSPS KAKENHI Grant No. 24890102, and Grants-in Aid for Scientific Research from the Japanese Ministry of Health, Labor, and Welfare (H26-kanjitu-kanen-wakate-008) and a grant from the National Institute of Biomedical Innovation.

References

1. Yamamoto T, Nakatani M, Narukawa K, Obika S (2011) Antisense drug discovery and development. Future Med Chem 3:339–365. https://doi.org/10.4155/fmc.11.2
2. Baumann V, Winkler J (2014) miRNA-based therapies: strategies and delivery platforms for oligonucleotide and non-oligonucleotide agents. Future Med Chem 6:1967–1984. https://doi.org/10.4155/fmc.14.116
3. Egli M, Manoharan M (2019) Re-engineering RNA molecules into therapeutic agents. Acc Chem Res 52:1036–1047. https://doi.org/10.1021/acs.accounts.8b00650
4. White PJ, Anastasopoulos F, Pouton CW, Boyd BJ (2009) Overcoming biological barriers to in vivo efficacy of antisense oligonucleotides. Expert Rev Mol Med 11:1–19. https://doi.org/10.1017/S1462399409001021
5. Crooke ST, Wang S, Vickers TA et al (2017) Cellular uptake and trafficking of antisense oligonucleotides. Nat Biotechnol 35:230–237. https://doi.org/10.1038/nbt.3779
6. Crooke ST (2017) Molecular mechanisms of antisense oligonucleotides. Nucleic Acid Ther 27:70–77. https://doi.org/10.1089/nat.2016.0656
7. Bijsterbosch MK, Manoharan M, Rump ET et al (1997) In vivo fate of phosphorothioate antisense oligodeoxynucleotides: predominant uptake by scavenger receptors on endothelial liver cells. Nucleic Acids Res 25:3290–3296
8. Kakiuchi-Kiyota S, Koza-Taylor PH, Mantena SR et al (2014) Comparison of hepatic transcription profiles of locked ribonucleic acid antisense oligonucleotides: evidence of distinct pathways contributing to non-target mediated toxicity in mice. Toxicol Sci 138:234–248
9. Burel SA, Hart CE, Cauntay P et al (2016) Hepatotoxicity of high affinity gapmer antisense oligonucleotides is mediated by RNase H1 dependent promiscuous reduction of very long pre-mRNA transcripts. Nucleic Acids Res 44:2093–2109
10. Burdick AD, Sciabola S, Mantena SR et al (2014) Sequence motifs associated with hepatotoxicity of locked nucleic acid—modified antisense oligonucleotides. Nucleic Acids Res 42:4882–4891
11. Hagedorn PH, Yakimov V, Ottosen S et al (2013) Hepatotoxic potential of therapeutic oligonucleotides can be predicted from their sequence and modification pattern. Nucleic Acid Ther 23:302–310. https://doi.org/10.1089/nat.2013.0436
12. Zhang Y, Qu Z, Kim S et al (2011) Down-modulation of cancer targets using locked nucleic acid (LNA)-based antisense oligonucleotides without transfection. Gene Ther 18:326–333. https://doi.org/10.1038/gt.2010.133
13. Nestle FO, Mitra RS, Bennett CF et al (1994) Cationic lipid is not required for uptake and selective inhibitory activity of ICAM-1 phosphorothioate antisense oligonucleotides in keratinocytes. J Invest Dermatol 103:569–575

14. Koller E, Vincent TM, Chappell A et al (2011) Mechanisms of single-stranded phosphorothioate modified antisense oligonucleotide accumulation in hepatocytes. Nucleic Acids Res 39:4795–4807. https://doi.org/10.1093/nar/gkr089
15. Stein CA, Hansen JB, Lai J et al (2010) Efficient gene silencing by delivery of locked nucleic acid antisense oligonucleotides, unassisted by transfection reagents. Nucleic Acids Res 38:e3–e3. https://doi.org/10.1093/nar/gkp841
16. Souleimanian N, Deleavey GF, Soifer H et al (2012) Antisense 2′-Deoxy, 2′-Fluroarabino nucleic acids (2′F-ANAs) oligonucleotides: in vitro Gymnotic silencers of gene expression whose potency is enhanced by fatty acids. Mol Ther Nucleic Acids 1:e43. https://doi.org/10.1038/mtna.2012.35
17. Hori SI, Yamamoto T, Waki R et al (2015) Ca^{2+} enrichment in culture medium potentiates effect of oligonucleotides. Nucleic Acids Res 43:e128. https://doi.org/10.1093/nar/gkv626
18. Shiraishi T, Pankratova S, Nielsen PE (2005) Calcium ions effectively enhance the effect of antisense peptide nucleic acids conjugated to cationic tat and oligoarginine peptides. Chem Biol 12:923–929. https://doi.org/10.1016/j.chembiol.2005.06.009
19. Sin Y, Makimura F, Saijo M, Obika S (2018) Generation of splice switching oligonucleotides targeting the Cockayne syndrome group B gene product in order to change the diseased cell state. Biochem Biophys Res Commun 500:163–169. https://doi.org/10.1016/j.bbrc.2018.04.015
20. Hande M, Saher O, Lundin EK et al (2019) Oligonucleotide–Palladacycle conjugates as splice-correcting agents. Molecules 24:1180
21. Rocha CSJ, Lundin KE, Behlke MA et al (2016) Four novel splice-switch reporter cell lines: distinct impact of oligonucleotide chemistry and delivery vector on biological activity. Nucleic Acid Ther 26:381–391. https://doi.org/10.1089/nat.2016.0631
22. Hayashi J, Nishigaki M, Ochi Y et al (2018) Effective gene silencing activity of prodrug-type 2′-O-methyldithiomethyl siRNA compared with non-prodrug-type 2′-O-methyl siRNA. Bioorg Med Chem Lett 28:2171–2174. https://doi.org/10.1016/j.bmcl.2018.05.016
23. Hayashi J, Samezawa Y, Ochi Y et al (2017) Syntheses of prodrug-type phosphotriester oligonucleotides responsive to intracellular reducing environment for improvement of cell membrane permeability and nuclease resistance. Bioorg Med Chem Lett 27:3135–3138. https://doi.org/10.1016/j.bmcl.2017.05.031
24. Mori S, Morihiro K, Okuda T et al (2018) Hydrogen peroxide-triggered gene silencing in mammalian cells through boronated antisense oligonucleotides. Chem Sci 9:1112–1118. https://doi.org/10.1039/C7SC04318J
25. Shimojo M, Kasahara Y, Inoue M et al (2019) A gapmer antisense oligonucleotide targeting SRRM4 is a novel therapeutic medicine for lung cancer. Sci Rep 9:7618. https://doi.org/10.1038/s41598-019-43100-1
26. Zammarchi F, de Stanchina E, Bournazou E et al (2011) Antitumorigenic potential of STAT3 alternative splicing modulation. Proc Natl Acad Sci U S A 108:17779–17784. https://doi.org/10.1073/pnas.1108482108
27. Straarup EM, Fisker N, Hedtjarn M et al (2010) Short locked nucleic acid antisense oligonucleotides potently reduce apolipoprotein B mRNA and serum cholesterol in mice and non-human primates. Nucleic Acids Res 38:7100–7111. https://doi.org/10.1093/nar/gkq457
28. Yamamoto T, Yasuhara H, Wada F et al (2012) Superior silencing by 2′,4′-BNANC-based short antisense oligonucleotides compared to 2′,4′-BNA/LNA-based apolipoprotein B antisense inhibitors. J Nucleic Acids 2012:707323. https://doi.org/10.1155/2012/707323
29. Lamers F, van der Ploeg I, Schild L et al (2011) Knockdown of survivin (BIRC5) causes apoptosis in neuroblastoma via mitotic catastrophe. Endocr Relat Cancer 18:657–668. https://doi.org/10.1530/ERC-11-0207

Chapter 11

Evaluating the Knockdown Activity of MALAT1 ENA Gapmers In Vitro

Shinzo Iwashita, Takao Shoji, and Makoto Koizumi

Abstract

Antisense oligonucleotides (ASOs) are widely used for the identification of gene functions and regulation of genes involved in different diseases for therapeutic purposes. For in vitro evaluation of the knockdown activity of gapmer ASOs, we often use lipofection or electroporation to deliver gapmer ASOs into the cells. Here, we describe a method for evaluating the knockdown activity of gapmer ASOs by a cell-free uptake mechanism, termed as gymnosis, using MALAT1 gapmer ASOs modified with 2′-*O*-methoxyethyl RNA (2′-MOE) or 2′-*O*,4′-*C*-ethylene–bridged nucleic acid (ENA). This method is robust because it does not involve the use of any transfection reagent and has minimal effects on cell growth. Further, we describe a convenient technique for performing one-step reverse transcription and real-time qPCR using cell lysates without RNA extraction. Data for up to 96 samples can be obtained following these methods.

Key words MALAT1, antisense oligonucleotide (ASO), Gapmer, Gymnosis, 2′-*O*-methoxyethyl RNA (2′-MOE), 2′-*O*-methyl RNA (2′-OMe), 2′-*O*,4′-*C*-ethylene–bridged nucleic acid (ENA)

1 Introduction

Antisense oligonucleotides (ASOs), which are single-strand nucleic acids having 15–20 nucleotide sequences complementary to the target mRNA, have been widely used in clinical applications in recent years and are attracting attention as a new pharmaceutical modality [1]. In order to protect ASOs from nuclease degradation and to improve the pharmacokinetics and biodistribution of ASOs, chemical modifications of ASOs are performed in many cases [2]. Furthermore, the positions and numbers of chemical modifications of ASOs are important for increasing their affinity with the target mRNA. ASOs of various designs having high efficacy have been studied [1–3]. One of the most effective ASOs is a gapmer, which is composed of a central DNA gap flanked with two wings at both the 3′- and 5′-ends of the gap region. These gapmer ASOs bind to the target mRNA to form ASO-mRNA duplexes that are substrates for RNase H1 cleavage. Mipomersen, inotersen, and

Toshifumi Yokota and Rika Maruyama (eds.), *Gapmers: Methods and Protocols*, Methods in Molecular Biology, vol. 2176, https://doi.org/10.1007/978-1-0716-0771-8_11,

volanesorsen, which are gapmers modified with 2′-*O*-methoxyethyl RNA (2′-MOE), have been approved by the FDA or EMA [4–6].

We reported oligonucleotides modified with 2′-*O*,4′-*C*-ethylene–bridged nucleic acids (ENAs), which restrict sugar puckering to the C3′-*endo* conformation [7]. Oligonucleotides containing ENA residues show two notable properties: (1) high affinity to the complementary single-strand RNA, and (2) dramatically high resistance against exonucleases and endonucleases [7, 8]. Based on these properties, ENA oligonucleotides are expected to function as nucleic acid therapeutics for target gene knockdown by RNase H1 cleavage [9, 10] and for splicing modulation in genetic diseases, such as DMD, by exon skipping [11, 12].

Here, we demonstrate the method of evaluating in vitro ASO knockdown activity using the MALAT1-ASO sequence with various chemical modifications, such as 2′-MOE, ENA, and 2′-*O*-methyl RNA (2′-OMe). Typically, lipofection or electroporation is used to deliver ASO to the cells; however, here we describe an evaluation method using the mechanism of cell-free uptake (called gymnosis); this is a robust method that does not require any transfection reagent or any additives to serum, thereby minimizing the effects on cell growth [13]. Also, we describe a convenient technique for performing one-step reverse transcription and real-time qPCR using cell lysates without RNA extraction. This will help in testing a large number of samples and performing concentration dependence tests easily.

As a positive control, we selected a MALAT1 MOE gapmer with high knockdown activity in the cell, as previously reported [14]. We designed two MALAT1 ENA gapmers 1 and 2, which were 16mers with the same sequence as that of cEt gapmer, and were 3-nucleotide-wings shorter than the 5-nucleotide-wings of the MALAT1 MOE gapmer, as previously reported [14]. Moreover, the middle nucleotide of the 3-nucleotide-wings of MALAT1 ENA gapmer 2 was substituted with 2′-OMe to fine-tune its binding affinity to the target mRNA. Using this method of evaluation, we were able to determine and compare the knockdown activities of these three gapmers in mouse Hepa1-6 hepatoma cells and Renca cells (mouse renal carcinoma cells). From these results, the knockdown activity of MALAT1 ENA gapmer 2 was found to be notably higher than that of the MALAT1 MOE gapmer (Figs. 1 and 2).

2 Materials

2.1 Cell Culture

1. Hepa1-6 cells.
2. Renca cells.
3. Cell culture medium: Dulbecco's Modified Eagle Medium (DMEM) containing 4.5 g/L D-glucose, 4.5 g/L L-glutamine,

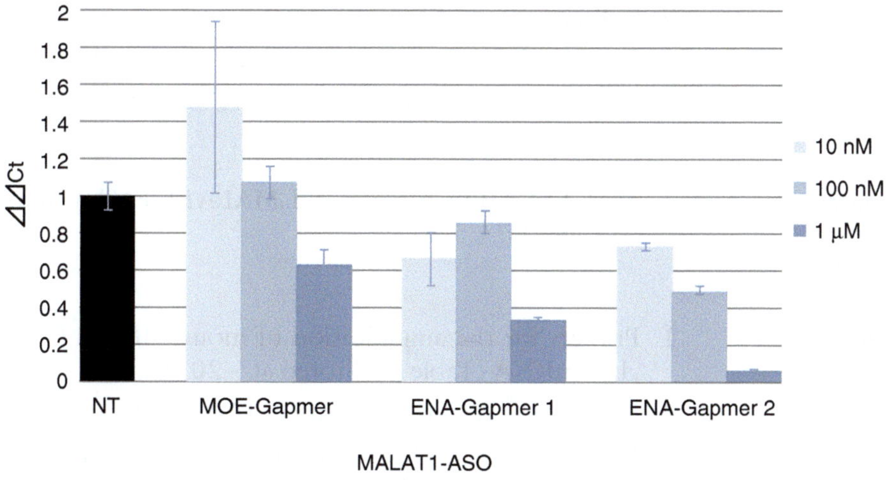

Fig. 1 Knockdown activity of *MALAT1* by chemically modified ASOs in Hepa1-6 cells (NT: no treatment)

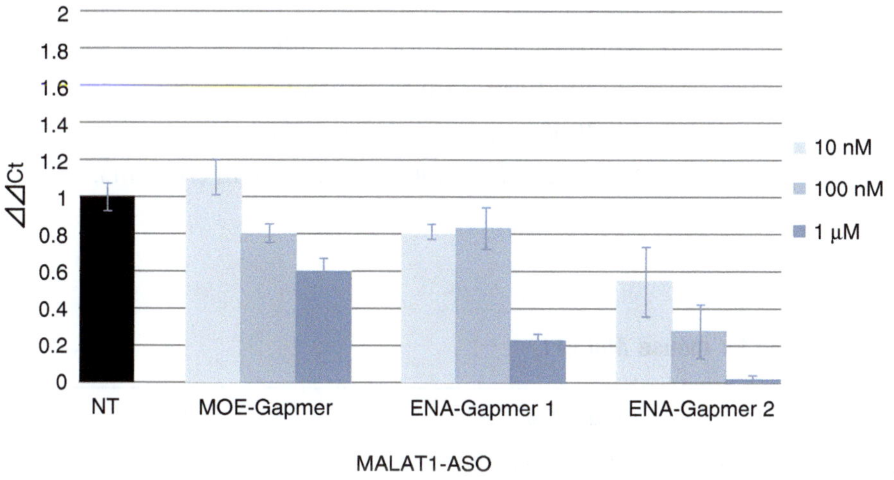

Fig. 2 Knockdown activity of *MALAT1* by chemically modified ASOs in Renca cells (NT: no treatment)

110 mg/L sodium pyruvate, 15 mg/L phenol red, and 10% fetal serum albumin for Hepa1-6 cells. RPMI1640 containing 4.5 g/L D-glucose, 2.383 g/L HEPES, 300 mg/L L-glutamine, 1.5 g/L sodium bicarbonate, 110 mg/L sodium pyruvate, 10% fetal serum albumin for Renca cells. Stored at 4 °C.

4. Dulbecco's phosphate-buffered saline (D-PBS): without calcium, magnesium, and phenol red. Stored at 4 °C.
5. 0.25% Trypsin-EDTA. Stored at −20 °C.
6. 96-well cell culture plate.
7. Cell culture T-75 flask.
8. 0.4 w/v% trypan blue solution.

9. Cell culture incubator.
10. Leica DMIRB Inverted microscope or equivalent.
11. Countess™ Automated Cell Counter (Invitrogen) or equivalent.

2.2 Chemically Modified MALAT1-ASOs

1. MALAT1 MOE gapmer and MALAT1 ENA gapmers 1 and 2 (Table 1).

2.3 qPCR

1. Primers for the amplification of mouse *MALAT1* and mouse *Actb* mRNA (Table 2). Stored at −20 °C.
2. SuperPrep® II Cell Lysis buffer (TOYOBO). Stored at −20 °C.
3. RNA-*direct*™ SYBR® Green Realtime PCR Master Mix (TOYOBO). Stored at −20 °C.
4. RNase-free grade distilled water.
5. Plate shaker.
6. Micro centrifuges for PCR tubes.
7. Mx3000P strip tubes (Agilent).
8. Mx3000P optical strip caps (Agilent).
9. Stratagene Mx3000P qPCR System (Agilent).

Table 1
Three ASOs targeted for mouse *MALAT1*

Name	Sequence	Length	Gapmer design
MALAT1 MOE gapmer[a]	5′-*gggtc*AGCTGCCAAT*gctag*-3′	20	5-10-5
MALAT1 ENA gapmer 1	5′-$\underline{\text{CTA}}$GTTCACTGAA$\underline{\text{TGC}}$-3′	16	3-10-3
MALAT1 ENA gapmer 2	5′-$\underline{\text{C}}$u$\underline{\text{A}}$GTTCACTGAA$\underline{\text{T}}$g$\underline{\text{C}}$-3′	16	3-10-3

Italic lowercase: 2′-MOE; uppercase: DNA; underlined uppercase: ENA; lowercase: 2′-OMe
All phosphate backbones represent phosphorothioate modification
[a]MALAT1 MOE gapmer is described previously [14]

Table 2
Primer sequences for the amplification of *Actb* mRNA and *MALAT1*

Name	Sequence
Actb F	Forward primer: 5′-GACTCATCGTACTCCTGCTTG-3′
Actb R	Reverse primer: 5′-GATTACTGCTCTGGCTCCTA-3′
Malat1 F	Forward primer: 5′-GTACTGTTCCAATCTGCTGCTA-3′
Malat1 R	Reverse primer: 5′-CTCCAACAACCACTACTCCAA-3′

3 Methods

Carry out all procedures at room temperature.

3.1 Cell Harvest

1. Culture cells in a T-75 flask until the cells reach confluence.
2. Wash the cells once with 20 mL D-PBS (*see* **Note 1**) and discard using an aspirator.
3. Add 3 mL trypsin-EDTA and incubate the cells at 37 °C for 5 min (or until the cells are detached from the flask) (*see* **Note 1**).
4. Add 7 mL of cell culture medium, followed by gentle pipetting (*see* **Note 2**).
5. Centrifuge the cells at 1000 × *g* for 5 min at room temperature.
6. Discard the supernatant using an aspirator and suspend the cells in 10 mL cell culture medium. Mix 10 μL cell suspension and 10 μL trypan blue solution; count the cells using an automated cell counter.
7. Adjust the cell suspension to 1.0×10^5 cells/mL with cell culture medium and harvest cells (100 μL per well) in a 96-well cell culture plate. Allow the harvested plates to sit undisturbed at room temperature for at least 30 min before transferring to a cell culture incubator (*see* **Note 3**).
8. Incubate the cells overnight at 37 °C in the presence of 5% CO_2.

3.2 Antisense Treatment with Gymnosis

1. Dilute MALAT1 ASOs with the cell culture medium.
2. Aspirate the cell culture medium (*see* **Note 4**) and add the cell culture medium-diluted MALAT1-ASOs.
3. Incubate the cells for 24 h at 37 °C in the presence of 5% CO_2 (*see* **Note 5**).

3.3 qPCR

1. Aspirate the cell culture supernatant (*see* **Note 4**) and wash cells twice with 200 μL D-PBS.
2. Aspirate D-PBS (*see* **Note 4**) and add 25 μL SuperPrep® II Cell Lysis Buffer directly. Shake the plates using a plate shaker for 10 min at room temperature, and store the lysate on ice (*see* **Notes 6** and 7).
3. Prepare PCR mixture using the following volumes per well: 10 μL RNA-*direct*™ SYBR® Green, 1 μL of 50 mM Mn $(OAc)_2$, 0.2 μL of 10 μM sense/antisense primers (final concentration of each primer, 0.2 μM), and 6.6 μL RNase-free grade distilled water.

4. Mix 18 μL PCR mixture and 2 μL cell lysate, and spin down with the help of a micro-centrifuge.
5. Commence thermal cycling using the following program:
   Reverse transcription.
   (a) Denaturation 1: 90 °C, 30 s.
   (b) Reverse transcription: 61 °C, 20 min.
   (c) Denaturation 2: 95 °C, 1 min.
   PCR cycle (40 cycles)
   (d) 95 °C, 15 s.
   (e) 55 °C, 15 s.
   (f) 74 °C, 45 s (data collection) (*see* **Notes 8** and **9**).

4 Notes

1. Pre-warm the cell culture medium, D-PBS, and trypsin-EDTA at 37 °C in a water bath before adding them to the cells.
2. To avoid cell debris, we often use a 100 μm cell-strainer and create a single-cell suspension if we are unable to remove the cell debris by pipetting.
3. We found that if we place the cells in the cell culture incubator immediately after cell harvesting, the cells are biased at the end of the well because of thermal convection. To disperse the cells, we usually set the plates undisturbed until the cells attach to the bottom of the plates.
4. To avoid sucking of the cells, tilt the plate so that it does not touch the bottom of the plate and soak the supernatant.
5. If the knockdown efficiency is not adequate, extend the incubation time up to 72 h.
6. Insoluble proteins may be precipitated owing to the volume of lysis buffer. In such cases, increase the volume of lysis buffer up to 50 μL, or transfer the lysate to a PCR tube and pellet the insoluble fraction down by centrifugation.
7. When storing the cell lysate, store at −80 °C and avoid freeze and thaw cycles.
8. If the amplification efficiency of the target gene and control gene is almost the same, quantitative C_t analysis can be performed. This method uses the theory that the detection of 1 cycle of PCR shows a twofold difference.

9. When SYBR Green is used for qPCR, the PCR product should be confirmed by the melting curve analysis. When using new primers, ensure that: the melting curve shows a single peak, the T_m value of the PCR product is correct, and the PCR product size is correct.

References

1. Crooke ST, Witztum JL, Bennett CF et al (2018) RNA-targeted therapeutics. Cell Metab 27:714–739
2. Geary RS, Norris D, Yu R, Bennett CF (2015) Pharmacokinetics, biodistribution and cell uptake of antisense oligonucleotides. Adv Drug Deliv Rev 87:46–51
3. Koizumi M (2007) True antisense oligonucleotides with modified nucleotides restricted in the N-conformation. Curr Top Med Chem 7:661–665
4. Raal FJ, Santos RD, Blom D et al (2010) Mipomersen, an apolipoprotein B synthesis inhibitor, for lowering of LDL cholesterol concentrations in patients with homozygous familial hypercholesterolaemia: a randomised, double-blind, placebo-controlled trial. Lancet 375:998–1006
5. Ackermann EJ, Guo S, Booten S et al (2012) Clinical development of an antisense therapy for the treatment of transthyretin-associated polyneuropathy. Amyloid 1:43–44
6. Gaudet D, Alexander VJ, Baker BF et al (2015) Antisense inhibition of apolipoprotein C-III in patients with hypertriglyceridemia. N Engl J Med 373:438–447
7. Morita K, Hasegawa C, Kaneko M et al (2002) 2'-*O*,4'-*C*-ethylene-bridged nucleic acids (ENA): highly nuclease-resistant and thermodynamically stable oligonucleotides for antisense drug. Bioorg Med Chem Lett 12:73–76
8. Morita K, Koizumi M (2018) Synthesis of ENA nucleotides and ENA oligonucleotides. Curr Protoc Nucleic Acid Chem 72:4.79.1–4.79.21
9. Takagi M, Morita K, Nakai D et al (2004) Enhancement of the inhibitory activity of oatp antisense oligonucleotides by incorporation of 2'-*O*,4'-*C*-ethylene-bridged nucleic acids (ENA) without a loss of subtype selectivity. Biochemistry 43:4501–4510
10. Koizumi M, Takagi-Sato M, Okuyama R et al (2006) Direct comparison of in vivo antisense activity of ENA oligonucleotides targeting PTP1B mRNA with that of 2'-*O*-(2-methoxy) ethyl-modified oligonucleotides. Oligonucleotides 16:253–262
11. Yagi M, Takeshima Y, Surono A et al (2004) Chimeric RNA and 2′-*O*,4′-*C*-ethylene-bridged nucleic acids have stronger activity than phosphorothioate oligodeoxynucleotides in induction of exon 19 skipping in dystrophin mRNA. Oligonucleotides 14:33–40
12. Lee T, Awano H, Yagi M et al (2017) 2'-*O*-methyl RNA/ethylene-bridged nucleic acid chimera antisense oligonucleotides to induce dystrophin exon 45 skipping. Genes (Basel) 8:E67
13. Stein CA, Hansen JB, Lai J et al (2010) Efficient gene silencing by delivery of locked nucleic acid antisense oligonucleotides, unassisted by transfection reagents. Nucleic Acids Res 38:e3
14. Hung G, Xiao X, Peralta R et al (2013) Characterization of target mRNA reduction through in situ RNA hybridization in multiple organ systems following systemic antisense treatment in animals. Nucleic Acid Ther 23:369–378

Chapter 12

Albumin-Binding Fatty Acid–Modified Gapmer Antisense Oligonucleotides for Modulation of Pharmacokinetics

Yunpeng Cai, Chenguang Lou, Jesper Wengel, and Kenneth A. Howard

Abstract

Prolonged circulation and modulation of the pharmacokinetic profile are important to improve the clinical potential of antisense oligonucleotides (ASOs). Gapmer ASOs demonstrate excellent nuclease stability and robust gene silencing activity without the requirement of transfection agents. A major challenge for in vivo applications, however, is the short blood circulatory half-life. This work describes utilization of the long circulation of serum albumin to increase the blood residence time of gapmer ASOs. The method introduces fatty acid modifications into the gapmer ASOs design to exploit the binding and transport property of serum albumin for endogenous ligands. The level of albumin–gapmer ASOs interaction, blood circulatory half-life and biodistribution was dependent on number, position, and fatty acid type (palmitic or myristic acid) within the gapmer ASO sequence and either phosphorothioate or phosphodiester backbone modifications. This work offers a strategy to optimize gapmer ASO pharmacokinetics by a proposed endogenous assembly process with serum albumin that can be tuned by gapmer ASO design modifications.

Key words Gapmer antisense oligonucleotides, Palmitic acid, Myristic acid, Serum albumin, Circulatory half-life, Pharmacokinetics

1 Introduction

Antisense oligonucleotides (ASOs) are synthetic single-stranded nucleic acids that bind to complementary mRNA through Watson–Crick base pairing and activate RNase H-mediated cleavage to act as gene silencing therapeutics [1, 2]. In combination with flanked modified oligonucleotides such as locked nucleic acid (LNA) [3], the so-called gapmer ASOs are highly stable against nucleases and offer flexible gene silencing approaches due to unassisted intracellular gymnotic delivery [4, 5]. However, the short circulatory half-life has remained one of the major challenges following systemic administration due to their small size (~6 kDa) that results in rapid renal clearance [6, 7]. Human serum albumin (HSA) has been exploited as a potent drug half-life extension strategy [8–10] due to its property as a natural transport protein

Toshifumi Yokota and Rika Maruyama (eds.), *Gapmers: Methods and Protocols*, Methods in Molecular Biology, vol. 2176, https://doi.org/10.1007/978-1-0716-0771-8_12,

facilitated by binding sites for endogenous and exogenous ligands and its long circulatory half-life by engagement with the cellular recycling neonatal Fc receptor (FcRn) [11, 12]. Multiple fatty acid binding sites on albumin [13, 14], therefore, can be utilized to extend the circulatory half-life of gapmer ASOs by including fatty acids into the nucleic acid design.

This chapter describes albumin interaction with ASOs with either palmitoyl or myristoyl modifications varying in position and number, with either a phosphorothioate or phosphodiester backbone and the effect on gapmer pharmacokinetics. The level of albumin-binding affinity had a significant impact on the circulatory half-life in mice, with a maximal 2.4-fold increase of circulatory half-life in mice (66 min) that correlated with the highest albumin-binding affinity of two-palmitoyl phosphorothioate gapmer ASO in vitro [15]. Fatty acid gapmer ASOs also show maintained unassisted in vitro gene silencing activity in different cell lines [16]. Systemic administration of fatty acid gapmer ASO that exhibit prolonged circulation may be a promising gene silencing strategy for therapeutic applications.

2 Materials

2.1 Oligonucleotide Synthesis

All basic chemical reagents used were purchased from Sigma-Aldrich (St. Louis, MO, USA) or Fluka (Honeywell lab, Morristown, NJ, USA) and used without further purification. Use ultrapure water (resistivity, 18.2 MΩ cm at 25 °C) and analytical grade reagents to prepare all solutions. Prepare and store all reagents at room temperature unless otherwise specified.

1. Standard DNA phosphoramidite monomers, solid supports and cyanine 5.5 (Cy5.5) phosphoramidite and additional reagents from Sigma-Aldrich, GE Healthcare (Chicago, IL, USA) or Glen Research (Sterling, VA, USA).
2. Locked nucleic acid (LNA) phosphoramidite monomers from Exiqon, now a QIAGEN company (Hilden, Germany).
3. Synthesize the 2′-*N*-palmitoyl-amino-LNA-T and 2′-*N*-myristoyl-amino-LNA-T phosphoramidite monomers according to literature procedure [17].
4. Anhydrous acetonitrile: activated molecular sieves (3 Å, 8–12 mesh) is added to acetonitrile (HPLC grade) and the mixture is kept for 24 h. The dryness of the solvent is confirmed on a Coulometric Karl-Fischer titrator (*see* **Note 1**).
5. Solution of nucleoside phosphoramidite monomers: activated molecular sieves (3 Å, 8–12 mesh) are added into the 0.1 M solutions of standard DNA and LNA monomers in anhydrous acetonitrile (≤15 ppm) and the solutions are dried for 24 h (*see* **Note 2**).

2.2 RP-HPLC, IE-HPLC, and MALDI-MS

1. A-buffer (0.05 M triethylammonium acetate, pH 7.4): (1) Triethylammonium acetate stock solution (1 M, pH 7.4): 1 mol triethylamine (139.88 mL) is emulsified in ultrapure water (600 mL) and the resulting mixture is cooled in a water–ice bath. To this 1 mol glacial acetic acid (57.25 mL) is added slowly and the solution is then warmed to room temperature. Triethylamine or acetic acid can be used to adjust pH to 7.4 if necessary. The solution is transferred into a 1 L volumetric flask, and ultrapure water is added to have the solution volume reach 1 L. (2) 1 M stock solution (250 mL, pH 7.4) is diluted to 5 L using ultrapure water (*see* **Note 3**). The resulting solution is filtered using a nylon membrane (Waters, 47 mm, 0.2 μm) before use.
2. B-buffer (A-buffer–acetonitrile = 1:3, v/v): Mix A-buffer (500 mL) and acetonitrile (1500 mL, HPLC grade) extensively (*see* **Note 3**). Filter the resulting solution using a nylon membrane (Waters, 47 mm, 0.2 μm) before use.
3. Solution of sodium acetate (3 M): Dissolve 0.75 mol sodium acetate (61.53 g) in ultrapure water (150 mL). Transfer the solution into a 250 mL volumetric flask and add ultrapure water to have the final volume at 250 mL.
4. Solution of sodium perchlorate (5 M): Dissolve 1.25 mol sodium perchlorate (153.05 g) in 150 mL ultrapure water. Transfer the solution into a 250 mL volumetric flask and add ultrapure water to have the final volume at 250 mL.
5. 3-Hydroxypicolinic acid (HPA) matrix: Dissolve 100 mg diammonium hydrogen citrate in 1 mL ultrapure water, of which 10 μL is taken and mixed with 10 mg HPA in 0.99 mL ultrapure water.
6. C-buffer (sodium perchlorate, 1.0 M): Dissolve 3 mol sodium perchlorate (367.32 g) is dissolved in 2400 mL ultrapure water. Transfer the solution to a 3000 mL volumetric flask and add ultrapure water to have the final volume at 3000 mL (*see* **Note 3**). Filter the resulting solution using a nylon membrane (Waters, 47 mm, 0.2 μm) before use.
7. D-buffer (0.25 M Tris–HCl, pH 8.0): Dissolve 0.75 mol trisaminomethane (90.84 g) in 2400 mL ultrapure water, to which add concentrated hydrochloric acid to adjust the pH to 8. Transfer the solution into a 3 L volumetric flask and add ultrapure water to have the final volume of 3 L (*see* **Note 3**). Filter the resulting solution using a nylon membrane (Waters, 47 mm, 0.2 μm) before use.
8. Reversed-phase HPLC (RP-HPLC, Waters System 600 HPLC equipment) with a Waters XBridge BEH C18-column (5 μm, 100 mm × 19 mm).

2.3 Electrophoretic Mobility Shift Assay

Prepare all solutions of oligonucleotides (RNA, DNA, gapmer ASOs) using nuclease-free water. Gapmer ASOs are stored at −20 °C and thawed on ice before use. All the chemical reagents are analytical grade and purchased from Sigma-Aldrich (St. Louis, MO, USA) unless noted.

1. Fatty acid gapmer ASOs (either palmitoyl or myristoyl modifications varying in one or two fatty acids, both at 3′ end or one either at the 3′ and 5′ end, with phosphodiester or phosphorothioate backbone).
2. Recombinant human serum albumin (rHSA)(Sigma-Aldrich, St. Louis, MO, USA).
3. Novex Gel Cassettes (8 cm × 8 cm × 1 mm).
4. ProtoGel (30%).
5. Ammonium persulfate: 10% (w/v) solution in Milli-Q water.
6. *N,N,N,N′*-tetramethyl-ethylenediamine (TEMED).
7. Running buffer: 1× TBE (tris–borate–ethylenediaminetetraacetic acid [EDTA]) buffer (from 10× stock; GIBCO, Life Technologies, Grand Island, NY, USA).
8. Loading buffer: 20 mg Orange G and 1.5 mL glycerol in 10 mL Milli-Q water.
9. Gapmer ASOs staining solution: Diluted in Milli-Q water from 10,000× SYBR Gold Nucleic Acid Gel Staining Concentrate in DMSO).
10. Coomassie blue staining solution: 45% methanol, 10% acetic acid, and 3 g/L Coomassie Brilliant Blue R in Milli-Q water.
11. Coomassie Blue destaining solution: 10% methanol and 7.5% acetic acid in Milli-Q water.
12. XCell SureLock® Mini-Cell, Gel Doc™ EZ Imager.

2.4 Nanoparticle Tracking Analysis

1. Palmitoyl gapmer ASOs and myristoyl gapmer ASOs (varying of 1 or 2 fatty acids, both at 3′ end or one either at the 3′ and 5′ end, with phosphodiester or phosphorothioate backbone).
2. Recombinant human serum albumin (rHSA) (Sigma-Aldrich, St. Louis, MO, USA).
3. Two-component 1 mL syringe (Frisenette ApS, Knebel, Denmark).
4. Nanoparticle tracking analysis (NanoSight LM10, 405 nm laser, Malvern Instruments SA, Worcestershire, UK), thermometer (OMEGA HH804, Engineering, Inc./Stamford, CT, USA).

2.5 In Vitro Cellular Gene Silencing

1. Complete cell culture medium dependent on cell lines.
2. RNA extraction: TRIzol, chloroform, isopropanol, absolute ethanol, glycogen (optional).
3. High-Capacity cDNA Reverse Transcription Kit (Applied Biosystems, Warrington, UK).
4. Real-Time Polymerase Chain Reaction (PCR): SYBR Green PCR Master Mix (Applied Biosystems, Warrington, UK). 384-well standard PCR Plate and Adhesive PCR Plate Seals.
5. Forward and reverse primers of target genes and housekeeping gene glyceraldehyde 3-phosphate dehydrogenase (GAPDH). GAPDH forward primer (Integrated DNA Technologies, Coralville, IA, USA): 5′-GTCAGCCGCATCTTCTTTTG-3′; reverse primer: 5′-GCGCCCAATACGACCAAATC-3′.
6. LightCycler 480 System (Roche Diagnostics, Basel, Switzerland).

2.6 In Vivo Pharmacokinetic Evaluation of Oligonucleotides

1. Cy5.5-labeled fatty acid gapmer ASOs (with one or two palmitoyl modifications both at 3′ end).
2. Seven- to eight-week-old female C57BL/6 mice (Janvier Labs, Berthevin, France).
3. Capillary tubes coated with heparin (Hirschmann® Laborgeräte GmbH & Co).
4. Low-fluorescent diet (AIN-76A, Research Diets Inc., New Brunswick, NJ).
5. In Vivo Imaging System (IVIS, PerkinElmer, USA).

3 Methods

3.1 Oligonucleotide Synthesis

All oligonucleotide preparation and analysis are carried out at room temperature unless otherwise specified.

1. Mount solutions (0.1 M) of standard DNA and LNA monomers in anhydrous acetonitrile onto the designated positions on the DNA synthesizer respectively (*see* **Note 4**).
2. Perform oligonucleotide synthesis on an ÄKTA oligopilot plus 10 system in 1.0 μmol scale (polystyrene support) or Uni80 support using the phosphoramidite approach. The coupling time for standard DNA and LNA monomers is 720 s and monitor stepwise coupling efficiencies in a continuous manner determined by the absorbance of the dimethoxytrityl cation at 500 nm (*see* **Note 5**).
3. Incorporate palmitoyl-amino-LNA-T phosphoramidite, myristoyl-amino-LNA-T phosphoramidite and Cy5.5 phosphoramidite (0.05 M, in anhydrous acetonitrile) into oligonucleotides

via a manual coupling procedure [18] using 5-[3,5-bis(trifluoromethyl)phenyl]-*H*-tetrazole (0.25 M, in anhydrous acetonitrile) as activator and extended coupling time (20 min) (*see* **Note 6**).

4. Cleavage of oligonucleotides from solid support and removal of nucleobase protecting groups are performed using 28% aqueous ammonia (1 mL) for 16 h at 55 °C (*see* **Note 7**).
5. Purify the resulting oligonucleotides by reversed-phase HPLC (RP-HPLC). Start elution with an isocratic hold of A-buffer for 2 min followed by a linear gradient to 70% B-buffer over 38 min at a flow rate of 2.5 mL/min. Omit reversed-phase HPLC purification when the purities of the crude oligonucleotides was sufficiently high (>80% by ion-exchange HPLC analysis).
6. Combine fractions of isolated oligonucleotides and evaporate under a flow of nitrogen to remove the solvents (*see* **Note 8**).
7. The purified oligonucleotides are subjected to standard detritylation (80% aqueous acetic acid at room temperature) and then precipitate from abs. Ethanol by first adding aqueous solutions of 3 M sodium acetate (15 μL) and 5 M sodium perchlorate (15 μL) followed by addition of abs. Ethanol (1 mL, −20 °C). Store the resulting suspensions at −20 °C for 1 h, and remove the supernatants after centrifugation (13,200 rpm (16,110 × *g*), 5 min, 4 °C), and wash the pellet with cold abs. Ethanol (2 × 1 mL; −20 °C) and dry for 20 min at 50 °C on a metal heat block under a flow of nitrogen.
8. Dissolve the pellets in ultrapure water (1 mL) and analyze by MALDI-TOF MS and analytical ion-exchange HPLC (IE-HPLC). Record mass spectra of oligonucleotides on a Bruker Daltonics Microflex LT MAIDI-TOF MS instrument in ES+ mode using HPA matrix. Record analytical IE-HPLC traces on a Merck-Hitachi Lachrom system equipped with a DNAPac PA100 analytical column (13 μm, 250 mm × 4 mm) heated to 60 °C. Perform elution with an isocratic hold of D-buffer (10%), starting from 2 min hold using 2% C-Buffer in ultrapure water, followed by a linear gradient to 30% C-buffer in 23 min at a flow rate of 1.1 mL/min.
9. Determine concentrations of purified oligonucleotides by UV absorption measurements at 260 nm.

3.2 Electrophoretic Mobility Shift Assay

The binding affinity between fatty acid gapmer ASOs and rHSA can be estimated by visualizing the assembled rHSA/gapmer ASOs constructs at different molar ratios using an electrophoretic mobility shift assay.

1. For preparation of an 8% native polyacrylamide gel: mix 2.7 mL of Protogel, 6.3 mL of Milli-Q water, and 1 mL of 10× TBE running buffer. Add 100 μL of 10% (w/v) APS solution and 10 μL of TEMED.
2. Cast gel solution into an empty gel cassette with a 10-well comb and wait for 40 min to allow for gelation.
3. Mix fatty acid-gapmer ASOs (either palmitoyl or myristoyl modifications varying in 1 or 2 fatty acids, both at 3′ end or one either at the 3′ and 5′ end, with phosphodiester or phosphorothioate backbone) at 1 μM final concentration with rHSA at molar ratios of 40, 20, 10, 5, 2.5, 1.25, 0.63, and 0 (rHSA: gapmer ASOs) in a final volume of 20 μL for each sample. Incubate for 30 min at room temperature before loading onto gels.
4. Mix samples with loading buffer at volume ratio 1:1, and load 14 μL for each well in the gel. Run the gel at 125 V for 45 min in an XCell SureLock® Mini-Cell.
5. Stain the fatty acid-gapmer ASOs with 50 mL of 1× SYBR Gold solution for 15 min and image the gel with a Gel Doc™ EZ Imager. To stain rHSA, incubate the gel with Coomassie blue staining solution for 15 min and destain in destaining solution overnight before imaging (*see* **Note 9**).

3.3 Nanoparticle Tracking Analysis

Fatty acid–modified gapmer ASOs may tend to aggregate in aqueous solutions. The addition of rHSA can solubilize the aggregated fatty acid gapmer ASOs and demonstrate the formation of rHSA/ fatty acid gapmer ASOs constructs by determining the nanoparticle number changes in solution using nanoparticle tracking analysis (NTA).

1. Incubate for 4 h at room temperature fatty acid gapmer ASOs (1 μM) either mixed with or without rHSA (40 uM). Dilute samples in 500 μL of Milli-Q water (*see* **Note 10**) are load into the sample chamber of NTA using a 1 mL syringe (*see* **Note 11**).
2. Set cameral level of NanoSight to 10–13 (*see* **Note 12**) and record 5 videos of samples for 60 s by advancing syringe of 100 μL each time. Temperature in sample chamber is recorded manually using a thermometer within the range of 22–24 °C.
3. Analyze videos of particle motions with threshold between 2 and 5 (*see* **Note 13**) to obtain nanoparticle sizes and concentrations.

3.4 In Vitro Cellular Gene Silencing

The unassisted gene silencing ability of gapmer ASOs offers great flexibility of formulation design and reduced cytotoxicity due to the absence of a transfection agent. Gapmer ASOs can be added directly into complete cell culture medium and silencing activity can be determined using real-time polymerase chain reaction (PCR).

1. Thaw cells and centrifuge to remove residual DMSO and grow in complete cell culture media for at least 2 weeks before determination of gene expression levels.
2. Seed cells onto plates when cells are 80–90% confluent in flasks. Optimize the number of cells seeded by determination of target gene expression levels to allow the amount of extracted RNA to be sufficient for Real-Time PCR measurement. For example, use 12-well plates for primary human chondrocytes with 1×10^5 cells/well in 1 mL complete medium and 24-well plates for Caco-2 cells with 5×10^4 cells/well in 0.5 mL complete medium.
3. Transfect cells with fatty acid gapmer ASOs when at 90% confluency in wells. On the day of transfection, refresh the complete cell culture media 1 h before transfection and add fatty acid gapmer ASO solutions directly to the media at final concentrations of 0.5–10 μM (*see* **Note 14**). Mismatch control groups include gapmer ASOs with and without the same fatty acid modification (*see* **Note 15**).
4. After 36–72 h incubation with fatty acid gapmer ASOs (*see* **Note 16**), extract total RNA using TRIzol reagent according to the manufacturer's protocol. After resuspension of RNA in nuclease-free water, reversely transcript RNA into complementary DNA (cDNA) using High-Capacity cDNA Reverse Transcription Kit according to the manufacturer's protocol. (*see* **Note 17**) cDNA can be stored at −20 °C for at least 1 month or at 4 °C overnight.
5. Use the cDNA for real-time polymerase chain reaction (PCR) with SYBR Green master Mix to determine the gene expression levels normalized to housekeeping gene GAPDH. Dilute cDNA samples in nuclease-free water for 5–20 times and mix with corresponding forward and reverse primers in 384-well PCR plate (*see* **Note 18**). Add SYBR green master mix (from 2× concentrated stock) to the samples (*see* **Note 19**) and seal plates with Adhesive PCR Plate Seals. Perform Real-time PCR using a LightCycler 480 System with thermal cycles following the manufacturer's protocol of SYBR green. Analyze real-time PCR data using LightCycler480 software (version 1.5) to obtain the cycle quantification (Cq) values and calculate gene expression levels using double delta analysis by normalization to the housekeeping gene expression levels.

3.5 In Vivo Pharmacokinetic Evaluation of Oligonucleotides

Circulatory half-life and biodistribution of fatty acid gapmer ASOs can be determined by detection of a far-red fluorophore Cy5.5 attached at the 5′ end of oligonucleotides. Fatty acid gapmer ASOs can display an extended circulatory half-life that correlates to the albumin-binding affinity.

1. Maintain Female C57BL/6 mice of 7- to 8-week-old for 2 weeks prior to experiments and feed with low-fluorescent diet. Divide mice into experimental groups (fatty acid gapmer ASOs) and control groups (PBS injection and non–fatty acid gapmer ASOs) with 4–6 mice per group.
2. On the day of pharmacokinetic evaluation, administer 100 μL of a 3.5 mg/kg Cy5.5-labeled fatty acid gapmer ASOs intravenously by tail vein injection from a 5 mg/mL stock diluted in PBS. After injection, collect 10 μL blood samples either by sublingual puncture at 1 min or by tail-nicking at other time points. Isolate plasma samples by centrifuge and kept at 4 °C before fluorescence intensity scan in the In Vivo Imaging System with the emission filter settings (720 nm, 740 nm, 760 nm, and 780 nm) for Cy5.5 fluorophore.
3. At 24 h, terminate the mice and harvest organs such as kidneys, liver, spleen, lung and heart for IVIS scanning as described above.
4. Analyze fluorescence intensities by quantitation of regions of interest (ROI) and normalize to the PBS control group (*see* **Note 20**) using Living Image Software (version 3.2). Calculate the circulatory half-life of gapmer ASOs using a one phase decay model in GraphPad Prism software (version 7.0).

4 Notes

1. The amount of water in acetonitrile should be lower than 15 ppm as measured on a coulometric Karl-Fischer titrator. If the water content is above this value, additional molecular sieves should be added for further water removal before remeasuring.
2. The molecular sieves are added after dissolving all the solid phosphoramidite monomers in anhydrous acetonitrile.
3. A homogeneous solution should be obtained after extensive mixing.
4. To prime all the positions on the DNA synthesizer, it is important to ensure that each tube is filled with the corresponding monomer solution.
5. Stepwise coupling efficiency in all cases should be >98.0%.

6. Each phosphoramidite monomer (0.1 mmol) is dissolved in anhydrous acetonitrile (1.0 mL), 0.5 mL of which is taken and gently mixed with the activator solution (0.5 mL, 5-[3,5-bis (trifluoromethyl)phenyl]-*H*-tetrazole in acetonitrile, 0.25 M) in a 1 mL syringe. The syringe is placed on one end of the synthesis column and the solution is gradually injected into the column during 20 min (starting with 0.2 mL, followed by 0.05 mL per minute).
7. The cleavage and deprotection should be carried out in a sealed tube, preventing the evaporation of ammonia.
8. RP-HPLC, IE-HPLC, and MALDI-TOF MS can be carried out on some fractions to guide the proper pooling of fractions.
9. Stain fatty acid-gapmer ASOs first and then stain rHSA for high quality gel pictures if both oligonucleotide and protein staining is required.
10. Optimal dilution factors for NanoSight analysis need to be determined by the number of particles with ideally 50–100 particles in the field of view.
11. Ensure no observable bubbles in NanoSight chamber when injecting samples with a syringe.
12. Higher camera levels of the NanoSight can maximize the visualization of small particles but can overexpose particles (overexposed particles will show purple color on screen).
13. When analyzing NanoSight videos of samples, keep threshold as low as possible to include as many particles as possible but number of blue crosses (background noise) should be below 5 per frame.
14. Optimal concentration of fatty acid-gapmer for cellular gene silencing is dependent on the cell type and gene expression level, corresponding siRNA formulated with transfection agents such as lipofectamine can be used as a positive control but the media needs to be changed 4 h after transfection to minimize cytotoxicity of transfection agents.
15. Blast the mismatch control sequence with the Nucleotides BLAST program (https://blast.ncbi.nlm.nih.gov/BlastAlign.cgi) to ensure no relevant off-target gene silencing effects.
16. Maximal gene silencing time point after addition of fatty acid-gapmers to media needs to be optimized based on cell types and gene targets.
17. Maximal 10 μg RNA can be transcripted per sample using cDNA reverse transcription kit.
18. Optimal concentration of cDNA and primers for real-time PCR need to be optimized but can start with a relatively high cDNA concentration (5–10 × dilution) and a final concentration of 250 nM with each primer.

19. SYBR green master mix can be premixed with primers to be added at the same time for real-time PCR.
20. Spectral Unmixing Mode is used to generate an overlapping emission spectrum from different emission results obtained at 720 nm, 740 nm, 760 nm, and 780 nm for Cy5.5 fluorescence intensity analysis using Average Radiant Efficiency.

Acknowledgments

This work was supported by CEMBID (Center for Multifunctional Biomolecular Drug Design, Grant Number: NNF17OC0028070). This work has received funding from the European Union's Horizon 2020 research and innovation program under grant agreement No 810685.

References

1. Rossor AM, Reilly MM, Sleigh JM (2018) Antisense oligonucleotides and other genetic therapies made simple. Pract Neurol 18 (2):126–131
2. Agrawal S, Zhao Q (1998) Antisense therapeutics. Curr Opin Chem Biol 2:519–528
3. Koshkin AA, Singh SK, Nielsen P et al (1998) LNA (locked nucleic acids): synthesis of the adenine, cytosine, guanine, 5-methylcytosine, thymine and uracil bicyclonucleoside monomers, oligomerisation, and unprecedented nucleic acid recognition. Tetrahedron 54:3607–3630
4. Stein CA, Hansen JB, Lai J et al (2009) Efficient gene silencing by delivery of locked nucleic acid antisense oligonucleotides, unassisted by transfection reagents. Nucleic Acids Res 38:e3–e3
5. Soifer HS, Koch T, Lai J et al (2012) Silencing of gene expression by gymnotic delivery of antisense oligonucleotides. Methods Mol Biol 815:333–346
6. Srinivasan SK, Iversen P (1995) Review of in vivo pharmacokinetics and toxicology of phosphorothioate oligonucleotides. J Clin Lab Anal 9:129–137
7. Geary RS, Watanabe TA, Truong L et al (2001) Pharmacokinetic properties of 2′-*O*-(2-methoxyethyl)-modified oligonucleotide analogs in rats. J Pharmacol Exp Ther 296:890–897
8. Sleep D, Cameron J, Evans LR (2013) Albumin as a versatile platform for drug half-life extension. Biochim Biophys Acta 1830:5526–5534
9. Larsen MT, Kuhlmann M, Hvam ML et al (2016) Albumin-based drug delivery: harnessing nature to cure disease. Mol Cell Ther. 4:3.
10. Bienk K, Hvam ML, Pakula MM et al (2016) An albumin-mediated cholesterol design-based strategy for tuning siRNA pharmacokinetics and gene silencing. J Control Release 232:143–151
11. Schmidt EGW, Hvam ML, Antunes F et al (2017) Direct demonstration of a neonatal Fc receptor (FcRn)-driven endosomal sorting pathway for cellular recycling of albumin. J Biol Chem 292:13312–13322
12. Larsen MT, Rawsthorne H, Schelde KK et al (2018) Cellular recycling-driven in vivo half-life extension using recombinant albumin fusions tuned for neonatal Fc receptor (FcRn) engagement. J Control Release 287:132–141
13. Curry S, Brick P, Franks NP (1999) Fatty acid binding to human serum albumin: new insights from crystallographic studies. Biochim Biophys Acta Mol Cell Biol Lipids 1441:131–140
14. Curry S, Mandelkow H, Brick P et al (1998) Crystal structure of human serum albumin complexed with fatty acid reveals an asymmetric distribution of binding sites. Nat Struct Mol Biol 5:827–835
15. Hvam ML, Cai Y, Dagnæs-Hansen F et al (2017) Fatty acid-modified Gapmer antisense oligonucleotide and serum albumin constructs for pharmacokinetic modulation. Mol Ther 25:1710–1717
16. Cai Y, Makarova A-M, Wengel J et al (2018) Palmitoylated phosphodiester gapmer designs

with albumin binding capacity and maintained in vitro gene silencing activity. J Gene Med 20: e3025

17. Johannsen MW, Crispino L, Wamberg MC et al (2011) Amino acids attached to 2′-amino-LNA: synthesis and excellent duplex stability. Org Biomol Chem 9:243–252

18. Kværnø L, Kumar R, Dahl BM et al (2000) Synthesis of abasic locked nucleic acid and two seco-LNA derivatives and evaluation of their hybridization properties compared with their more flexible DNA counterparts. J Org Chem 65:5167–5176

Part III

In Vitro/In Vivo Evaluation of Gapmers

Chapter 13

The Use of Gapmers for In Vivo Suppression of Hepatic mRNA Targets

David S. Greenberg, Yonat Tzur, and Hermona Soreq

Abstract

This chapter describes the use of locked nucleic acid (LNA) GapmeRs for the in vivo knockdown of specific mRNAs in the mouse liver and phenotype analysis. LNA GapmeRs may be tested for efficacy by transfection in cultured cells. They are delivered into mice in vivo by intravenous tail injection.

Key words LNA gapmers, In vivo knockdown, Liver mRNA, Steatosis, Diet-induced obese mice

1 Introduction

GapmeRs are potent single-stranded, antisense oligonucleotide research tools which enable the efficient knockdown, a critical component of loss-of-function studies, of mRNA or long noncoding RNA (lncRNA). GapmeRs make use of RNase H-dependent catalysis for performing the degradation of complementary RNA targets. Here, we report the use of GapmeRs consisting of 16 nucleotide long oligonucleotides protected with locked nucleic acid (LNA) residues in the flanking regions and the LNA-free central gap of which consists of DNA. LNAs create high-affinity RNA analogs in which the ribose ring is "locked" in the ideal conformation for Watson–Crick binding. As a result, LNA-protected oligonucleotides exhibit unprecedented thermal stability when hybridized to a complementary DNA or RNA strand. Furthermore, when the GapmeR is hybridized to its target RNA, the central DNA gap can induce RNase H cleavage of the opposing RNA strand. The first GapmeR patents were filed over 30 years ago, but they have been used to target specific tissues in an in vivo setting only more recently. In this chapter, we describe a protocol for the efficient in vivo knockdown of important regulatory hepatic transcripts involved in liver steatosis, the first step in nonalcoholic

Toshifumi Yokota and Rika Maruyama (eds.), *Gapmers: Methods and Protocols*, Methods in Molecular Biology, vol. 2176, https://doi.org/10.1007/978-1-0716-0771-8_13, © Springer Science+Business Media, LLC, part of Springer Nature 2020

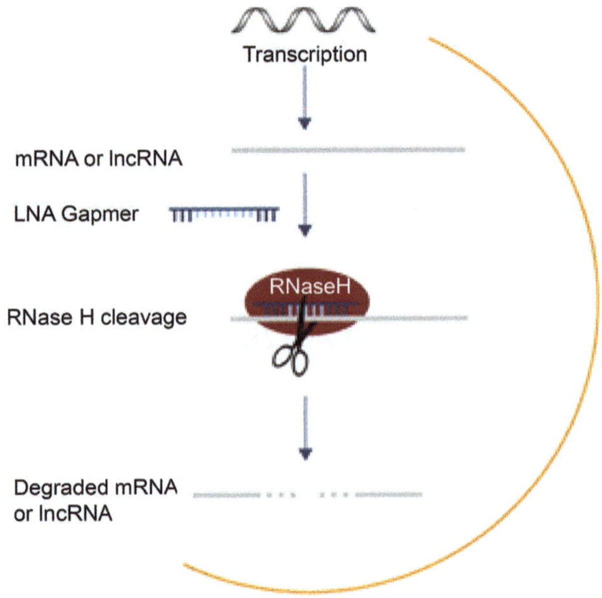

Fig. 1 Schematic diagram of mRNA knockdown by LNA GapmeRs

fatty liver disease. Figure 1 shows the schematic function of GapmeRs.

We have recently discovered that miR-132 elevation plays a causative role in liver steatosis and that miR-132 knockdown is able to reverse steatosis in diet-induced obese mice [1]. We intravenously injected C57BL/6J male mice with diet-induced obesity (DIO), which were fed a high-fat diet for 11 weeks starting at the age of 6–7 weeks, with a 16-mer LNA-protected AS oligonucleotide complementary to miR-132 (AM132) or with a 16-mer LNA-based control. By 7 days of AM132 treatment, DIO mice showed a striking reduction in liver miR-132 levels and a concomitant decline in liver steatosis. We have further found that the hepatic levels of a number of miR-132 targets and of metabolic transcripts were altered. They included FoxO3, Pten, Sirt1, and Cyp2e1, which were all upregulated. To establish the relative contribution of the upregulation of specific proteins with the elimination of steatosis, we used in vivo GapmeR treatments to target each of these specific proteins individually and measure the physiological effects on the mice.

2 Materials

1. C2C12 cells purchased from ATCC (Manassas, Virginia, USA).
2. DMEM supplemented with 10% fetal bovine serum (FBS), 2 mM L-glutamine, 1000 units/mL penicillin, 0.1 mg/mL

streptomycin sulfate, and 0.25 μm/mL amphotericin B (Biological Industries, Beit-Haemek, Israel).

3. LNA GapmeRs (Exiqon, Vedbaek, Denmark). Resuspend in DDW and use at a concentration of 17 pmolL/μL.
4. HiPerfect transfection reagent (Qiagen, CA, USA).
5. Six-well tissue culture plates (Thermo Scientific, USA).
6. Qiagen miReasy (Sigma-Aldrich, St. Louis, MO, USA).

3 Methods

To confirm the efficacy of tested GapmeRs and identify the most potent GapmeR for each target, we screened five different GapmeRs targeting each mRNA in cultured C2C12 myoblast mouse cells.

1. Grow C2C12 cells in a humidified atmosphere at 37 °C, 5% CO_2, in DMEM supplemented with FBS, glutamine, penicillin, streptomycin sulfate, amphotericin B.
2. LNA GapmeRs are designed to complement each of the tested mRNA targets. Transfection using HiPerfect transfection reagent, typically 50 pmoL GapmeR and 25 μL HiPerfect per single 6-well plate (*see* **Note 1**).
3. For the in vivo experiments, 16-mer LNA GapmeRs are used which are complementary to mature mRNA sequences. Those were injected intravenously by tail injection using a 30G needle and 1 mL syringe. Localization of the tail vein is easier if the tail is prewarmed under a heat lamp. The injections can be done for three successive days, at 3.3 mg/kg per day, or once, at 10 mg/kg. Table 1 shows the sequences of the GapmeRs used to target our genes of interest (*see* **Notes 2–6** on factors affecting in vivo efficacy).
4. For mRNA quantification, RNA is extracted using Qiagen miReasy kit according to the manufacturer's protocol, followed by RNA concentration determination (NanoDrop, Thermo, DE, USA). RNA should then be run on a 1% agarose gel to check RNA quality. mRNA levels are determined by quantitative reverse transcription PCR (RT-qPCR) using Promega (WI, USA) reagents, iTaq™ Universal SYBR® Green Supermix (Bio-Rad, CA, USA) according to the manufacturer's instructions and an ABI prism 7900HT PCR machine (Applied Biosystems, CA, USA). All measurements are normalized to a housekeeping gene. Primer sequences used in the qPCR analysis are listed in Table 2.

Table 1
The sequences of the oligonucleotides used in the in vivo experiments

Name	Chemistry	Sequence	mer
Anti sirt1	LNA GapmeR	5′-CTCAGGTGGAGGAATT-3′	16
Anti Pten	LNA GapmeR	5′-TTAAATTTGGCGGTGT-3′	16
Anti P300	LNA GapmeR	5′-TGGTAGAACAGTCTT-3′	15
Anti FoxO3	LNA GapmeR	5′-CTGTGGCTGAGTGAGT-3′	16
Negative control A	LNA GapmeR	5′-AACACGTCTATACGC-3′	15

The negative control oligonucleotide was designed against the primate-specific miR-608 [1], following a careful screen ensuring that it has no complementary targets in the mouse

Table 2
The sequences of the primers used for qPCR validation in the in vivo experiments

Primer	Sequence (5′ → 3′)
mP300 fwd	CAAAGGAGGCTAAAGGTGAGGA
mP300 rev	CAGAGGTGCTTGGCTGTTCT
mSirt1 fwd	GAGCTGGGGTTTCTGTCTCC
mSirt1 rev	AACATGGCTTGAGGGTCTGG
mCyp2e1 fwd	AGGAGTACAAGAACAAGGGGAT
mCyp2e1 rev	CACCAGGAAGTGTGCCTCTC
mPTEN fwd	AGAACAAGATGCTCAAAAAGGACA
mPTEN rev	AGGGTGAGTACAAGATACTCCT
mFOXO3 fwd	AGGGAGGAGGAGGAATGTGG
mFOXO3 rev	GCTCCAGCTCGGCTCCTTC
mNdufc fwd	GTAGTGCTGCGCTCGTTTT
mNdufc rev	GTTAGGTTTGGCATTGACTG

4 Notes

1. We found that the predicted GapmeR sequences showed large efficacy differences when experimentally tested (Fig. 2), highlighting the importance of these tests before using such agents in in vivo experiments. In other studies, we found yet larger variability in neuronal-originated cell cultures [2], extending this rule to other systems than the liver.
2. Injected oligonucleotide accumulates in the liver for biological reasons. This may predict greater difficulties when planning gapmer tests for nonliver tissues and cell types.

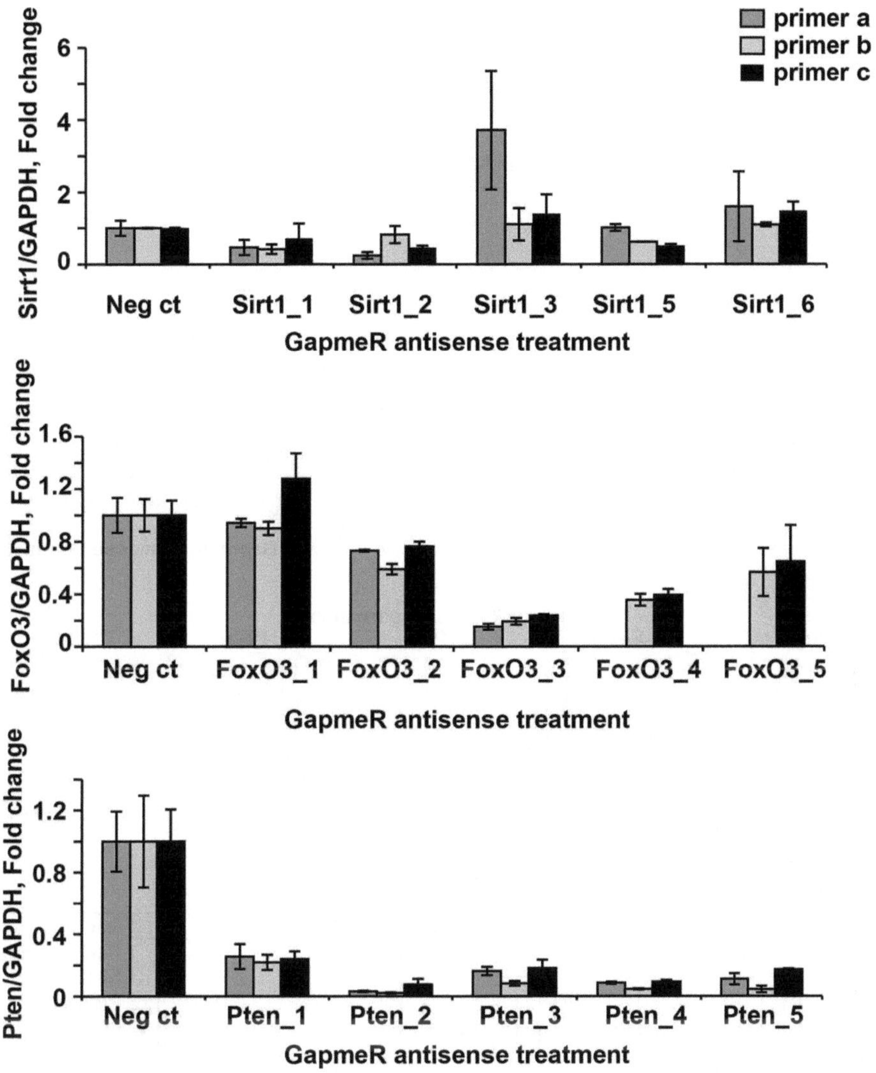

Fig. 2 Cell culture validation of targets knockdown in cells. In vitro validation of knockdown efficiency of miR-132 target transcripts in the C2C12 mouse myoblast cell line: Cells were transfected with the five GapmeRs for each target or under control conditions, using HiPerfect transfection reagent (Qiagen), and RNA was extracted 48 h posttransfection. qRT-PCR quantification was performed using different primer pairs located outside of the GapmeR area. Knockdown results are shown for Sirt1, FoxO3, and Pten. Note the variability in the efficacy of knockdown between the different GapmeRs

3. Suppressing several individual miR-132 targets failed to mimic the phenotype caused by miR-132 elevation; however, their cumulative impact exceeded that of the miR-132 transgene, highlighting the complexity aspect of GapmeR experiments. This might reflect a mirror image of the complexity mode of action of microRNAs at large [3, 4].
4. The varied physiological effects of the different tested GapmeRs (Fig. 3) are likely due to a synergistic impact reflecting

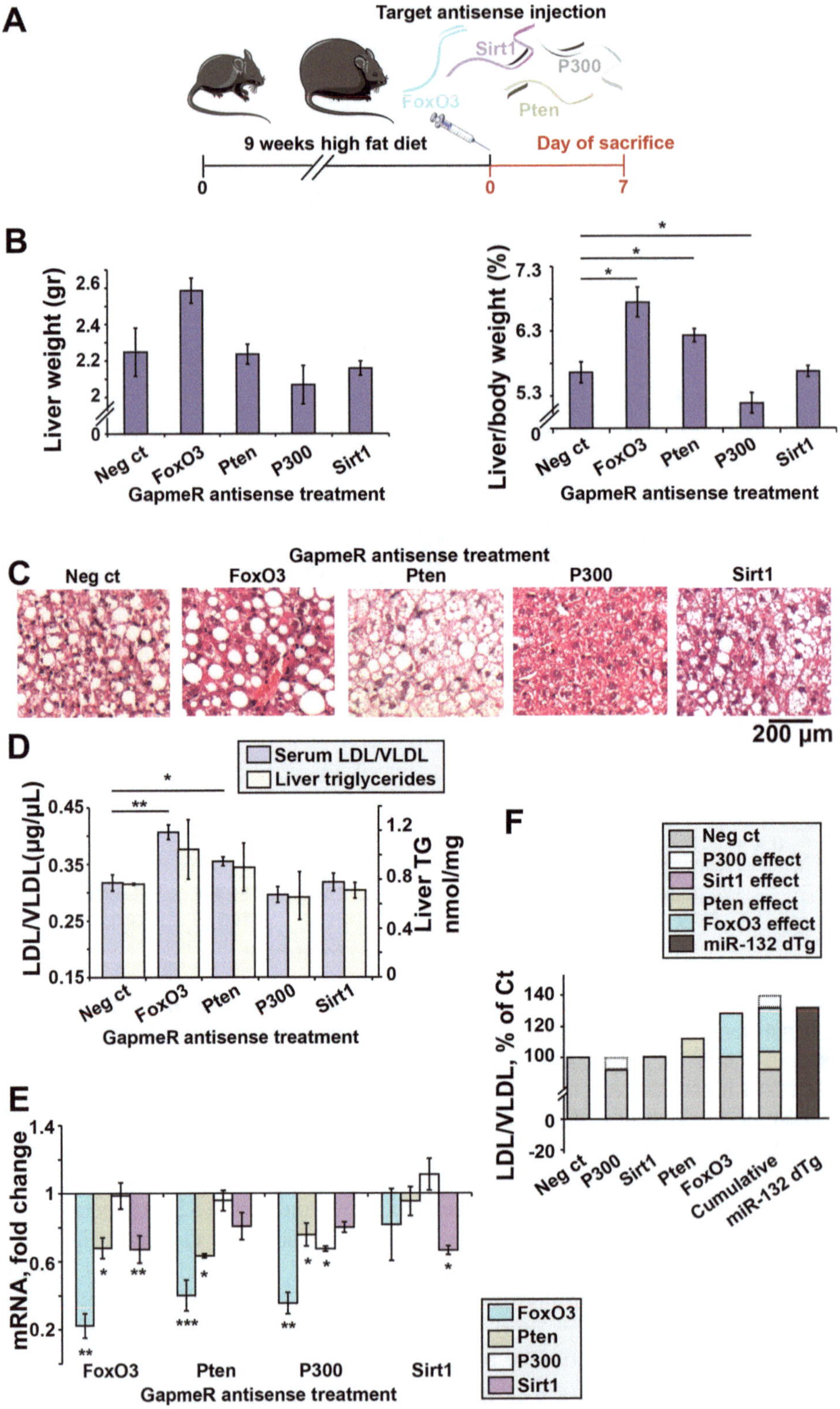

Fig. 3 GapmeRs knockdown of hepatic genes in vivo. (**a**) Experimental design: C57Bl/6J mice were fed with a high-fat diet for 9 weeks, followed by a single intravenous injection of 10 mg/kg antisense GapmeRs for

the cumulative changes in the levels of different targets of a microRNA targeting numerous hepatic transcripts. An additional compounding problem may be the hepatotoxicity of GapmeRs which has been shown is due to nonspecific RNase H activity which is triggered by the GapmeRs [5, 6].

5. The recently reported inverse mode of action of several transcripts which target their complementary microRNAs and induce their destruction [7–9] might add one more layer of difficulty to GapmeR studies. Specifically, miR-132 has been shown to be destroyed by synthetic sequences complementary to its sequence [7]. While none of the currently validated targets of this microRNA shows the necessary level of complementarity for inducing such destruction, this phenomenon should be taken into consideration whenever planning such GapmeR studies.
6. Particular genomic changes in the targeted and/or other affected genes might modify the efficacy of GapmeRs in vitro and in vivo, with implications for their future use as therapeutic targets.

Fig. 3 (continued) Fox03, Pten, P300, Sirt1 or a negative control oligonucleotide. Mice were sacrificed 7 days posttreatment. (**b**) Elevated liver weight and liver/body weight in Fox03- and Pten-AS treated mice, and reduced liver weight and liver/body weight in P300-AS treated mice ($n = 4$). (**c**) Hematoxylin and eosin staining of liver sections. Note variable size of fat vacuoles. (**d**) Serum LDL/VLDL, liver triglycerides (TG) and HDL in mice treated with target antisense GapmeRs ($n = 4$). (**e**) Single target mRNA levels, obtained from Fluidigm analysis, normalized to a number of housekeeping genes and then to negative control-treated mice ($n = 4$). Note cross-target effects of Fox03- and Pten-AS oligonucleotides. (**f**) Individual and cumulative effects of each antisense GapmeR on the levels of LDL/VLDL (as shown in **d**) compared to miR-132 dTg mice. $*p < 0.05$, $**p < 0.01$, $***p < 0.001$ (two-way ANOVA). Bars show mean + SEM

Acknowledgments

This study has been supported by the Nofar program of the Israel Inovation Authority (grant no 65812) to D. S. Greenberg as well as by the Israel Science Fund (No 1016/18) to H. Soreq.

References

1. Hanin G, Yayon N, Tzur Y, Haviv R, Bennett ER, Udi S, Krishnamoorthy YR, Kotsiliti E, Zangen R, Efron B, Tam J, Pappo O, Shteyer E, Pikarsky E, Heikenwalder M, Greenberg DS, Soreq H (2018) miRNA-132 induces hepatic steatosis and hyperlipidaemia by synergistic multitarget suppression. Gut 67 (6):1124–1134
2. Simchovitz A, Hanan M, Niederhoffer N, Madrer N, Yayon N, Bennett ER, Greenberg DS, Kadener S, Soreq H (2019) NEAT1 is overexpressed in Parkinson's disease substantia nigra and confers drug-inducible neuroprotection from oxidative stress. FASEB J 33 (10):11223–11234
3. Meydan C, Shenhar-Tsarfaty S, Soreq H (2016) MicroRNA regulators of anxiety and metabolic disorders. Trends Mol Med 22(9):798–812
4. Soreq H, Wolf Y (2011) NeurimmiRs: microRNAs in the neuroimmune interface. Trends Mol Med 17(10):548–555
5. Burel SA, Hart CE, Cauntay P, Hsiao J, Machemer T, Katz M, Watt A, Bui HH, Younis H, Sabripour M, Freier SM, Hung G, Dan A, Prakash TP, Seth PP, Swayze EE, Bennett CF, Crooke ST, Henry SP (2016) Hepatotoxicity of high affinity gapmer antisense oligonucleotides is mediated by RNase H1 dependent promiscuous reduction of very long pre-mRNA transcripts. Nucleic Acids Res 44 (5):2093–2109
6. Kasuya T, Hori S, Watanabe A, Nakajima M, Gahara Y, Rokushima M, Yanagimoto T, Kugimiya A (2016) Ribonuclease H1-dependent hepatotoxicity caused by locked nucleic acid-modified gapmer antisense oligonucleotides. Sci Rep 6:30377
7. Bitetti A, Mallory AC, Golini E, Carrieri C, Carreno Gutierrez H, Perlas E, Perez-Rico YA, Tocchini-Valentini GP, Enright AJ, Norton WHJ, Mandillo S, O'Carroll D, Shkumatava A (2018) MicroRNA degradation by a conserved target RNA regulates animal behavior. Nat Struct Mol Biol 25(3):244–251
8. de la Mata M, Gaidatzis D, Vitanescu M, Stadler MB, Wentzel C, Scheiffele P, Filipowicz W, Grosshans H (2015) Potent degradation of neuronal miRNAs induced by highly complementary targets. EMBO Rep 16(4):500–511
9. de la Mata M, Grosshans H (2018) Turning the table on miRNAs. Nat Struct Mol Biol 25 (3):195–197

Chapter 14

Development of LNA Gapmer Oligonucleotide-Based Therapy for ALS/FTD Caused by the C9orf72 Repeat Expansion

Chaitra Sathyaprakash, Raquel Manzano, Miguel A. Varela, Yasumasa Hashimoto, Matthew J. A. Wood, Kevin Talbot, and Yoshitsugu Aoki

Abstract

Several neurodegenerative diseases, including amyotrophic lateral sclerosis (ALS), have a complex genetic background, in addition to cases where the disease appears to manifest sporadically. The recent discovery of the hexanucleotide repeat expansion in the C9orf72 gene as the causative agent of ALS (C9ALS) gives rise to the opportunity to develop new therapies directed at this mutation, which is responsible for a large proportion of ALS and/or frontotemporal dementia cases. Mammalian models conscientiously replicating the late-onset motor defects and cellular pathologies seen in human patients do not exist. In this context, patient-derived cells give us a platform to test potential antisense oligonucleotide therapies, which could be the key to treat this subtype of motor neuron disease. Recently, we described that locked nucleic acid gapmer oligonucleotide-based treatment targeting C9orf72 repeat expanded transcripts resulted in recovery from the disease-related phenotypes in patient-derived fibroblasts. Our findings highlight the therapeutic potential of C9ALS using this gapmer oligonucleotide-based approach.

Key words Amyotrophic lateral sclerosis (ALS), C9orf72, C9ALS, Antisense oligonucleotide (ASO), Locked nucleic acid (LNA), Gapmer

1 Introduction

The C9orf72 hexanucleotide repeat expansion is associated with the most common subtype of inherited cases of amyotrophic lateral sclerosis (C9ALS) [1, 2]. It consists of a $[GGGGCC]_n$ repeat region in intron 1a of the C9orf72 gene [3]. The presence of this repeat results in the transcription and accumulation of repeat-containing transcripts in the nucleus of several neuronal subtypes in the central nervous system, including hippocampal, cerebellar, frontal cortex, and spinal motor neurons; the most significantly affected cell type in C9ALS [1, 4, 5]. This accumulation has been

Toshifumi Yokota and Rika Maruyama (eds.), *Gapmers: Methods and Protocols*, Methods in Molecular Biology, vol. 2176, https://doi.org/10.1007/978-1-0716-0771-8_14, © Springer Science+Business Media, LLC, part of Springer Nature 2020

identified using fluorescence in situ hybridization (FISH), revealing RNA *foci* to be comprised of both the sense and antisense transcripts [6].

The exact threshold for the number of repeats leading to an ALS pathology is currently unknown, though it is usual to find hundreds to thousands of repeats in affected patients [7, 8]. It should also be noted that frontotemporal dementia (FTD), the second most common form of dementia, also shares this genetic and molecular pathology, affecting neurons in the frontal lobe of the brain [1, 3, 9, 10]. The repeat causes the formation of robust G-quadruplex structures [11, 12] in the nucleus, with non-specific interactions and sequestration of several different RNA-binding proteins to these immature transcripts [13–15]. The sequestration of RNA binding proteins has been hypothesized to have a toxic gain-of-function effect on downstream pathways in a neuron-specific manner. However, the non-specific nature of these interactions has made our ability to identify key cellular pathways to target significantly challenging. Furthermore, the presence of dipeptide repeat proteins, arising from repeat-associated non-ATG translation in several regions of the brain and spinal cord have been reported to interfere with pathways that may lead to further gain-of-function disease phenotypes [16–18]. These include interference in nucleocytoplasmic transport and triggering stress-induced pathways of translation [19–22]. As a result, the most efficient strategy to treat C9ALS is often contested.

Targeted intervention of C9orf72 repeat-containing cells, via antisense oligonucleotide (ASO) knockdown, has been shown in several studies to ameliorate the pathology of RNA *foci* in the nucleus of affected cells [5, 13, 23, 24], and reverse other reported cellular and behavioral pathologies in a C9orf72 bacterial artificial chromosome mouse model (though no motor deficits were reported) [25]. This successful treatment was administered to the mice via an intracerebroventricular stereotactic injection; the ASOs were acquired from Ionis Pharmaceuticals. Furthermore, ASO treatments in other neurodegenerative and neuromuscular diseases, including spinal muscular atrophy and Duchenne muscular dystrophy, have shown significant success with this strategy for therapeutic interventions [26, 27].

One of the most significant challenges of studying a human neurodegenerative disease like ALS is to identify the molecular mechanisms that lead to the decline of affected neurons in disease. This is partially due to their inaccessibility for biopsies, as well as the fact that post-mortem examination shows major atrophy of motor neurons, leaving little clue as to the cause of this loss. Approximately 10% of ALS cases are thought to be inherited, with specific mutations, including in the C9orf72 gene. These mutations are present in every cell therefore, we may use more accessible fibroblasts obtained from skin biopsies to study the endogenous function of C9orf72, and related cellular pathologies.

The function of endogenous C9orf72 protein has been challenging to identify. Sequence motif scanning was used to predict the presence of a differentially expressed in normal and neoplasia (DENN)-like domain in the secondary and tertiary structures of the protein [28]. This linked it to a potential role as a guanine exchange factor (GEF), well-known to interact with the Rab family of GTPases, which have numerous roles in vesicular and intracellular transport at and between membrane-bound organelles [29]. Several studies have established the interaction between C9orf72 with various Rab proteins, using immunoprecipitation, proximity ligation assays and immunostaining [30–32). These studies have implicated them to have roles in endosomal trafficking, vesicular secretion and autophagy, exhibiting robust pathological phenotypes in C9ALS models. Significantly, in more than one study wildtype C9orf72 has been shown to have an association with Rab1a, ULK1 and SMCR8, as a regulator in autophagy (33–36). The precise nature of this mechanism is yet to be determined, due to an inconsistency in the models and experimental conditions used across studies, though the implication remains that C9orf72 is likely to have diverse cellular functions, making the targeted knockdown of mutant transcripts a promising form of clinical treatment. We have previously identified a reduction in the number of extracellular vesicles (EV) secreted by C9ALS patient fibroblasts and SH-SY5Y cells where C9orf72 or Rab7L1 have been knocked down [31]. This pathological phenotype was rescued by the knockdown of mutant repeat-containing C9orf72 transcripts [31]. This suggested a loss-of-function role for C9orf72 protein in disease. Furthermore, previous studies have also confirmed a reduction in the overall number of C9orf72 transcripts in brain tissue affected by the mutation, potentially suggesting a partial contribution to pathology from this phenotype [1–3]. This, in combination with the toxic gain-of-function phenotypes, may suggest ASO knockdown as a potential therapy for C9ALS whereby the nuclear RNA *foci* may be targeted and restore endogenous transcript levels.

A number of barriers are presented in this therapeutic strategy, including the potential half-life of an ASO and ability of the molecule to cross the blood–brain barrier and cell membrane of motor neurons, and finally to cross the nucleocytoplasmic membrane and interact with the *foci*, finally targeting the molecules toward degradation pathways. Several chemistries with which to modify the ASOs exist and must be carefully implemented to optimize this kind of therapy.

Locked nucleic acid (LNA) gapmers are a hybrid molecule designed as ASOs (Fig. 1a, b) [37–39], to interact with a complementary sequence, which is the target for knockdown: in this case, intron 1a of the C9orf72 transcript, where the repeat exists. LNA gapmers contain flanking regions with the altered locked chemistry, and a middle region containing artificially modified

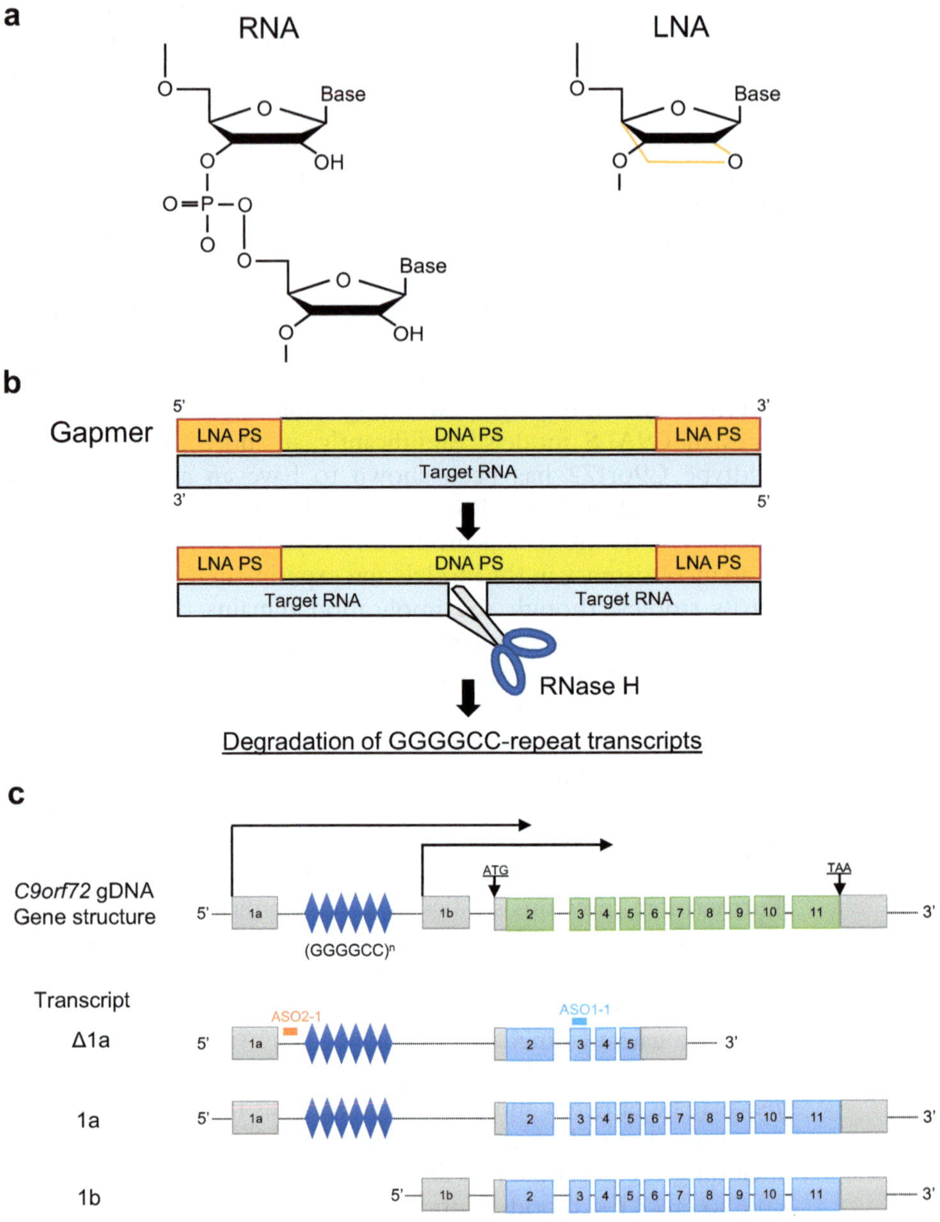

ASO 1-1	+C*+A*+C*C*A*C*T*C*T*C*T*G*C*+A*+T*+T
ASO 2-1	+G*+C*+G*A*C*T*C*C*T*G*A*G*T*+T*+C*+C
ASO Scramble	+C*+A*+G*T*G*T*G*C*T*C*A*G*T*+C*+A*+A

+, Locked nucleic acid (LNA)
*, phosphorothioate linkage.

Brain2017 Table S2

Fig. 1 LNA Gapmer design, targeting the hexanucleotide expansion in intron 1a of the C9orf72 gene. (**a**) Locked nucleic acid (LNA) is modified with an extra bridge (connecting the 2′ oxygen and 4′ carbon) and has high binding affinity to mRNA and high nuclease resistance. (**b**) Mechanism of LNA-based knockdown treatment. "Gapmer" structures were developed where LNA phosphorothioate (PS) residues are present either side of DNA PS backbone. The external LNA PS residues thus enhance RNA binding while the internal DNA PS

deoxyribonucleotide or ribonucleotide monomers, which, if bound to an mRNA molecule will firstly inhibit transport or translation, through steric hindrance, and secondly direct the DNA–RNA double-stranded molecule toward the ribonuclease H pathway (Fig. 1b) [40, 41].

Here we present our strategy for designing and targeting of the C9orf72 repeat expansion in C9ALS patient-derived fibroblasts using LNA–DNA mixmers, which has shown successful knock-down of the repeat-containing transcripts and successful reversal of an EV-related pathology [31].

2 Materials

2.1 Design of antisense LNA Gapmers

1. The website of the UCSC Genome Browser was used to identify the mRNA sequence and various known transcripts of C9orf72 (https://genome.ucsc.edu/index.html).
2. The ASO sequences targeting all C9orf72 transcripts, intron 1a of the C9orf72 open reading frame (hexanucleotide expansion-containing transcripts) or a scrambled control are: +C*+A* +C*C*A*C*T* C*T*C*T*G*C*+A*+T*+T (ASO 1-1; 16mer), +G*+C*+G*A*C*T*C*C* T*G*A*G* T*+T*+C*+C (ASO 2-1; 16mer) and +C*+A*+G*T*G* T*G*C*T*C*A*G*T*+C*+A*+A (ASO Scramble) (Fig. 1c). Phosphorothioated DNA base: G*, A*, T*, C*. Phosphorothioated LNA base: +G*, +A*, +T*, +C*. *See* Fig. 1c for region of alignment at the *C9orf72* locus. All LNA–DNA *mixmers* were synthesized by Exiqon. Each mixmer sequence was examined by the BLAST online software to check for any potential negative off-target effects.

2.2 Obtaining Human Skin Biopsies

1. Fibroblast media; Minimum Essential Medium (MEM) with Earle's Salts and L-glutamine PAA (Sigma-Aldrich) (500 mL; stored at 4 °C in clean tissue culture grade environment), 10% fetal bovine serum (FBS; stored as 50 mL aliquots; stored at −20 °C in clean tissue culture grade environment), 1% 100× MEM non-essential amino acids (Lonza 11140-035; 100 mL; stored at 4 °C), 1% 100× MEM vitamins (Lonza 11120-037, stored at 4 °C), 100× penicillin–streptomycin solution (containing 10,000 IU penicillin and 10,000 μg/mL streptomycin

Fig. 1 (continued) residues still allow RNase H cleavage. (**c**) Two alternative first exons are used, 1a and 1b, and both of these lies upstream of the translation start site. Between exons 1a and 1b lie the GGGGCC repeat expansion region. Using an antisense oligonucleotide (ASO), ASO2-1, targeting the repeat expansion can lead to selective knocking down of the mutated-transcripts which could otherwise cause the formation of RNA *foci*. It also preserves the expression of the normal transcript that should have an essential function for cell survival. ASO is indicated by a blue or orange line, dot line indicates introns and squares indicate exons

(Lonza) stored as 5 mL aliquots; stored at −20 °C), 100 mM sodium pyruvate (Lonza; stored at 4 °C) and 50 mg/mL uridine (Sigma-Aldrich U-3003-5G; stored at room temperature in chemical store; solution prepared by dissolving 300 mg of uridine in 6 mL of phosphate-buffered saline (PBS)). Pass through a sterile 0.2 or 0.1 μM filter to sterilize. Aliquot 0.6 mL and stored at −20 °C. Prior to preparation, freshly thaw all reagents stored at −20 °C at room temperature, or 4 °C overnight. To 500 mL of MEM, add: 50 mL FBS, 5 mL penicillin–streptomycin solution, 5 mL MEM vitamin solution, 5 mL MEM non-essential amino acids, 5 mL Na pyruvate 100 mM stock, 0.5 mL uridine stock. Gently swirl 500 mL bottle after all the supplements have been added and store at 4 °C.

2. 70% ethanol in a spray bottle
3. Mycoplasma testing kit.

2.3 Sterile Cell Culture of Patient-Derived Fibroblasts and SH-SY5Y Cells

2.3.1 Obtaining Fibroblasts from Skin Biopsies

1. Biopsy sample(s).
2. Fibroblast media (Subheading 2.2).
3. FBS gold (PAA; A15-751).
4. Scalpel with a sterile blade.
5. Pipette buoy.
6. 5 mL stripettes in individual casings.
7. T25 flasks tissue culture.
8. Tissue culture grade petri dishes (10 cm).
9. 70% ethanol in a spray bottle.
10. Waste beaker in a sterile hood with Virkon® solution or similar disinfectant.

2.3.2 Expanding Human-Derived Fibroblasts

1. Fibroblast media (Subheading 2.2).
2. PBS pH 7.4 tissue culture grade.
3. T25, T72, or T125 flasks.
4. 15 mL and 50 mL tissue culture sterile falcon tubes.
5. 1× trypsin–EDTA.
6. Centrifuge (15 and 50 mL falcons).
7. Tissue culture 37 °C incubator, 5% CO_2.
8. Pipette buoy.
9. 5, 10, and 25 mL stripettes in individual casings.
10. Waste beaker in a sterile hood with Virkon® solution or similar disinfectant.
11. 70% ethanol in a spray bottle.
12. Mycoplasma testing kit.

2.3.3 Freezing and Thawing Human-Derived Fibroblasts

1. Early freezing media; 8 mL fibroblast media, 1 mL tissue culture grade DMSO, 1 mL fetal calf serum (PAA) (for early passage fibroblasts).
2. Established freezing media; high glucose Dulbecco's modified Eagle's medium (Gibco), 20% FBS, 10% DMSO (for established fibroblasts of later passages and SH-SY5Y cells).
3. Fibroblast media (Subheading 2.2).
4. Feeding media; high glucose Dulbecco's modified Eagle's medium (Gibco) supplemented with 10% FBS and 1% penicillin–streptomycin solution.
5. Mr. Frosty box with isopropanol.
6. T75 flasks of expanded fibroblasts (80% confluency).
7. Tissue culture grade 1× trypsin–EDTA.
8. Tissue culture 37 °C incubator, 5% CO_2.
9. Centrifuge (15 and 50 mL falcons).
10. Pipette buoy.
11. 5, 10, and 25 mL stripettes in individual casings.

2.3.4 Culture of SH-SY5Y Neuroblastoma Cell Line

1. SH-SY5Y cell line (ATCC; CRL-2266™).
2. Feeding media; Dulbecco's modified Eagle's medium F12 (DMEM-F12) (Gibco), 10% FBS, 1% penicillin–streptomycin.
3. T75 and T125 tissue culture flasks.

2.4 Transfection of Fibroblasts with LNA Gapmers

1. Fibroblast cultures at 80–90% confluency for passaging.
2. Tissue culture grade 6-well plates.
3. Lipofectamine reagent RNAiMAX (Thermo Fisher; catalog no. 13778030).
4. LNA gapmers working stock 10 μM in RNase–DNase-free water (short term storage at 20 °C) (*see* **Note 1**).
5. Feeding media; high glucose Dulbecco's modified Eagle's medium (Gibco), 10% FBS, 1% penicillin–streptomycin.
6. Opti-MEM serum-free media (Gibco).
7. PBS pH 7.4 tissue culture grade.
8. 15 mL or 50 mL tissue culture grade falcons.
9. RNase–DNase-free 1.5 mL tubes.
10. Gilson pipettes (P1000, P200, P20, and P2) and sterile, filtered tips.
11. 70% Ethanol in a spray bottle.

2.5 Transfection of SH-SY5Y Cells with Expression Plasmids

1. SH-SY5Y cells seeded at 40% density and cultured overnight (Subheading 3.5).
2. 6-well plates (tissue culture grade).
3. Lipofectamine 3000 (Thermo Fisher, catalog no. L3000015).
4. Feeding media; high glucose Dulbecco's modified Eagle's medium, 10% FBS, 1% penicillin–streptomycin.
5. GST:RAB-GTPase expression plasmid (4 μg per reaction) (−20 °C storage) (Gifted from Prof. Mitsunori Fukuda, Tohoku University, Japan).
6. C9orf72 expression plasmid (4 μg per reaction) (−20 °C storage) (Gifted from Prof. Mitsunori Fukuda, Tohoku University, Japan).
7. Opti-MEM serum-free media (Gibco).
8. 70% ethanol in a spray bottle.

2.6 Harvesting of Patient Fibroblasts for RNA or Protein Analysis

2.6.1 Harvesting Protein

1. Fibroblast cultures seeded in 6-well plates with desired treatments.
2. Chilled PBS pH 7.4, tissue culture grade.
3. Protein sample buffer; 70 mM Tris–HCl pH 6.7, 10% SDS, 5 mM EDTA, 5% 2-mercaptoethanol, 1× Complete™ Protease Inhibitor Cocktail (Roche). For 10 mL of protein sample buffer, mix 1 g of SDS, 100 μL of 0.5 M EDTA-2Na, and 700 μL of 1 M Tris–HCl (pH 6.7), and add distilled water to a total volume of 9.1 mL. Make up aliquots of buffer (455 μL) and store at 20–25 °C. On the day of the experiment, add 25 μL of 2-mercaptoethanol and 20 μL of 25× protease inhibitor cocktail to the aliquot (500 μL protein sample buffer).
4. Autoclaved 1.5 mL Eppendorf tubes.
5. Ice bucket and fresh ice.
6. RNase–DNase-free tissue culture cell scrapers.
7. Tabletop centrifuge.
8. Tissue culture sterile Gilson pipettes and tips.
9. Tissue homogenizer.

2.6.2 Harvesting RNA

1. Fibroblast cultures seeded in 6-well plates with desired treatments.
2. Chilled PBS pH 7.4, tissue culture grade.
3. QIAGEN RNeasy Mini Kit, stored at room temperature for RNA extraction.
4. RNaseZap™ solution spray (Thermo Fisher).
5. RNase–DNase-free tissue culture cell scrapers.

6. Gilson pipettes and RNase–DNase-free filtered tips.
7. RNase–DNase-free 1.5 mL Eppendorf tubes.
8. Ice bucket and fresh ice.
9. Rocker.
10. Tabletop centrifuge.
11. NanoDrop.

2.7 Reverse Transcription PCR and Semi-quantitative PCR

1. cDNA synthesis kit (High Capacity cDNA reverse transcription kit, Applied Biosystems),
2. TaqMan primers for C9orf72 isoforms (Hs00376619_m1) and Beta-actin (ACTB) (Hs99999903_m1) as a housekeeping gene.
3. TaqMan master mix (Applied Biosystems).
4. RNase–DNase-free DEPC-treated water (molecular grade).
5. RNase–DNase-free 96-well plates for qPCR, plus sealing foils.
6. Centrifuge with an attachment for 96-well plate.
7. qPCR machine.

2.8 Immunoblotting

1. BCA Protein assay kit.
2. Distilled water.
3. Protein sample buffer (Subheading 2.6).
4. Heat block.
5. NuPAGE® Tris-Acetate SDS Running Buffer (20×) (Thermo Fisher).
6. 12% Tris–glycine SDS–polyacrylamide gel.
7. NuPAGE® LDS Sample Buffer (4×) (Thermo Fisher).
8. NuPAGE® Sample Reducing Agent (10×) (Thermo Fisher).
9. Mini-PROTEAN Tetra cell (Bio-Rad).
10. HiMark™ Pre-Stained Protein Standard (Thermo Fisher).
11. Primary antibodies: rabbit polyclonal anti-C9orf72 (Santa Cruz S-14138763) (1:1000); rabbit polyclonal anti-VPS 26 (GTX106297) (1:1000); mouse monoclonal anti-beta-actin (Abcam ab6276) (1: 5000).
12. Fluorescence-conjugated secondary antibodies: IRDye 800CW goat anti-mouse IgG (1: 5000); Licor IRDye 680RD goat anti-rabbit IgG (1:2000).
13. Sodium dodecyl sulfate (SDS).
14. 2-Mercaptoethanol.
15. TBS-T; 25 mM Tris–HCl, pH 7.4, 137 mM NaCl, 2.7 mM KCl, 0.08% Tween-20.

16. Polyvinylidene fluoride (PVDF) membrane (pore size: 0.45 μm) (Millipore).
17. Extra thick paper, 2.5 mm thickness, 8.0 cm × 13.5 cm (Thermo Fisher).
18. Methanol.
19. Concentrated anode buffer; 0.3 M Tris, 20% methanol.
20. Anode buffer; 0.03 M Tris, 20% methanol.
21. Cathode buffer; 25 mM Tris, 20% methanol, 40 mM 6-amino-*n*-hexanoic acid, 0.01% SDS.
22. PBS-T; PBS, 0.1% Tween 20.
23. Amersham ECL Prime Blocking Reagent (GE Healthcare, Little Chalfont, UK).
24. Amersham ECL Select Western Blotting Detection Reagent (GE Healthcare).
25. NCL-DYS1 (Leica Biosystems, Newcastle Upon Tyne, UK).
26. Licor Odyssey CLx infrared imaging system.

2.9 Purification of Extracellular Vesicles and Nanoparticle Tracking Analysis

1. T75 or T125 flasks of fibroblasts or SH-SY5Y cells with desired treatments or conditions.
2. Sterile stripettes, pipettes, and filtered tips.
3. Serum-free media; Opti-MEM (Gibco) or Earle's Balanced Salt Solution (EBSS) (Gibco).
4. 50 mL falcon tubes.
5. Ultracentrifuge.
6. Syringe and 0.22 μm syringe filters.
7. PBS.
8. Sterile 1.5 mL Eppendorf tubes.
9. NanoSight NS500 instrument (Malvern).

3 Methods

3.1 Design of Antisense LNA Gapmers

1. Select the target, the retained mutant intron 1a of the C9orf72 mRNA transcript.
2. Search for the region of oligonucleotide expansion $[GGGGCC]_n$ and scan flanking regions as potential oligonucleotide binding sites.
3. Select the target region based on coverage of transcripts of interest, GC content between 30% and 60%, lack of self-complementarity, and cross-hybridization to other oligonucleotides; avoid the hexanucleotide repeat expansion in target

site; avoid blast hits in other transcripts; and avoid stretches of more than three guanines and cytosines (preferably <3).

4. Choose specific nucleotides for LNA and other chemical modifications. In general, a ~16-mer with a full phosphorothioate backbone and up to three LNA modifications at each end of the oligonucleotide is preferable for RNase H–mediated cleavage (*see* **Note 2**). When steric blocking is required LNAs can be placed with two unmodified bases between LNAs and, in any case, avoiding stretches of more than four LNA bases, except when designing short oligos <10-mer (e.g., to block entire miRNA families). The design can be optimized by varying LNA content and length of the oligonucleotide to achieve good mismatch discrimination and avoid off-target effects (*see* **Note 3**).
5. This evaluation can be performed in https://www.ensembl.org/index.html (see for example: https://www.ensembl.org/Multi/Tools/Blast). Access to additional tools and design algorithms can be found in https://www.qiagen.com (although sequences are only provided upon purchase).

3.2 Obtaining Human Skin Biopsies from ALS Patients to Generate Fibroblast Cultures

The *standard operating procedure* (SOP) used to obtain skin biopsy samples for this study was conducted according to a multicenter project agreement in the UK, including the University of Oxford. All tissue culture procedures were carried out in standard sterile conditions in sterile working hoods. All reagent containers, pipettes, pipettes buoys, pipette boxes, stripette casings, and so on were always thoroughly cleaned before use in the hoods with 70% ethanol solution.

All patient skin biopsies were conducted by qualified individuals at the Oxford Motor Neuron Disease Centre, University of Oxford, John Radcliffe Hospital, Oxford. The punch biopsy procedure or scalpel method was used to obtain around 2 cm^2 of flat skin and transferred to a sterile 25 cm tube with around 20 mL of fibroblast media to keep the biopsy tissue viable. All samples were tested for mycoplasma according to the standard instructions of kits purchased by the group/institution. Samples are stored at 4–8 °C prior to plating and brought to room temperature upon use.

3.3 Sterile Cell Culture of Patient-Derived Fibroblast Cultures and SH-SY5Y Cells

3.3.1 Obtaining Fibroblasts from Skin Biopsies

1. On day 1 (**steps 1–6**), warm the fibroblast media and fresh FBS to 37 °C just before use. Thaw frozen stocks of FBS slowly overnight and warm to 37 °C on the day of culture. Transfer all materials into hood following thorough cleaning with 70% ethanol.
2. Pipet 4 mL of FBS into the petri dish and discard the stripette into the waste beaker. Pour off most of the fibroblast media in the sample into the waste beaker, leaving the biopsy in about 5 mL of media.

3. Take a sterile scalpel and use to carefully lift skin biopsy from its tube and transfer to the petri dish containing FBS. The bottom side of the sample may contain fatty, translucent residue. This should be cleanly cut off using the scalpel and carefully discarded. The skin sample should then be placed top side up and cut into 6–8 segments, keeping them in the FBS, without allowing them to dry out.
4. Meanwhile, take a T25 flask and coat with up to 5 mL of FBS by pipetting. Swill the flask to coat the entire surface of the flask bottom. Aspirate the FBS and repeat the coating process five times.
5. Use a fresh, sterile scalpel to careful transfer each piece of skin sample into the FBS-coated T25 flask (top side up), placing them equally far apart from each other. Using a pipette buoy and stripettes, carefully pipet 5 mL of fibroblast media into the original 25 mL sample tube, gently mix and take 2.5 mL of media. Pipet around 0.25–0.5 mL of media on each piece of skin in the T25 flask.
6. Carefully put the T25 flask in an incubator and ensure it is not agitated. Clear up hood and clean thoroughly with 70% ethanol.
7. On day 2 (**steps 7–10**), warm FBS and fibroblast media to 37 °C just before use. Combine 1:1 FBS and media in a sterile tube and mix (volume as needed).
8. Switch on and sterilize tissue culture hood and clean using 70% ethanol; place a waste beaker containing disinfectant in the hood.
9. Carefully retrieve T25 culture flask from the incubator and examine skin biopsy samples for dry patches. Use pipette to replenish 0.25–0.5 mL of FBS–media mixture to skin pieces in flask, without dislodging the biopsies from the flask.
10. Check the T25 flask under a light microscope for cell growth around skin biopsies. Replace the flask in incubator and clean up as usual with 70% ethanol.
11. On subsequent days, until fibroblasts begin to grow and establish themselves in culture, continue to replenish FBS–media mixture on top of the skin biopsies (*see* **Note 4**).
12. When around 30–50 fibroblasts per skin piece can be seen, the cells can be fed with 100% fibroblast media and may be left in this media for up to 3 days without replacement. Top up the T25 flask with 5 mL of media rather than feeding spots of skin biopsies dropwise.
13. If significant floating debris begins to be seen, aspirate 50% of the media in sterile conditions and replace with fresh,

pre-warmed fibroblast media by gentle pipetting using a 5 mL stripette onto the side of the T25 flask.

14. Once areas around skin pieces become more established (50–70% confluent) with fibroblasts, total media maybe be aspirated and replaced with 5 mL of fresh media (*see* **Note 4**).

3.3.2 Expanding Human-Derived Fibroblasts

When the area around the skin biopsies is fully confluent with fibroblasts, they should be passaged and also checked for mycoplasma contamination. Any samples positive for mycoplasma should be discarded.

1. Collect the conditioned media from confluent culture and store in a sterile 15 mL falcon tube. If testing for mycoplasma, collect 2 mL of conditioned media and store at 4 °C until used.
2. Wash the cells by pipetting 2 mL of PBS on to the sides of the flask (not directly onto the cells) and swirling gently. Repeat this step once.
3. By similar pipetting, add 1 mL of trypsin–EDTA on to the cells and place in the incubator for 2 min. Check the cells under a microscope and if they have become rounded in appearance, knock the side of the flask gently, to dislodge them. The flask may be placed back in the incubator for a further 1–2 min if cells are firmly stuck to the bottom surface. The skin biopsy pieces may remain in the flask.
4. Add 2 mL of previously saved conditioned media into the flask of cells with trypsin to neutralize the enzyme.
5. Collect the dislodged cell suspension and gently transfer to a sterile 15 or 50 mL falcon tube. Centrifuge at 400 × *g* for 4 min (brake 9, acceleration 9).
6. Aspirate and discard supernatant media without dislodging the cell pellet.
7. Gently resuspend fibroblast pellet in 2 mL of fresh media into a single cell suspension by pipetting up and down using a stripette.
8. Transfer this suspension to a fresh T25 flask. Swirl from side to side to ensure uniform distribution of cells and place in the incubator.

   When the fibroblasts have become totally confluent within the T25 flask, they may be further passaged by the same method into further T25 flasks, and eventually T75 and T125 flasks. 10–12 mL of media should be used to resuspend and seed cells into a T75 flask and 20–25 mL of media should be used to resuspend and seed cells into a T125 flask. 2 mL and 4 mL of trypsin should be used to dislodge cells from T75 and T125 flasks, respectively (*see* **Note 4**).

3.3.3 Freezing and Thawing Human-Derived Fibroblasts

Cells should be frozen down for storage from 80% to 90% confluent T75 cell culture flasks. Early established fibroblasts, following stable growth from skin biopsies should be expanded and frozen down at an early passage (P1-4) for long-term storage in early freezing media. Later passages may be grown and cryogenically frozen in established freezing media.

1. Aspirate media from T75 flask of confluent fibroblasts. Add 5 mL of PBS and swirl to wash. Aspirate and discard PBS. Repeat if additional floating debris can be observed.
2. Pipet 2 mL of trypsin–EDTA into flasks and place in the incubator for 2 min, making sure the liquid covers the bottom of the flask.
3. Check cells under a microscope. If cells have a rounded shape, lightly tap the sides of the flask to detach the cells, and add 4 mL of media to inhibit trypsin activity.
4. Collect media into 15 mL falcon tube and centrifuge at 400 × *g*, 3 min, room temperature (brake 9, acceleration 9). Aspirate and discard the supernatant.
5. Using a Gilson pipette, gently and briefly resuspend the cell pellet in 1 mL of freezing media and quickly transfer to a cryotube (prelabeled with patient cell line details, date and passage).
6. Cryotubes should be placed in a Mr. Frosty box and immediately stored in the −80 °C freezer, overnight. Tubes may then be transferred to liquid nitrogen for long term storage or dry ice for transport.
7. Cells may be recovered from frozen suspension by rapidly thawing the cells in a 37 °C water bath for 1 min, diluting the suspension in 10 mL of pre-warmed fibroblast media or feeding media and centrifuging at 400 × *g*, 3 min.
8. Supernatant containing DMSO should be aspirated in full and the recovered cell pellet should be gently resuspended in 10 mL fresh fibroblast or feeding media.
9. Transfer resuspended cells to a sterile T75 flask and place in 37 °C incubator. The media should be changed after overnight incubation.

3.3.4 Culture of SH-SY5Y Neuroblastoma Cell Line

1. SH-SY5Y cells are thawed from commercial frozen stocks as described for fibroblasts in Subheading 3.3.3, using feeding media (Subheading 2.3.4).
2. Culture media is changed at least every 48 h, or more often judging by the color of the phenol indicator in the media (*see* **Note 5**).

3. SH-SY5Y cultures are passaged according to the same protocol as fibroblasts, though harsher passages are well tolerated by this line. Up to a 1 in 10 split can be done.

3.4 Transfection of C9orf72 Patient-Derived Fibroblasts with LNA Gapmers

1. LNA gapmers and Lipofectamine RNAiMAX reagent are kept on ice during the experiment and brought into the tissue culture hood during pipetting following sterilization of their tubes with 70% ethanol.
2. The LNA gapmer (Scrambled and C9orf72) master mix is calculated for transfection at 50 nM final concentration in 2 mL of media per well. 10 μL of 10 μM stock is mixed with 70 μL Opti-MEM media (80 μL total) for one well. Use 80 μL of Opti-MEM for non-treated conditions.
3. The Lipofectamine RNAiMAX master mix consists of 6 μL of reagent per well and 74 μL of Opti-MEM. These reagents are mixed separately in 1.5 mL tubes, and each incubated for 5 min at room temperature. The two solutions are then mixed together in a 1:1 ratio by pipetting, and incubated for 15 min at room temperature.
4. LNA gapmer–Lipofectamine RNAiMAX mixture is then added to the center of wells where cells would be seeded onto. These are kept at room temperature until cells are seeded.
5. Fibroblasts from a P75 flask (70–80% confluent) are detached by adding 2 mL of trypsin–EDTA and incubating at 37 °C for 2 min. When cells are dislodged, add feeding media to halt trypsinization.
6. Collect the cell suspension and centrifuge at 400 × *g*, 3 min. Discard the supernatant and resuspend the cell pellet in 10 mL media.
7. 10 μL of the suspension is taken for cell counting via a hemocytometer and cell suspension volumes were adjusted using dilutions to seed fibroblasts in 6-well plates at 40% density (4.8×10^5 cells/well) in 2 mL of media if desired for RNA or protein analysis. Resuspended cells are pipetted dropwise on top of the LNA gapmer mixtures and transferred carefully to the incubator (*see* **Note 6**).
8. After a 24-h incubation the media containing transfection reagents is aspirated. The cells are briefly washed with PBS and the feeding media is replaced.
9. Forty-eight hours after transfection the cells are harvested for RNA or protein (Subheading 3.6).

3.5 Transfection of SH-SY5Y Cells with Expression Plasmids

1. SH-SY5Y cells from a P75 flask (70–80%) confluent are detached by adding 2 mL of trypsin–EDTA, spun down (400 × *g*, 3 min) (*see* Subheading 3.4) and the cell pellet is resuspended in 10 mL media.
2. Ten microliters of the suspension is taken for cell counting via a hemocytometer and cell suspension volumes are adjusted using dilutions to seed fibroblasts in 6-well plates at 25% density (3×10^5 cells/well) in 2 mL of media, if desired for RNA or protein analysis.
3. The plates are kept in the incubator overnight prior to transfection treatment.
4. Lipofectamine 3000 reagent and expression vectors are kept on ice during the experiment and brought into the tissue culture hood for pipetting, following sterilization of their tubes with 70% ethanol.
5. The plasmid (GST:RAB-GTPase or C9orf72 expression plasmid) master mix is calculated for transfection at 4 μg in 2 mL of media per well. The stock should be mixed with 70 μL Opti-MEM media (80 μL total) for one well. Use 80 μL of Opti-MEM for non-treated conditions.
6. The Lipofectamine 3000 master mix consists of 6 μL of reagent per well and 74 μL of Opti-MEM. These reagents are mixed separately in 1.5 mL tubes, and each incubated for 5 min at room temperature. The two solutions are then mixed together in a 1:1 ratio by pipetting, and incubated for 15 min at room temperature.
7. LNA gapmer–Lipofectamine 3000 mixture is then added dropwise into the appropriate wells and the plate is returned to the incubator.
8. After a 24-h incubation the media containing transfection reagents are aspirated. The cells are briefly washed with PBS and the media is replaced.
9. Forty-eight hours after transfection, the cells are harvested for RNA or protein (Subheading 3.6).

3.6 Harvesting of Patient Fibroblasts for RNA or Protein Analysis

3.6.1 Harvesting Protein

1. Aspirate media and wash cells twice with chilled PBS.
2. Scrape cells from wells in 500 μL of fresh chilled PBS and carefully pipet into a chilled 1.5 mL tube, without pipetting the cells up and down (*see* **Note** 7). Place tubes immediately on ice.
3. Centrifuge tubes at 1000 × *g*, 4 °C and discard the supernatant (*see* **Note 8**).
4. Prepare protein sample buffer and resuspend cell pellets for protein in 150 μL of sample buffer.

5. Using harsh pipetting or a homogenizer, briefly homogenize the protein samples on ice. Centrifuge at 16,500 × *g* for 15 min, and retain the supernatant. Samples may be kept on ice until prepared for SDS-PAGE (same day) or stored at −80 °C.

3.6.2 Harvesting RNA

1. Spray working environment, including pipettes and tip boxes, with RNaseZap™ to minimize contamination.
2. Aspirate media from 6-well plates and wash cells by pipetting PBS to the sides of the well and gently swirling the plate. Repeat this step twice.
3. Perform on-dish lysis by adding 300 μL RLT buffer from QIAGEN kit to each well.
4. Transfer to ice and place on a rocker for 15 min (alternatively place plate on a rocker in a cold room) (*see* **Note 9**).
5. Detach lysed cells full by rigorously pipetting buffer in the well and transfer to RNase–DNase-free 1.5 mL tubes. Place tubes on ice.
6. Centrifuge tubes at maximum speed, 4 °C, for 3 min to pellet cell debris.
7. Retain the supernatant and mix with 70% ethanol (from kit).
8. Follow the manufacturers protocol for column RNA extraction and elute RNA in the final step in 30 μL of RNase–DNase-free water.
9. Quantify total RNA concentration using a NanoDrop (UV-Vis spectrophotometry at 260 nm), using RNase–DNase-free water as a background control.
10. Store RNA at −80 °C and slowly thaw on ice when reusing.

3.7 Reverse Transcription PCR of RNA and Semi-quantitative Detection of PCR Product

1. Thaw RNA to be used in experiment on ice and vortex contents of tube to help denature non-specific interactions.
2. Set up cDNA synthesis reaction. For 1 reaction, mix 2 μL 10× RT Random Primers, 0.8 μL 25× dNTP Mix, 2 μL 10× RT Buffer, and 1 μL Reverse Transcriptase from cDNA Reverse Transcription Kit with 100 ng total RNA. Add RNase-free water to bring the final volume to 20 μL.
3. Set the conditions of the thermal cycler as follows: 25 °C for 10 min, 37 °C for 120 min, and 85 °C for 5 min. Store the reverse transcripts at 4 °C (short-term) or −20 °C (long-term).
4. Set up the RT-PCR reaction. For 1 reaction, mix 14.3 μL water, 0.5 μL 10 μM forward primer, 0.5 μL 10 μM reverse primer, 1.6 μL dNTPs, 2 μL 10× Ex Taq Buffer, and 0.1 μL Ex Taq HS from the Ex Taq Hot Start Version kit with 1 μL cDNA

template. Reactions are performed in triplicate for each sample and the average Ct value is considered for quantification.

5. Set the conditions of the thermal cycler as follows: 1 cycle of 95 °C for 4 min; 35 cycles of 95 °C for 30 s, 61 °C for 30 s, and 72 °C for 30 s; 1 cycle of 72 °C for 7 min.
6. Perform analysis by using housekeeping genes as a control for normalization (*see* **Note 10**) (Fig. 2a).

3.8 Immunoblotting

1. Determine the protein concentration by BCA Protein Assay Kit according to the manufacturer's protocol. Store protein samples at −80 °C.
2. Add 10–20 μg total protein in a 0.5 mL tube, and dilute by adding protein sample buffer to a total volume of 10 μL (run 12 μg total protein per lane).
3. Prepare master mix by adding 50 μL NuPAGE® LDS Sample Buffer to 20 μL NuPAGE® Sample Reducing Agent (for 10 samples).
4. Add 7 μL master mix and 3 μL distilled water to each sample (1:1). Heat samples at 100 °C for 5 min before loading.
5. Prepare 1× running buffer according to the manufacturer's protocol.
6. Load the sample (20 μL) into the 12% Tris–glycine SDS–polyacrylamide gel, and perform electrophoresis at 100 V for 60–70 min until the loading dye has run to the bottom of the tank.
7. Soak PVDF membrane for 20 s in methanol and then in the anode buffer until use (at least 10 min).
8. Soak blotting papers in each transfer buffer for at least for 30 min.
9. Soak the gel after electrophoresis in cathode buffer for 5–10 min.
10. Place blotting papers, PVDF membrane, and gel on the transfer machine.
11. Transfer at 100 V for 60–70 min.
12. Rinse the membrane PBS-T twice.
13. Block membrane in 1% (w/v) skimmed milk in TBS-T for 1 h at room temperature or overnight at 4 °C.
14. Incubate the membrane with primary antibodies in blocking solution overnight at 4 °C.
15. Rinse the membrane twice and wash three times for 10 min each with TBS-T at 20–25 °C.
16. Incubate the membrane with secondary antibodies in blocking solution for 1 h at 20–25 °C in the dark.

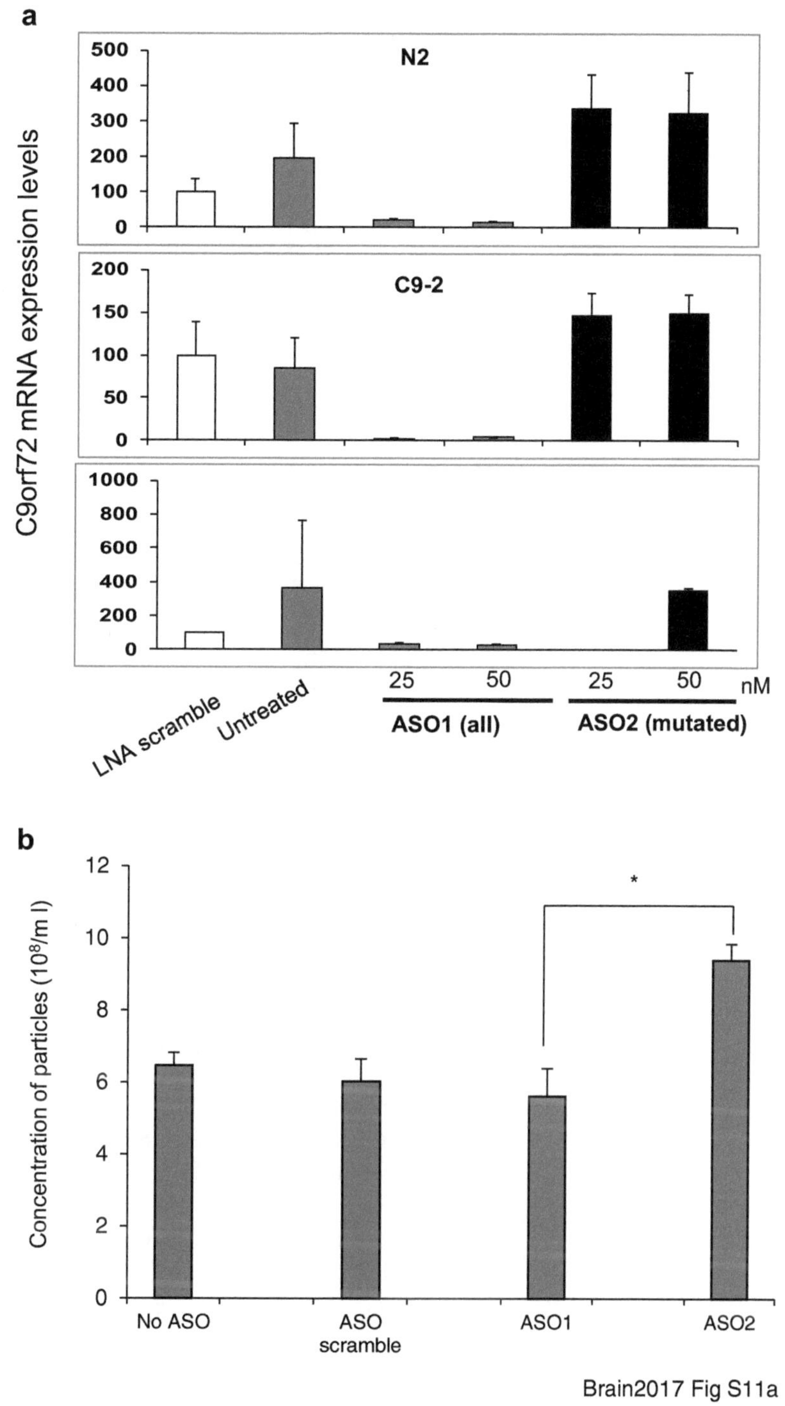

Fig. 2 qPCR validation of LNA gapmer based knockdown of repeat-containing C9orf72 transcripts in human-derived fibroblasts. (**a**) LNA gapmer ASOs-based knockdown in control (N2) and C9ALS/FTD-patient fibroblasts (C9-2). Cells were incubated for 48 h with ASO1-1 or ASO2-1 and the mRNA expression (%) of C9orf72 expression levels were evaluated by semi-quantitative PCR. The data are presented as mean $\pm$ SEM. (**b**) Nanoparticle Tracking Analysis (NTA) quantitative analysis of particle concentration ($\times$10e8) in the media from C9ALS/FTD fibroblasts treated with ASO1-1 and ASO2-1 and scramble LNA gapmers. Asterisks denote one-way ANOVA test significant *P*-value ($*P < 0.05$). Error bars represent $\pm$ SEM

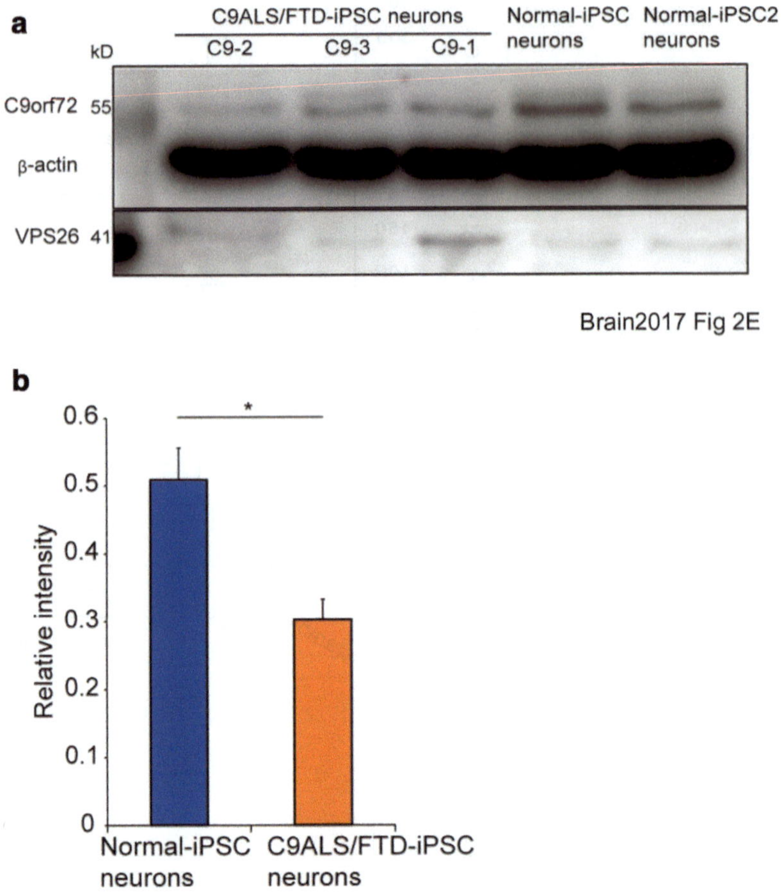

Fig. 3 Immunoblotting shows an overall decrease in C9orf72 protein in human-derived iPSC-motor neurons. (**a**) Protein expression level of C9orf72 in iPSC-neurons from C9ALS/FTD patient was significantly downregulated compared to those from normal individual. (**b**) Semi-quantitative analysis of C9orf72 expression from (**a**). Error bars represent SEM. Asterisks denote one-way ANOVA test significant *P*-value ($*P < 0.05$)

17. Rinse the membrane twice and wash three times for 10 min with PBS-T at 20–25 °C, in the dark.
18. Visualize bands in the Licor Odyssey CLx infrared imaging system (Fig. 3).

3.9 Purification of Extracellular Vesicles and Nanoparticle Tracking Analysis

1. Patient and control fibroblasts or SH-SY5Y cells with desired treatments are grown to 100% confluency and carefully washed with PBS twice. Cells are fed with serum-free media (Opti-MEM or EBSS) 48 h prior to harvesting.
2. Culture media is collected in 50 mL falcons in sterile tissue culture conditions and centrifuged at room temperature at 300 × *g* for 5 min and then 1200 × *g* for 10 min to pellet out all cell matter. Supernatant is retained (*see* **Note 11**).

3. Spun media is passed through a 0.22 μm filter and ultracentrifuged at 120,000 × *g* for 70 min at 4 °C.
4. Supernatant is carefully aspirated, and the EV pellet is washed with 5 mL of PBS and the ultracentrifugation step is repeated (*see* **Note 12**).
5. The EV pellet is thoroughly resuspended in 50 μL of PBS and transferred to an Eppendorf tube (*see* **Note 13**).
6. EV samples stocks from the ultracentrifugation step should be diluted in cold PBS at least 1:100 prior to Nanoparticle tracking analysis (*see* **Note 14**).
7. The NanoSight instrument should be flushed with PBS to clean before and between loading samples. PBS is also used as the control background against all samples (*see* **Note 15**).
8. Samples are loaded onto the instrument from 1.5 mL Eppendorf tubes and the camera focus is adjusted to optimize imaging of the maximum number of particles. Where individual particles are not discernible by the software, samples are accordingly diluted in PBS to improve the acquisition of data.
9. Analysis should be run with scripts for tracking particles in the range 40–100 nm. Measurements are analyzed using the batch process function. Raw data is extracted and analyzed in Microsoft Excel (Fig. 2b). At least five different fields per sample should be imaged and analyzed for sufficient statistical power.

4 Notes

1. Keep stocks of LNA gapmer preparations at 100 μM in RNase–DNase-free water at −80 °C if storing for more than 6 months.
2. Introduction of a single 2′-*O*-methyl (2′-OMe) modification at gap position 2 decreased protein binding and hepatotoxicity, and improved the therapeutic index with minimal impairment of antisense activity [42].
3. Replacing one phosphorothioate backbone at position 2 or 3 from the 5′-side of the DNA gap with alkylphosphonates reduced or eliminated toxicity of several hepatotoxic gapmer ASOs [43].
4. It should be noted that the expansion time for lines from different individuals can vary significantly. This also includes samples from skin biopsies from different regions (arm, shoulder etc.). Usual duration for cultures to become established may take days to weeks. Until this occurs, the skin biopsies should be patiently kept moist by rehydrating with FBS–fibroblast media daily.

5. SH-SY5Y cells grow much faster than fibroblasts. Please note that media will need to be changed more often and cultures will require passaging more often.
6. Media may be gently swirled by moving the plate from side to side, and up and down to ensure equal mixing of LNA gapmer–lipofectamine complexes.
7. Do not pipet scraped cells up and down while harvesting as the cells will lyse and die due to the force. This will significantly reduce the yield of material to harvest protein or RNA.
8. Cell pellets can be frozen and stored at −80 °C at this stage. Prior to purifying protein, thaw pellets slowly on ice.
9. RNA extraction can also be performed by scraping cells using RNase–DNase-free scrapers, transferring to sterile tubes and lysing cells using 300 μL RLT buffer.
10. If using more than one housekeeping gene for normalization (recommended), calculate the geometric of the observed C_t values from the qRT-PCR raw data before normalizing against the C_t values of the C9orf72 transcript.
11. After harvesting conditioned media from cells, samples should be kept on ice or 4 °C at all times in the duration of the experiment to prevent denaturing of EVs.
12. Following ultracentrifugation the samples should be collected immediately to prevent EV pellets from slipping. We also recommending marking the tubes as to the orientation during spinning so that the region where the pellet should be is known. Washing pellet step with PBS may also be omitted if the number of cells from cultures of conditioned media was low.
13. The storage of EV preparations is not recommended due to the possibility of denaturation. We recommend proceeding directly to Nanoparticles tracking analysis.
14. Dilutions between 1:10 and 1:500 were used in our experiments, aiming to reach particle concentrations between 2×10^8 to 1×10^9 particles per mL.
15. For all recordings, the camera level was adjusted to 13 or 15. The default functions for all post-acquisition settings were used. Blur and minimum expected particle size, except for the detection threshold, was fixed at 5.

Acknowledgments

This work was supported by a Japan Society for the Promotion of Science Grant-in-Aid for Scientific Research (C) [grant number 18 K11067 to N.M and 18 K07544 to Y.A.], Grants-in-Aid for

Research on Nervous and Mental Disorders [grant number 28-6 to Y.A.], and the Japan Agency for Medical Research and Development [grant numbers 19ek0109239h0003 and 19lm0203086 h0001 to Y.A.].

References

1. DeJesus-Hernandez M, Mackenzie IR, Boeve BF et al (2011) Expanded GGGGCC Hexanucleotide repeat in noncoding region of C9ORF72 causes chromosome 9p-linked FTD and ALS. Neuron 72:245–256
2. Renton AE, Majounie E, Waite A et al (2011) A hexanucleotide repeat expansion in C9ORF72 is the cause of chromosome 9p21-linked ALS-FTD. Neuron 72:257–268
3. Gijselinck I, Van Langenhove T, van der Zee J et al (2012) A C9orf72 promoter repeat expansion in a Flanders-Belgian cohort with disorders of the frontotemporal lobar degeneration-amyotrophic lateral sclerosis spectrum: a gene identification study. Lancet Neurol 11:54–65
4. Gendron TF, Bieniek KF, Zhang YJ et al (2013) Antisense transcripts of the expanded C9ORF72 hexanucleotide repeat form nuclear RNA foci and undergo repeat-associated non--ATG translation in c9FTD/ALS. Acta Neuropathol 126:829–844
5. Lagier-Tourenne C, Baughn M, Rigo F et al (2013) Targeted degradation of sense and antisense C9orf72 RNA foci as therapy for ALS and frontotemporal degeneration. Proc Natl Acad Sci U S A 110(47):E4530–E4539
6. Cooper-Knock J, Higginbottom A, Stopford MJ et al (2015) Antisense RNA foci in the motor neurons of C9ORF72-ALS patients are associated with TDP-43 proteinopathy. Acta Neuropathol 130:63–75
7. van Blitterswijk M, Dejesus-hernandez M, Niemantsverdriet E et al (2013) Association between repeat sizes and clinical and pathological characteristics in carriers of C9ORF72 repeat expansions (Xpansize-72): a cross-sectional cohort study. Lancet Neurol 12:978–988
8. van Blitterswijk M, Gendron TF, Baker MC et al (2015) Novel clinical associations with specific C9ORF72 transcripts in patients with repeat expansions in C9ORF72. Acta Neuropathol 130:863–876
9. Cruts M, Gijselinck I, Van Langenhove T et al (2013) Current insights into the C9orf72 repeat expansion diseases of the FTLD/ALS spectrum. Trends Neurosci 36:450–459
10. Almeida S, Gascon E, Tran H et al (2013) Modeling key pathological features of frontotemporal dementia with C9ORF72 repeat expansion in iPSC-derived human neurons. Acta Neuropathol 126:385–399
11. Haeusler AR, Donnelly CJ, Periz G et al (2014) C9orf72 nucleotide repeat structures initiate molecular cascades of disease. Nature 507:195–200
12. Zhou B, Liu C, Geng Y et al (2015) Topology of a G-quadruplex DNA formed by C9orf72 hexanucleotide repeats associated with ALS and FTD. Sci Rep 5:16673
13. Sareen D, O'Rourke JG, Meera P et al (2013) Targeting RNA foci in iPSC-derived motor neurons from ALS patients with a C9ORF72 repeat expansion. Sci Transl med 5:208ra149. Sci Transl Med 5(208):208ra149
14. Cooper-Knock J, Walsh MJ, Higginbottom A et al (2014) Sequestration of multiple RNA recognition motif-containing proteins by C9orf72 repeat expansions. Brain 137 (Pt 7):2040–2051
15. Lee YB, Chen HJ, Peres J et al (2013) Hexanucleotide repeats in ALS/FTD form length-dependent RNA foci, sequester RNA binding proteins, and are neurotoxic. Cell Rep 5:1178–1186
16. Mann DM, Robinson A, Rollinson S et al (2013) Dipeptide repeat proteins are present in the p62 positive inclusions in patients with frontotemporal lobar degeneration and motor neurone disease associated with expansions in C9ORF72. Acta Neuropathol Commun 1:1
17. Mori K, Weng S, Arzberger T et al (2013) The C9orf72 GGGGCC repeat is translated into aggregating dipeptide-repeat proteins in FTLD/ALS. Science 339:1335–1339
18. Ash PEA, Bieniek KF, Gendron TF et al (2013) Unconventional translation of C9ORF72 GGGGCC expansion generates insoluble polypeptides specific to c9FTD/ALS. Neuron 77:639–646
19. Jovičić A, Mertens J, Boeynaems S et al (2015) Modifiers of C9orf72 dipeptide repeat toxicity connect nucleocytoplasmic transport defects to FTD/ALS. Nat Neurosci 18:1226–1229
20. Zhang K, Donnelly CJ, Haeusler AR et al (2015) The C9orf72 repeat expansion disrupts nucleocytoplasmic transport. Nature 525 (7567):56–61

21. Freibaum BD, Lu Y, Lopez-Gonzalez R et al (2015) GGGGCC repeat expansion in C9orf72 compromises nucleocytoplasmic transport. Nature 525(7567):129–133
22. Moens TG, Niccoli T, Wilson KM et al (2019) C9orf72 arginine-rich dipeptide proteins interact with ribosomal proteins in vivo to induce a toxic translational arrest that is rescued by eIF1A. Acta Neuropathol 137(3):487–500
23. Donnelly CJ, Zhang PW, Pham JT et al (2013) RNA toxicity from the ALS / FTD C9ORF72 expansion is mitigated by antisense intervention. Neuron 80:415–428
24. O'Rourke JG, Bogdanik L, Muhammad AKMG et al (2015) C9orf72 BAC transgenic mice display typical pathologic features of ALS/FTD. Neuron 88:892–901
25. Jiang J, Zhu Q, Gendron TF et al (2016) Gain of toxicity from ALS/FTD-linked repeat expansions in C9ORF72 is alleviated by antisense oligonucleotides targeting GGGGCC-containing RNAs. Neuron 90:535–550
26. Mendell JR, Al-Zaidy S, Shell R et al (2017) Single-dose gene-replacement therapy for spinal muscular atrophy. N Engl J Med 377:1713–1722
27. Shimizu-Motohashi Y, Miyatake S, Komaki H et al (2016) Recent advances in innovative therapeutic approaches for Duchenne muscular dystrophy: from discovery to clinical trials. Am J Transl Res 8:2471–2489
28. Levine TP, Daniels RD, Gatta AT et al (2013) The product of C9orf72, a gene strongly implicated in neurodegeneration, is structurally related to DENN Rab-GEFs. 29:499–503
29. Marat AL, Dokainish H, McPherson PS (2011) DENN domain proteins: Regulators of Rab GTPases. J Biol Chem 286:13791–13800
30. Farg MA, Sundaramoorthy V, Sultana JM et al (2014) C9ORF72, implicated in amytrophic lateral sclerosis and frontotemporal dementia, regulates endosomal trafficking. Hum Mol Genet 23:3579–3595
31. Aoki Y, Manzano R, Lee Y, et al (2017) C9orf72 and RAB7L1 regulate vesicle trafficking in amyotrophic lateral sclerosis and frontotemporal dementia. 140:887–897
32. Corbier C, Sellier C (2017) C9ORF72 is a GDP/GTP exchange factor for Rab8 and Rab39 and regulates autophagy. Small GTPases 8:181–186
33. Webster CP, Smith EF, Bauer CS, et al (2016) The C9orf72 protein interacts with Rab1a and the ULK1 complex to regulate initiation of autophagy. EMBO J 1–21
34. Yang M, Liang C, Swaminathan K et al (2016) A C9ORF72/SMCR8-containing complex regulates ULK1 and plays a dual role in autophagy. Neurosci Lett 1–17
35. Amick J, Roczniak-Ferguson A, Ferguson SM (2016) C9orf72 binds SMCR8, localizes to lysosomes, and regulates mTORC1 signaling. Mol Biol Cell 27:3040–3051
36. Sullivan PM, Zhou X, Robins AM et al (2016) The ALS/FTLD associated protein C9orf72 associates with SMCR8 and WDR41 to regulate the autophagy-lysosome pathway. Acta Neuropathol Commun 4:51
37. Obika S, Nanbu D, Hari Y et al (1997) Synthesis of 2′-O,4′-*C*-methyleneuridine and -cytidine. Novel bicyclic nucleosides having a fixed c3-endo sugar puckering. Tetrahedron Lett 38:8735–8738
38. Inoue H, Hayase Y, Iwai S et al (1987) Sequence-dependent hydrolysis of RNA using modified oligonucleotide splints and RNase H. FEBS Lett 215:327–330
39. Kurreck J, Wyszko E, Gillen C et al (2002) Design of antisense oligonucleotides stabilized by locked nucleic acids. Nucleic Acids Res 30:1911–1918
40. Fedoroff OY, Salazar M, Reid BR (1993) Structure of a DNA : RNA hybrid duplex. J Mol Biol 233(3):509–523
41. Crooke ST (1999) Molecular mechanisms of action of antisense drugs. Biochim Biophys Acta 1489:31–43
42. Shen W, De Hoyos CL, Migawa MT et al (2019) Chemical modification of PS-ASO therapeutics reduces cellular protein-binding and improves the therapeutic index. Nat Biotechnol 37:640–650
43. Migawa MT, Shen W, Wan WB et al (2019) Site-specific replacement of phosphorothioate with alkyl phosphonate linkages enhances the therapeutic profile of gapmer ASOs by modulating interactions with cellular proteins. Nucleic Acids Res 47:5465–5479

Chapter 15

Targeted Gene Silencing in Malignant Hematolymphoid Cells Using GapmeR

Atish Kizhakeyil, Mobashar Hussain Urf Turabe Fazil, and Navin Kumar Verma

Abstract

Delivery of conventional antisense oligonucleotides or small interfering RNA (siRNA) molecules into hematolymphoid cells for targeted gene silencing has been proven to be difficult. Here, we describe a simple protocol to knockdown specific gene(s) in malignant hematolymphoid cells using "GapmeR." This protocol could be applicable to a wide range of cell-types and thus solves an important problem for researchers working with cell lines or primary cells derived from patients with hematolymphoid malignancies.

Key words GapmeR, RNAi, Gene knockdown, Lymphoma cells, HuT78, STAT3

1 Introduction

Hematolymphoid malignancies represent a spectrum of cancers of the hematopoietic and lymphoid tissues, mainly affecting blood cells, bone marrow, lymph nodes and other parts of the lymphatic system. This group of cancer includes lymphomas, leukemias, myeloproliferative neoplasms, mast cell neoplasms, plasma cell neoplasms, histiocytic tumors, and dendritic cell neoplasms. The "Lymphoma and Leukemia Society" estimates that hematolymphoid malignancies would account for 9.4% of the 0.6 million cancer deaths in the USA in 2019 [1]. Approximately every 3 min one person is diagnosed with a hematolymphoid cancer and more than 6 people die every hour due to these diseases in the USA alone [1].

As with most cancer types, hematolymphoid malignancies result mainly from genetic alterations that drive the transformation of somatic cell clones [2, 3]. To better understand specific role(s) of individual genes in the development and progression of lymphoid tumors, it becomes imperative to perform targeted silencing of

Toshifumi Yokota and Rika Maruyama (eds.), *Gapmers: Methods and Protocols*, Methods in Molecular Biology, vol. 2176, https://doi.org/10.1007/978-1-0716-0771-8_15,

genes of interest in these cells under various experimental conditions. In addition, a number of drugs utilizing the RNA interference (RNAi) mechanism, such as short interfering RNAs (siRNA) and antisense oligonucleotides, are being developed for precision therapeutics to treat a wide range of diseases. In fact, targeted gene silencing approach has shown successful outcomes in clinical trials against many diseases and several antisense oligonucleotide/siRNA molecules have recently been approved by the United States Food and Drug Administration (FDA). For example, Vitravene (also known as Fomivirsen) was the first antisense oligonucleotide drug approved in 1998 by the FDA to treat cytomegalovirus retinitis [4, 5]. Macugen (formerly pegaptanib) was approved by the FDA to treat age-related macular degeneration (AMD) of the retina [6]. Kynamro, also known as Mipomersen, a second-generation antisense oligonucleotide inhibitor of apolipoprotein B-100, was approved for homozygous familial hypercholesterolemia (HoFH), a disease associated with very high plasma concentration of low-density lipoprotein (LDL) [7, 8]. In addition, several other antisense oligonucleotide molecules, including Custirsen [9], EZN-2938 [10], and Cetuximab [11], are undergoing clinical trials for various cancer types. The STAT3-targeted antisense oligonucleotide AZD9150 is being evaluated for lymphoma [12].

Despite promising potential of gene silencing-based techniques in biomedical and clinical research and for drug development, there has been limited success in targeting hematolymphoid cells. Conventionally, researchers use viral vectors and the electroporation or nucleofection systems to deliver RNA interfering molecules into cells to perform gene silencing in the laboratory. However, there are a number of challenges associated with the use conventional gene silencing techniques in lymphoid cells. These include suboptimal delivery of oligonucleotides into lymphoid cells, poor efficiency, and inconvenience of use. Lymphoid cell lines or primary cells are very difficult to transfect [13, 14]. Although viral vector systems have been applied for gene transfer into lymphoid cells with 30–80% transduction efficiency [15], they have drawbacks, mostly related to their immunogenicity [16, 17] and limitation in the use of target molecules to be transferred. Electroporation or nucleofection of RNAi molecules in lymphoid cells results in significant cell death and perturbs important cellular functions [18]. Chemical transfection methods, such as the use of cationic lipids or peptides and lipofection or polyfection, often fail to deliver oligonucleotides into hematolymphoid cells [19]. In addition, use of siRNA along with vehicles may trigger unwanted signaling in the cells derived from a hematopoietic lineage [17].

GapmeR is a chimeric locked nucleic acid (LNA)-conjugated single-strand antisense DNA molecule and has emerged as an alternate approach to gene silencing [20, 21]. It can permeate into mammalian cells, including T lymphocytes, and silence the target gene with precise specificity [18, 22]. Typically, a GapmeR molecule contains 2–5 chemically modified LNA "wings" at each terminus flanking a central "gap" of 5–10 base single strand antisense DNA [20]. The combination of both chemistry and structural modifications provides GapmeR with high binding affinity for the target mRNA, resistance to nucleases, improved stability in biological serum or cell culture medium, and facilitates unassisted cellular uptake or "gymnosis" [20, 21].

Herein, we provide an easy-to-follow, stepwise method to knockdown specific genes of interest in cultured hematolymphoid cancer cells. Although the protocol described here was optimized for targeting STAT3 in the cutaneous T cell lymphoma (CTCL) cell line HuT78, it can be used as a starting point for gene silencing in various other types of malignant hematolymphoid cells.

2 Materials

2.1 GapmeR Molecules

1. GapmeR molecules targeted against gene(s) of interest and nontargeting GapmeR as control (*see* **Note 1**).
2. Stock solutions of GapmeR molecules (50 mM each) are prepared in nuclease free water, aliquoted and stored at −80 °C until use (*see* **Note 2**).

2.2 Cells and Cell Culture Materials

1. Primary cells or cell lines derived from hematolymphoid malignancies (e.g., the CTCL line HuT78) (*see* **Note 3**).
2. Cell culture medium: Roswell Park Memorial Institute (RPMI) 1640 medium supplemented with 10% fetal bovine serum (FBS), 1 mM sodium pyruvate, and 10,000 U/mL penicillin–streptomycin.

2.3 Quantitative RT-PCR (qRT-PCR) Analysis

1. RNA isolation kit.
2. 5× qScript cDNA super mix.
3. Gene-specific primers for target amplification.
4. 2× SYBR green qPCR Master Mix.
5. RNase-free water.
6. 8-Well strip tubes for qPCR.
7. Real-time PCR thermocycler.

2.4 Western Immunoblot Analysis

1. Cell lysis buffer: 50 mM HEPES (pH 7.4), 150 mM NaCl, 1.5 mM $MgCl_2$, 1 mM EGTA, 10 mM sodium pyrophosphate, 50 mM sodium fluoride, 50 mM β-glycerophosphate, 1 mM Na_3VO_4, 1% Triton X-100, 2 mM phenylmethylsulfonyl fluoride, 10 μg/mL leupeptin, and 10 μg/mL aprotinin [23] (*see* **Note 4**).

2. 5× Laemmli sample buffer.
3. Gel running buffer: 3 g tris base, 14.2 g glycine, 1 g sodium dodecyl sulfate (SDS), and distilled water to 1 L.
4. Western transfer buffer: 5.8 g tris base, 29 g glycine, 1 g SDS, 200 mL methanol, and distilled water to 1 L.
5. Polyvinylidene difluoride (PVDF) or nitrocellulose membrane.
6. Phosphate buffered saline (PBS), pH 7.2.
7. PBST: PBS containing 0.05% Tween 20.
8. Membrane blocking solution: 5% nonfat dry milk in PBST.
9. Protein quantification assay kit.
10. Appropriate primary antibodies and horseradish peroxidase (HRP)-labeled secondary antibodies at dilutions recommended for probing the immunoblot.
11. Chemiluminescence reagents.
12. X-ray films.

2.5 Equipment

1. PCR thermocycler.
2. Real-time PCR system.
3. Refrigerated centrifuge.
4. Polyacrylamide gel electrophoresis system.
5. Western transfer apparatus.
6. Chemiluminescence-based blot imaging or X-ray film processing system.
7. X-ray/blot scanner.

3 Methods

3.1 Treatment of HuT78 Cells with GapmeR

1. Maintain HuT78 cells in RPMI 1640 medium (*see* **Note 3**).
2. Pipette out cells in a 15 mL centrifuge tube and wash once with cell culture medium by centrifugation.
3. Count cell number with a hemocytometer (*see* **Note 5**).
4. Dispense 2×10^5 cells in 2 mL culture medium per well into a 6-well tissue culture plate (*see* **Note 6**).
5. Add required amount of target gene-specific GapmeR molecules to the cells in culture in different wells of the 6-well plates (Fig. 1). Nontargeting GapmeR molecules can be used as control. Keep one well with equivalent number of cells without GapmeR as experimental control (*see* **Note 7**).
6. Shake plates gently to evenly mix GapmeR molecules in the culture medium.

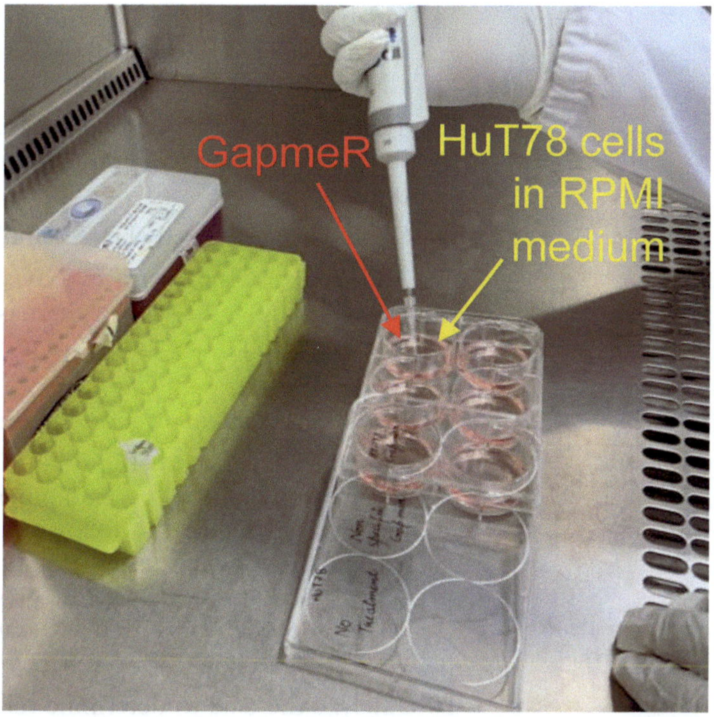

Fig. 1 An illustration of in vitro "Add and Analyse" protocol for GapmeR-mediated gene silencing in hematolymphoid malignancies. Required amount of GapmeR molecules are directly added to cells. Treated cells are incubated for 24 h to several days and analyzed for silencing of target gene(s) by molecular and functional assays

7. Place the plate containing GapmeR-treated cells in the tissue culture incubator for 48–96 h (*see* **Note 8**).
8. After the GapmeR incubation period (48–96 h), pipette out cells gently and place in 1.5 mL microfuge tubes (1 mL in each tube).
9. Pellet cells by centrifugation at 1000 × *g* for 5 min in a refrigerated centrifuge.
10. Wash cells at least one time with PBS.
11. Harvest cell pellets for either RNA or protein analysis.

3.2 Real-Time Polymerase Chain Reaction (RT-qPCR) Analysis of GapmeR-Mediated Knockdown

1. Harvest cells by centrifugation at 8000 × *g* for 10 min in a refrigerated centrifuge.
2. Extract total RNA from the cells using RNA extraction kits according to the manufacturer's instructions (*see* **Note 9**).
3. Add 4 μL 5× cDNA synthesis master-mix solution to 1 μg RNA in a total volume of 20 μL in a microfuge tube for cDNA synthesis according to the manufacturer's protocol (*see* **Note 10**).

4. Incubate the above RNA mixture for 5 min at 25 °C in a thermocycler, followed by 30 min at 50 °C to complete cDNA synthesis.
5. Terminate the reaction by increasing the incubation temperature to 85 °C for 5 min in the thermocycler to inactivate reverse transcriptase.
6. Store the synthesized cDNA samples at −20 °C until use.
7. Add 2 μL of the above cDNA samples to 18 μL premixed cocktail of 2× SYBR green master-mix and primers (e.g., STAT3, GAPDH as a control) prepared in nuclease free water (*see* **Note 11**).
8. Transfer the above reaction mixture to 8-well strip tubes.
9. Run the samples into the RT-qPCR system to quantify gene specific amplicons as per equipment's manual. Typically, thermal cycle is set to the following: Holding stage (95 °C for 10 min) → Cycling stage (95 °C for 15 s; 60 °C for 1 min; 95 °C for 15 s) × 40 cycles → Melt curve stage (95 °C for 15 s → 60 °C for 1 min → gradual increase at the rate of +0.3 °C/s to 95 °C for 5 min) (*see* **Note 12**).
10. Analyze the data to determine relative quantities of mRNA of interest (*see* **Note 13**).

3.3 Western Immunoblot Analysis of GapmeR-Treated Cells

1. Harvest cells by centrifugation at 8000 × *g* for 10 min in a refrigerated centrifuge.
2. Lyse the cells in 50 μL cell lysis buffer containing protease and phosphatase inhibitor cocktails on ice for 1 h.
3. After 1 h, centrifuge cellular lysates at 13,000 × *g* for 15 min at 4 °C and collect the supernatant (*see* **Note 14**).
4. Determine the amount of protein in cellular lysates using a protein quantification assay kit as per manufacturer's instructions (*see* **Note 15**).
5. Mix 20 μg equivalent protein from each cellular lysate samples with required amount of 5× Laemmli sample buffer and then heat at −95 °C for 5 min.
6. Resolve the protein samples by SDS–polyacrylamide gel electrophoresis (SDS-PAGE) using the gel running buffer (*see* **Note 16**).
7. Preactivate PVDF membrane by soaking in methanol and then washing once in transfer buffer. Alternatively, nitrocellulose membrane can be used for Western transfer purpose. Use the preactivated PVDF/nitrocellulose membrane in trans-blot apparatus for transfer of proteins from polyacrylamide gel, according to manufacturer's instructions.

8. Electrophoretically transfer separated proteins (from the gel) to PVDF/nitrocellulose membrane by using a Western transfer system for 60 min at 4 °C.
9. Block the membrane using blocking solution for 1 h at room temperature.
10. Incubate the membrane with relevant primary antibody (e.g., anti-STAT3 or anti-GAPDH using an appropriate concentration/dilution as recommended by the manufacturer) overnight at 4 °C with gentle agitation.
11. After washing the membrane three times in PBST, incubate it with corresponding HRP-conjugated secondary antibody (e.g., HRP-linked anti-rabbit IgG using an appropriate concentration/dilution as recommended by the manufacturer) for 1 h at room temperature.
12. Visualize immunoreactive bands using chemiluminescence reagent and subsequent exposure to a light sensitive X-ray film. Alternatively, blots can be visualized by a chemiluminescence-based blot imaging system (Fig. 2) (*see* **Note 17**).
13. Scan X-ray films using a scanner at 600 dpi (dot per inch) resolution in tiff image format and save file in a data storage drive.
14. Quantify the intensity of individual immunoreactive protein bands using the ImageJ software. Normalize the intensity values of protein of interest using reference housekeeping proteins (GAPDH, actin, etc.) to get relative expression of targets (*see* **Note 18**).

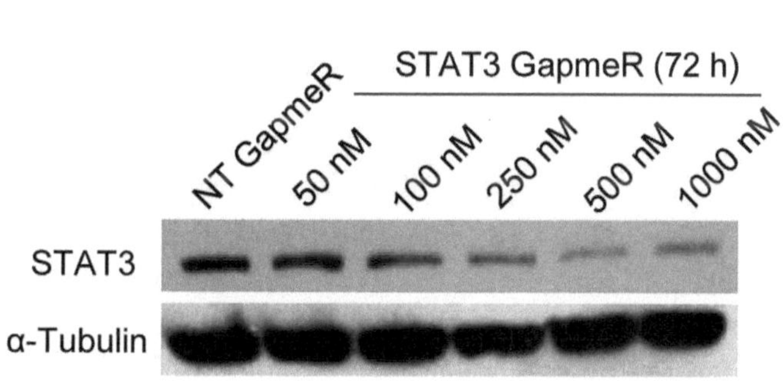

Fig. 2 Dose-dependent effect of GapmeR targeting STAT3 in hematolymphoid cancer cells. HuT78 cells were incubated with increasing amounts of STAT3 targeting GapmeR (50 nM to 1000 nM) for 72 and lysed. Nontargeting (NT) GapmeR was used as a control. Cellular lysates were analyzed for the knockdown of STAT3 expression by Western immunoblotting. The blot was reprobed with GAPDH to confirm equal protein loading

4 Notes

1. GapmeR design is a vital factor for the success of knockdown of the target gene. GapmeR molecules targeted against specific gene(s) of interest can be designed using Exiqon's GapmeR "Design-Tool," which ranks various possible sequences based on multiple parameters, including potential off-target effects. It is important to design, synthesize and screen multiple sequences (typically top 3–5 sequences) to determine the best GapmeR against an individual target.
2. Centrifuge the GapmeR-containing vial after adding nuclease free water and let it settle for at least 15 min before gentle mixing. Aliquot GapmeR stock into smaller microfuge tubes and store at −80 °C until use.
3. Current protocol uses HuT78 as a cell culture model. However, similar methodologies are applicable across various other primary and cultured cells of hematolymphoid origin, such as HH, MJ, MyLA, NKYS, SNK1, Raji, and Jurkat.
4. Commercially available cell lysis buffers can also be used for Western immunoblot analysis. Optimal cell lysis is essential for generating good quantity of protein. It is essential to add phosphatase inhibitors, especially if downstream process involves probing for phosphorylated forms of proteins.
5. Automated cell counters are available from commercial sources and can be used for accurate cell counting.
6. The effective outcome of GapmeR-mediated gene silencing is dependent on seeding density of the cells. For optimal results, seeding density should be empirically determined for each GapmeR design. The pH of the medium, cell cycle and stability of GapmeR depending on handling are also major concerns.
7. Final concentrations of GapmeR molecules in each well can vary between 0.1 and 1 μM depending on individual target, cell type, and protein turnover. In our experience with STAT3-targeted GapmeR, we observed optimal gene silencing effect at 500 nM GapmeR.
8. The duration of GapmeR-treatment varies depending on specific gene of interest, which should be optimized for each target. This, in turn, is dependent on various factors such as the half-life of mRNA or protein apart from macromolecular replenishment by the cellular machinery. For STAT3-targeted GapmeR in HuT78 cells, we observed that optimal mRNA depletion occurred at 48 h while maximal depletion of protein (>80% knockdown) detected at 96 h. Cells should be monitored periodically for signs of contamination.

9. Various types of RNA extraction kits are available commercially that can be used, including column- and TRIzol-based extraction methods. Certain RNA extraction modules also recommend DNase treatment to enhance the quality of RNA, which ensures reproducible and consistent results.
10. Different manufacturers may have their own specific protocols for mixing that need to be followed. cDNA synthesis is essential step for quantification of mRNA levels and can be performed with a commercially available kit or using cocktail of individual components. We have used both commercial kits and lab-made cocktail and our experience suggests that there is no major difference in the quality of cDNA.
11. The reaction mixtures should be prepared in triplicates to minimize potential technical errors. SYBR Green probe for detection of amplicon is generally used. However, there are options (e.g., Taqman, Molecular Beacons) available for similar use.
12. The thermal cycle protocol can vary depending on specific master-mix reagents, fluorescent tags, gene targets, and the equipment.
13. The $2^{-\Delta\Delta CT}$ method proposed by Livak and Schmittgen is frequently applied to calculate relative gene expression in target and reference samples using a reference gene (or internal control gene) as the normalizer [24]. CT (the threshold cycle) is the cycle at which the fluorescence level reaches the threshold. This method is the most popular software package for RT-qPCR experiments. Housekeeping genes, such as GAPDH and β-actin, are commonly used as reference genes.
14. Centrifugation of cell lysates is done to remove unwanted cellular debris.
15. Bradford assay is a simple method applied to quantify protein concentrations in cell lysates. Alternatively, bicinchoninic acid (BCA) protein assay kit is frequently used due to high sensitivity and compatibility to measure broad range of protein concentrations. Both types of assay kits are available commercially from various manufacturers.
16. Lysis of 1×10^6 lymphoma cells typically yields 200–300 μg total protein. In our case, the detection of STAT3 in HuT78 cells requires about 20 μg lysate protein. However, the amount of cellular lysates to be used for each assay depends on cell types and the target protein to be detected.
17. Traditionally, X-ray films are used for capturing chemiluminescent signals. Several digital imaging systems are available nowadays that can be used to visualize Western blots.
18. Various image analysis software packages, both commercial and free for use, are available for densitometry applications.

Acknowledgments

The authors acknowledge funding support from the Singapore Ministry of Education (MOE) under its MOE Academic Research Fund (AcRF) Tier 2 Grant (MOE2017-T2-2-004), the Skin Research Institute of Singapore (SRIS), Skin Innovation Grant (SIG18015) and the Agency for Science, Technology and Research (A*STAR) under its Wound Care Innovation for the Tropics (WCIT) Industry Alignment Fund Pre-Positioning (IAF-PP) grant (H17/01/a0/0K9) Singapore. A.K. is provided with PhD fellowship by Lee Kong Chian School of Medicine, Nanyang Technological University Singapore.

References

1. Facts and Statistics (2019) Lymphoma and Leukemia Society. Access 3 October 2019
2. Love C, Sun Z, Jima D, Li G, Zhang J, Miles R, Richards KL, Dunphy CH, Choi WW, Srivastava G, Lugar PL, Rizzieri DA, Lagoo AS, Bernal-Mizrachi L, Mann KP, Flowers CR, Naresh KN, Evens AM, Chadburn A, Gordon LI, Czader MB, Gill JI, Hsi ED, Greenough A, Moffitt AB, McKinney M, Banerjee A, Grubor V, Levy S, Dunson DB, Dave SS (2012) The genetic landscape of mutations in Burkitt lymphoma. Nat Genet 44:1321–1325
3. Leeksma OC, de Miranda NF, Veelken H (2017) Germline mutations predisposing to diffuse large B-cell lymphoma. Blood Cancer J 7:e532
4. Roehr B (1998) Fomivirsen approved for CMV retinitis. J Int Assoc Phys AIDS Care 4:14–16
5. Vitravene Study Group (2002) A randomized controlled clinical trial of intravitreous fomivirsen for treatment of newly diagnosed peripheral cytomegalovirus retinitis in patients with AIDS. Am J Ophthalmol 133:467–474
6. Ng EW, Shima DT, Calias P, Cunningham JET, Guyer DR, Adamis AP (2006) Pegaptanib, a targeted anti-VEGF aptamer for ocular vascular disease. Nat Rev Drug Discov 5:123–132
7. Hair P, Cameron F, McKeage K (2013) Mipomersen sodium: first global approval. Drugs 73:487–493
8. Raal FJ, Santos RD, Blom DJ, Marais AD, Charng MJ, Cromwell WC, Lachmann RH, Gaudet D, Tan JL, Chasan-Taber S, Tribble DL, Flaim JD, Crooke ST (2010) Mipomersen, an apolipoprotein B synthesis inhibitor, for lowering of LDL cholesterol concentrations in patients with homozygous familial hypercholesterolaemia: a randomised, double-blind, placebo-controlled trial. Lancet 375:998–1006
9. Chi KN, Higano CS, Blumenstein B, Ferrero JM, Reeves J, Feyerabend S, Gravis G, Merseburger AS, Stenzl A, Bergman AM, Mukherjee SD, Zalewski P, Saad F, Jacobs C, Gleave M, de Bono JS (2017) Custirsen in combination with docetaxel and prednisone for patients with metastatic castration-resistant prostate cancer (SYNERGY trial): a phase 3, multicentre, open-label, randomised trial. Lancet Oncol 18:473–485
10. Jeong W, Rapisarda A, Park SR, Kinders RJ, Chen A, Melillo G, Turkbey B, Steinberg SM, Choyke P, Doroshow JH, Kummar S (2014) Pilot trial of EZN-2968, an antisense oligonucleotide inhibitor of hypoxia-inducible factor-1 alpha (HIF-1α), in patients with refractory solid tumors. Cancer Chemother Pharmacol 73:343–348
11. Shapiro JD, Thavaneswaran S, Underhill CR, Robledo KP, Karapetis CS, Day FL, Nott LM, Jefford M, Chantrill LA, Pavlakis N, Tebbutt NC, Price TJ, Khasraw M, Van Hazel GA, Waring PM, Tejpar S, Simes J, Gebski VJ, Desai J, Segelov E (2018) Cetuximab Alone or With Irinotecan for Resistant KRAS-, NRAS-, BRAF- and PIK3CA-wild-type Metastatic Colorectal Cancer: The AGITG Randomized Phase II ICECREAM Study. Clin Colorectal Cancer 17:313–319
12. Reilley MJ, McCoon P, Cook C, Lyne P, Kurzrock R, Kim Y, Woessner R, Younes A, Nemunaitis J, Fowler N, Curran M, Liu Q, Zhou T, Schmidt J, Jo M, Lee SJ, Yamashita M, Hughes SG, Fayad L, Piha-Paul S, Nadella MVP, Xiao X, Hsu J, Revenko A, Monia BP, MacLeod AR, Hong DS (2018) STAT3 antisense oligonucleotide

AZD9150 in a subset of patients with heavily pretreated lymphoma: results of a phase 1b trial. J Immunother Cancer 6:119

13. Zhao N, Qi J, Zeng Z, Parekh P, Chang CC, Tung CH, Zu Y (2012) Transfecting the hard-to-transfect lymphoma/leukemia cells using a simple cationic polymer nanocomplex. J Control Release 159:104–110
14. Yin J, Ma Z, Selliah N, Shivers DK, Cron RQ, Finkel TH (2006) Effective gene suppression using small interfering RNA in hard-to-transfect human T cells. J Immunol Methods 312:1–11
15. Milone MC, Fish JD, Carpenito C, Carroll RG, Binder GK, Teachey D, Samanta M, Lakhal M, Gloss B, Danet-Desnoyers G, Campana D, Riley JL, Grupp SA, June CH (2009) Chimeric receptors containing CD137 signal transduction domains mediate enhanced survival of T cells and increased antileukemic efficacy in vivo. Mol Ther 17:1453–1464
16. Lamers CH, Willemsen R, van Elzakker P, van Steenbergen-Langeveld S, Broertjes M, Oosterwijk-Wakka J, Oosterwijk E, Sleijfer S, Debets R, Gratama JW (2011) Immune responses to transgene and retroviral vector in patients treated with ex vivo-engineered T cells. Blood 117:72–82
17. Whitehead KA, Dahlman JE, Langer RS, Anderson DG (2011) Silencing or stimulation? siRNA delivery and the immune system. Annu Rev Chem Biomol Eng 2:77–96
18. Fazil MH, Ong ST, Chalasani ML, Low JH, Kizhakeyil A, Mamidi A, , Lim CF, Wright GD, Lakshminarayanan R, Kelleher D, Verma NK (2016) GapmeR cellular internalization by macropinocytosis induces sequence-specific gene silencing in human primary T-cells. Sci Rep 6:37721
19. McManus MT (2002) Small interfering RNA-mediated gene silencing in T lymphocytes. J Immunol 169:5754–5760
20. Stein CA, Hansen JB, Lai J, Wu S, Voskresenskiy A, Høg A, Worm J, Hedtjärn M, Souleimanian N, Miller P, Soifer HS, Castanotto D, Benimetskaya L, Ørum H, Koch T (2010) Efficient gene silencing by delivery of locked nucleic acid antisense oligonucleotides, unassisted by transfection reagents. Nucleic Acids Res 38:e3
21. Castanotto D, Lin M, Kowolik C, Wang L, Ren XQ, Soifer HS, Koch T, Hansen BR, Oerum H, Armstrong B, Wang Z, Bauer P, Rossi J, Stein CA (2015) A cytoplasmic pathway for gapmer antisense oligonucleotide-mediated gene silencing in mammalian cells. Nucleic Acids Res 43:9350–9361
22. Fazil MHUT, Ong ST, Chalasani MLS, Kizhakeyil A, Verma NK (2019) GapmeR-mediated gene silencing in motile T-cells. Methods Mol Biol 1930:67–73
23. Verma NK, Dempsey E, Conroy J, Olwell P, Mcelligott AM, Davies AM, Kelleher D, Butini S, Campiani G, Williams DC, Zisterer DM, Lawler M, Volkov Y (2008) A new microtubule-targeting compound PBOX-15 inhibits T-cell migration via post-translational modifications of tubulin. J Mol Med (Berl) 86:457–469
24. Livak KJ, Schmittgen TD (2001) Analysis of relative gene expression data using real-time quantitative PCR and the $2^{-\Delta\Delta CT}$ method. Methods 25:402–408

Chapter 16

Gapmer Antisense Oligonucleotides to Selectively Suppress the Mutant Allele in *COL6A* Genes in Dominant Ullrich Congenital Muscular Dystrophy

Sara Aguti, Elena Marrosu, Francesco Muntoni, and Haiyan Zhou

Abstract

Allele-specific gene silencing by antisense oligonucleotide (ASO) or small interference RNA (siRNA) has been used as a therapeutic approach for conditions caused by dominant gain-of-function mutations. We here present an antisense approach using gapmer ASO to diminish the dominant-negative effect in Ullrich congenital muscular dystrophy (UCMD) caused by dominant mutation in one of the *COL6A* genes. We provide the details of methods that our lab has used. The methods comprise the design of gapmer ASOs and the in vitro evaluation of gapmer ASOs on the specific silencing of the mutant allele at mRNA levels, and functional assessment at protein levels. A fibroblast cell line cultured from a UCMD patient carrying a dominant mutation in one of the *COL6A* genes is used as a cellular model.

Key words Antisense oligonucleotides, Gapmer, Allele-specific silencing, COL6, Congenital muscular dystrophy, Dominant mutation, Fibroblasts

1 Introduction

Antisense oligonucleotide (ASO) has been widely used to correct the effects of specific mutations in different disorders. This approach is under active development for a number of genetic conditions including neuromuscular and neurodegenerative diseases [1–3].

Depending on their design and chemical modifications, ASOs can modify gene expression in different ways. One of the strategies is the use of gapmer ASOs to specifically silence a target mRNA. Gapmer ASOs are deigned to exploit the activity of a ubiquitous cellular enzyme called RNase H. RNase H recognizes RNA–DNA hybrids and cleaves the RNA strand by releasing the DNA fragment. It is essential for the elimination of any hybrids that are

Sara Aguti and Elena Marrosu contributed equally to this work.

Toshifumi Yokota and Rika Maruyama (eds.), *Gapmers: Methods and Protocols*, Methods in Molecular Biology, vol. 2176, https://doi.org/10.1007/978-1-0716-0771-8_16, © Springer Science+Business Media, LLC, part of Springer Nature 2020

produced during the replication and the repair processes as the presence of RNA–DNA hybrids could lead to DNA instability [4]. This is the underlying molecular mechanism of gapmer ASO on gene silencing. Gapmer ASO is designed as chimeric oligonucleotides. It consists of a central block of phosphorothioate (PS)-modified DNA (DNA gap) flanked by modified ribonucleotides (RNA wings) (Fig. 1a). The central DNA gap is essential to activate endogenous RNase H. Once ASO is bound to the mRNA target, RNase H cleaves the target mRNA and release the ASO. The PS linkages inside the DNA gap are essential to protect the central DNA sequence from the degradation by nucleases. While the chemically modified RNA wings at each end do not elicit RNase H activation, they do play a crucial role in increasing the binding affinity with the target mRNA and nuclease resistance. Several chemical modifications are commonly used in the RNA wings which include 2′-*O*-methyl (2′-OMe), 2′-*O*-methoxy-ethyl (MOE), and locked nucleic acid (LNA) [5].

Gapmer ASOs can be used as a therapeutic approach for disorders caused by toxic gain-of-function mutations by downregulating the target mRNA. It is possible to design gapmer ASOs to target both alleles, or the mutant allele specifically. The downregulation of both alleles may result in therapeutic benefit in disorders that are caused by the accumulation of a toxic protein. For instance in familial amyloid polyneuropathy, a disease caused by mutations in the transthyretin (*TTR*) gene that result in the accumulation of amyloid deposition in multiple tissues. In this condition, a gapmer ASO (Inotersen), which received approval from the Food and Drug Administration (FDA) in 2018, is used to reduce the *TTR* mRNA by targeting both the wild type and mutant *TTR* transcripts [6]. In other conditions, it is necessary to selectively silence the mutant allele only. It is possible to discriminate the wild type allele from the mutant allele by differences in the nucleic acid sequences whether this is an insertion/deletion or a single nucleotide variation (Fig. 1b). The allele-specific silencing approach has shown promise in preclinical studies in disorders such as Huntington disease and Ullrich congenital muscular dystrophy (UCMD) [7–10].

UCMD is the severe form of collagen VI–related congenital muscular dystrophy (COL6-CMD). It is caused by recessive or dominant mutations in one of the three genes encoding for the α-chains of collagen 6 protein (*COL6A1*, *COL6A2*, and *COL6A3*). Collagen 6 protein (COL6) is the major component of the extracellular matrix (ECM) in muscle. It is involved in maintaining muscle integrity by providing a structural link between basal lamina and the surrounding interstitial matrix. COL6 undergoes an extensive intracellular assembly process in which the three α-chains combine together to form COL6 tetramers that are secreted into ECM [11].

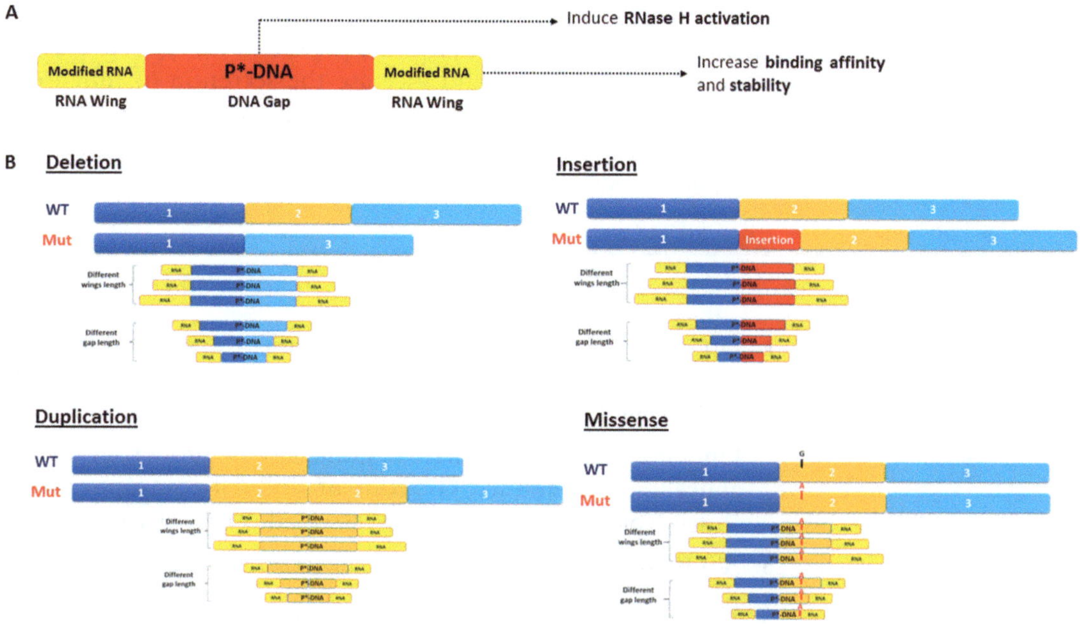

Fig. 1 The design of gapmer ASO for allele-specific gene silencing. (**a**) Gapmer ASO consists of a central phosphorothioate DNA (P∗-DNA) called "DNA Gap" that is essential to induce RNase H activation. The DNA Gap is flanked by two modified RNA sequences called "RNA wings". (**b**) When Gapmer ASO is used to selectively silence the mutant allele, the ASOs are specifically designed to recognize only the mutant transcript without effect on the wild type transcript. Different examples are shown according to the type of mutations to target. It is recommended to design ASOs with different RNA wings and DNA gap length in order to find the optimal ASO sequence

Patients carrying dominant mutations in one of the three *COL6A* genes are characterized by a reduction of COL6 protein in ECM, which is often accompanied with intracellular accumulation of this protein. This occurs for the presence of one mutated α-chain, which has a dominant-negative effect during tetramer formation, impairs the intracellular COL6 assembly process. Interestingly, complete deletion of one copy of the *COL6A* gene is not associated with any clinical features, as indicated by the parents of children affected by recessive variant, which carry null mutations but remain completely asymptomatic [12]. This suggests that suppression of the mutant allele of the *COL6A* gene by allele-specific ASOs may correct the abnormal collagen VI expression in CMD caused by dominant mutations. The fact that dominant-negative mutations account for 50–75% of UCMD patients, and nearly 100% of Bethlem myopathy (BM) cases, makes the allele-specific silencing approach an interesting therapeutic strategy for a large proportion of COL6-CMD patients [13, 14].

We recently reported an ASO approach to correct a dominant mutation in *COL6A3* gene [8]. The gapmer ASOs can specifically silence the mutant allele and increase the secretion of functional COL6 in ECM. This study has shown for the first time that gapmer ASO could be used as a therapeutic approach to treat dominant mutations-associated COL6-CMD.

2 Materials

2.1 ASO Design and Synthesis

1. Online *RNAstructure* software to predict the secondary structures of ASOs and the mRNA target region (http://rna.urmc.rochester.edu/RNAstructure.html).
2. Online software to calculate oligonucleotide properties such as annealing temperature, GC content, and self-complementarity (http://biotools.nubic.northwestern.edu/OligoCalc.html).
3. ASOs are synthesized commercially (e.g., Eurogentec Ltd).
4. RNase-free water (*see* **Notes 1** and **2**).

2.2 Fibroblast Culture

1. Growth medium: Dulbecco's modified eagle medium (DMEM), 10% fetal bovine serum (FBS), 1% GlutaMAX.
2. Trypsin–EDTA.
3. Phosphate-buffered saline (PBS).
4. Incubator set at 37 °C and 5% CO_2.

2.3 Fibroblast Transfection

1. Transfection reagent.
2. Low-serum transfection medium.
3. 6-well plate or 35 mm diameter culture dish.
4. Sterile 1.5 ml Eppendorf tubes.

2.4 RNA Extraction

1. RNA isolation kit.
2. β-mercaptoethanol.
3. 70% ethanol (molecular grade).
4. RNase-free water.
5. 1.5 ml RNase-free Eppendorf tubes.
6. NanoDrop spectrophotometer.

2.5 cDNA Synthesis

1. Reagents for reverse transcription.
2. Thermocycler.
3. 0.2 ml PCR tubes.

2.6 Primers Design

1. Design a pair of primers to amplify both wild type and the mutant transcripts.
2. Design two pairs of primers to specifically amplify the wild type or the mutant transcripts, respectively.
3. Use Primer 3 software for primers design (http://primer3.ut.ee/), avoiding regions with numerous SNPs and repeat regions.
4. Check the specificity of the primers with in silico PCR (https://genome.ucsc.edu/cgi-bin/hgPcr).
5. The primers length should be between 18 and 22 base pairs (bp), temperature (T_m) between 52 and 60 °C without secondary structures.

2.7 Polymerase Chain Reaction (PCR)

1. cDNA template from Subheading 2.5.
2. *Taq* Polymerase.
3. Primers from Subheading 2.6 (10 μM forward primer and 10 μM reverse primer) (*see* **Note 3**).
4. 10× PCR buffer, 10 mM dNTPs, 50 mM $MgCl_2$, Taq DNA polymerase (5 U/μl), and nuclease-free water.
5. 0.2 ml PCR tubes.
6. Thermocycler.
7. Tris–borate–EDTA 1× (TBE) buffer.
8. Agarose.
9. DNA gel stain.
10. Loading buffer.
11. DNA ladder.
12. Gel imaging system.

2.8 Quantitative Real Time PCR

1. cDNA template from Subheading 2.5.
2. qPCR primers from Subheading 2.6 (10 pmol/μl each) (*see* **Note 4**).
3. Universal SYBR Green Master Mix.
4. 96-well real-time PCR plate.
5. Sealing film.
6. Real-time PCR thermal cycler.

2.9 Immuno-fluorescence Staining for Extracellular and Intracellular COL6

1. 12-mm diameter glass coverslip precoated with collagen.
2. Ascorbic acid.
3. PBS.
4. Paraformaldehyde.
5. Blocking solution: PBS, 5% goat serum, 0.1% Triton X-100.

6. Mouse anti-collagen type VI antibody (e.g., MAB1944: 1:2000 dilution in blocking buffer).
7. Goat anti-mouse secondary antibody (e.g., A11029; goat anti-mouse Alexa 488; 1:500 dilution in blocking buffer).
8. Hoechst 33258.
9. Hydromount.
10. Fluorescence microscope.

3 Methods

3.1 Gapmer Antisense Oligonucleotide Design and Preparation

1. Design specific ASOs targeting the mutant RNA with different sequences and length in RNA wings and DNA gap (Fig. 1). The length of gapmer is usually between 13 and 22 mers and the DNA gap between 5 and 12 mers.
2. The central DNA gap is phosphorothioate (PS)-modified DNA, while the flanking RNA wings can be modified with different chemistries (e.g., 2′-OMe, MOE, and LNA).
3. The length of the flanking wings depends on the chemistry used (e.g., between 4 and 8 mers for 2′-OMe or 3 and 5 mers for MOE or LNA).
4. Dissolve ASOs in RNase free water at concentration of 100 μM as stocking solution.
5. ASO solution is aliquoted and stored at −20 °C for further usage.

3.2 Fibroblast Transfection

1. Seed the cells in 6-well plate at a concentration of 2×10^5 cells per well, which gives 90% confluence on the next day (Fig. 2a).
2. For each experiment, ideally four controls are needed. These include (1) blank control—untreated patient's cell line; (2) mock control—patient's cell line only treated with Lipofectamine; (3) scrambled control—patient's cell line treated with an ASO with scrambled sequence that does not bind to any mRNA; (4) an untreated healthy cell line (*see* **Notes 5** and **6**).
3. Cells are cultured in 2 ml of growth medium for 24 h.
4. 24 h later, change the growth medium to 1 ml Opti-MEM and leave the cells in the incubator during the preparation of transfections mixes.
5. Prepare the transfection reagents mix in a sterile 1.5 ml tubes. For each sample, prepare two mixes: the first mix (Mix A) contains 100 μl Opti-MEM and 1 μl ASO at desired concentration (e.g., 1 μl ASO at 100 μM to get a 100 nM final concentration). While for the mock control add only 100 μl

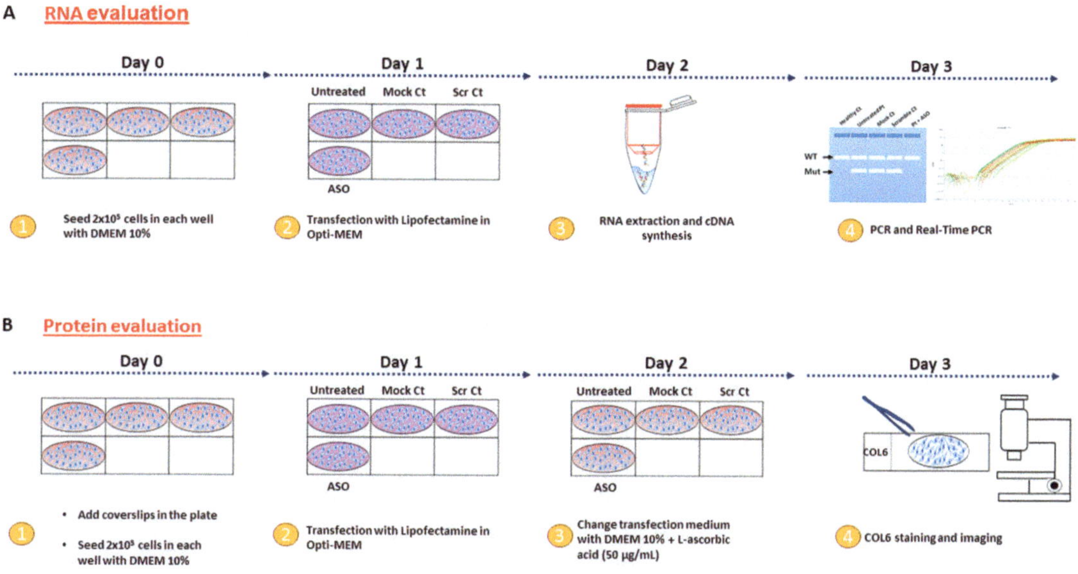

Fig. 2 The evaluation of ASOs on allele-specific silencing at mRNA and protein levels in cultured patient's skin fibroblasts. (**a**) For the evaluation of ASO at the mRNA levels, 2×10^5 cells per well are seeded in the six-well plate in DMEM with 10% FBS (Day 0). On the next day (Day 1), transfection experiment is performed in Opti-MEM using lipofectamine as transfection reagent. On day 2, RNA extraction and cDNA synthesis are carried out. This is followed by PCR and Real-Time PCR to evaluate the efficiency and the specificity of ASO treatment (Day 3). (**b**) For the evaluation of ASO at the protein levels, 2×10^5 cells are seeded on coverslips in the six-well plate (Day 0). Transfection experiment is performed on day 1. After 24 h, the transfection medium is replaced with DMEM 10% + L-ascorbic acid (50 μg/ml) (Day 2). This is followed by collagen VI immunostaining and imaging on day 3

Opti-MEM. The second mix (Mix B) contains 100 μl Opti-MEM and 5 μl Lipofectamine 2000.

6. Mix the ASO-containing tube (Mix A) with the lipofectamine-containing tube (Mix B) at a ratio of 1:1 (100 μl + 100 μl).
7. Incubate the transfection mix for 20 min at room temperature (RT).
8. Add 800 μl Opti-MEM in the transfection mix to top it up to a final volume of 1 ml.
9. Remove Opti-MEM from the 6 well plate and replace with 1 ml transfection mix in each well.
10. Incubate the plate for 24 h at 37 °C with 5% CO_2.

3.3 RNA Extraction and cDNA Synthesis

1. Harvest the cells after 24 h of transfection.
2. Extract total RNA with the RNeasy kit according to the manufacturer's instruction.
3. Measure the concentration and quality of extracted total RNA using NanoDrop.
4. Perform first-strand cDNA synthesis using 500 ng RNA.

3.4 PCR and Semiquantification

1. Perform PCR according to standard procedures. Add 1 μl cDNA in a 25 μl PCR reaction with 500 pmol of each primer, 200 μM of dNTPs, 1.5 mM $MgCl_2$, 1 unit of Taq polymerase, and 1x PCR buffer. The PCR amplification program is as follows: 1 cycle with 3 min at 94 °C (initial denaturation), 25–30 subsequent cycles of 30 s at 94 °C (denaturation), 30 s at 60 °C (annealing), and 30 s at 72 °C (extension) and a final 10-min extension at 72 °C.
2. Check the PCR product by electrophoresis in agarose gel with SYBR Safe DNA stain and visualize under an UV transillumi-nator. The gel concentration depends on the base pairs of PCR product (e.g., 2% gel for a 100-bp PCR product).
3. Digital pictures are captured. Quantify the intensity of each PCR band using ImageJ software (https://imagej.nih.gov). The ratio of densities of the wild-type band to mutant band is calculated.
4. Sanger sequencing may be used to analyze the PCR product to see how much mutant transcripts have been suppressed (*see* **Note 7**).

3.5 Quantitative Real-Time PCR

1. Use the first-strand cDNA synthesized in Subheading 3.3 above.
2. Perform a quantitative real-time PCR using specific primers to discriminate the wild type and mutant transcripts; cDNA and 1× PCR Mastermix are mixed in a 20 μl PCR reaction. The program includes activation at 95 °C for 5 min, 40 cycles of 95 °C for 3 s, and 60 °C for 1 min. The cycle at which the amount of fluorescence is above the threshold (C_t) is detected.
3. For quantification, it is possible to use the standard curves obtained from serial dilutions of cDNA from untreated patient's cells, or the $\Delta\Delta C_t$ method.
4. Normalize the ratios of wild-type and mutant *COL6A* to a reference gene (e.g., *18S* or *GAPDH*) of the treated samples, by taking the ratio in the scrambled-ASO treated sample as 1.0 (*see* **Note 8**).

3.6 Immuno-fluorescence Study

1. Place 12-mm diameter glass coverslip, precoated with collagen, into the 6-well plate (Fig. 2b).
2. Seed the cells at a concentration of 2×10^5 cells per well.
3. Transfect the cells next day as described in Subheading 3.2.
4. 24 h after transfection, change the transfection medium to growth medium containing 50 μg/ml L-ascorbic acid (*see* **Note 9**).
5. After 24 h treatment of L-ascorbic acid, remove the medium and wash once with PBS.

6. Fix the cells on the coverslip with 4% prewarmed paraformaldehyde for 10 min at RT. Rinse coverslips thoroughly with PBS (*see* **Notes 10** and **11**).
7. Incubate the cells for 1 h at RT with blocking buffer (*see* **Note 12**).
8. Incubate the cells with primary antibody for 1 h at RT (*see* **Notes 13** and **14**).
9. Rinse the coverslips three times in PBS for 5 min in each wash, and then incubate the cells with secondary antibody diluted in blocking buffer for 1 h at RT.
10. Wash the coverslips three times for 5 min with PBS and stain the nuclei with Hoechst 33258 for 3 min at RT.
11. Mount the coverslips with Hydromount.
12. Leave the slides to dry for at least 1 h.
13. Imaging under a fluorescence microscope.

4 Notes

1. ASOs are shipped in powder at room temperature. DNase and RNase free water is recommended to use to dissolve ASOs. However, DPEC-treated RNase-free water should be avoided to dissolve ASOs.
2. Dissolved ASOs should be aliquoted and stored at −20 °C, and avoid repeated freeze–thaw cycles.
3. The primers are designed to amplify both wild-type and mutant transcripts in the same PCR product.
4. The primers for real-time PCR are specifically designed to amplify either the wild-type or mutant transcripts, respectively.
5. Appropriate controls should be included in each experiment. In addition to an untreated healthy control cell line, the blank control represents the untreated disease condition, the mock control is used to evaluate lipofectamine's effect while the scrambled control is to evaluate the effect of both ASO chemistry and lipofectamine on target gene expression.
6. A standard concentration at 100 nM is recommended for ASO screening. The concentration may be further reduced in order to achieve the most efficient one (e.g., from 100 nM to 20 nM).
7. For Sanger sequencing, it is possible to use the same forward and reverse primers as used in PCR.
8. For real-time PCR data obtained in ASO treated samples, it is expected that the wild type and mutant transcripts should be similar among the untreated disease control, mock control and scrambled control.

9. A second transfection or longer than 24 h incubation period with ascorbic acid may be needed in some experiments.
10. For immunofluorescence staining, it is recommended to use 4% PFA prewarmed at 37 °C in order to preserve an intact morphology of the cells.
11. Fixed coverslips can be kept in PBS in fridge for a few weeks until the following staining.
12. Add 0.1% Triton in the blocking buffer only to visualize the intracellular COL6. Otherwise, stain the extracellular COL6 use the blocking buffer without Triton.
13. Collagen VI chain-specific antibodies are available. These include anti-collagen type VI α-1 chain antibody (17023-1-AP, Proteintech) and anti-collagen type VI α-3 chain antibody, clone 3C4 (MAB1944, Sigma-Aldrich). Anti-collagen type VI Antibody, clone VI-26 (MAB3303, Sigma-Aldrich), is to stain the generic collagen VI and recognize all α-1, 2, and 3 chains.
14. Primary antibodies should be stored at −20 °C, avoiding repeated freeze–thaw cycles. It is recommended to make small aliquots from the stock.

References

1. Benson MD et al (2018) Inotersen treatment for patients with hereditary transthyretin amyloidosis. N Engl J Med 379:22–31
2. Aartsma-Rus A (2017) FDA approval of Nusinersen for spinal muscular atrophy makes 2016 the year of splice modulating oligonucleotides. Nucleic Acid Ther 27:67–69
3. Aartsma-rus A, Krieg AM (2017) FDA approves eteplirsen for duchenne muscular dystrophy: the next chapter in the Eteplirsen saga. Nucleic Acid Ther 27:1–3
4. Cerritelli SM, Crouch RJ (2009) Ribonuclease H: the enzymes in eukaryotes. FEBS J 276:1494–1505
5. Moniass BP et al (1993) Evaluation of 2′ modified oligonucleotides containing 2′-deoxy gaps as antisense inhibitors of gene expression. J Biol Chem 268:14514–14522
6. Merrill D et al (2018) Inotersen treatment for patients with hereditary transthyretin Amylodosis. N Engl J Med 379:22–31
7. Carroll JB et al (2011) Potent and selective antisense oligonucleotides targeting single-nucleotide polymorphisms in the Huntington disease gene/allele-specific silencing of mutant huntingtin. Mol Ther 19:2178–2185
8. Marrosu E, Ala P, Muntoni F, Zhou H (2017) Gapmer antisense oligonucleotides suppress the mutant allele of COL6A3 and restore functional protein in Ullrich muscular dystrophy. Mol Ther Nucleic Acids 8:416–427
9. Gualandi F et al (2012) Antisense-induced messenger depletion corrects a COL6A2 dominant mutation in Ullrich myopathy. Hum Gene Ther 23:1313–1318
10. Noguchi S, Ogawa M, Kawahara G, Malicdan MC, Nishino I (2014) Allele-specific gene silencing of mutant mRNA restores cellular function in Ullrich congenital muscular dystrophy fibroblasts. Mol Ther 3:1–7
11. Cescon M, Gattazzo F, Chen P, Bonaldo P (2015) Collagen VI at a glance. J Cell Sci 128:3525–3531
12. Foley AR (2011) Large genomic deletions: a novel cause of Ullrich congenital muscular dystrophy. Ann Neurol 69:206–211
13. Brin L et al (2010) Early onset collagen VI myopathies: genetic and clinical correlations. Ann Neurol 68:511–520
14. Hicks D, Fiorillo C, Sutton R, Lochmu H (2008) A refined diagnostic algorithm for Bethlem myopathy. Neurology 70:1192–1199

Part IV

Detection of Gapmers

Chapter 17

Detection of Locked Nucleic Acid Gapmers from Mouse Muscle Samples Using ELISA

Kenji Rowel Q. Lim, Quynh Nguyen, and Toshifumi Yokota

Abstract

Antisense oligonucleotide (ASO)-mediated therapy is promising for the treatment of a variety of genetic disorders, such as Duchenne muscular dystrophy. As more ASOs advance in therapeutic development and enter clinical trials, it becomes necessary to have a means of quantifying their amounts in biological samples post-treatment. This information will be valuable for evaluating the safety and pharmacokinetic profiles of ASOs, and in deciding how the efficacy of these drugs can be improved. Gapmers are a class of ASOs characterized by having a central DNA portion that is surrounded by chemically modified nucleotides on both ends. While relatively simple and accessible methods to quantify other ASOs such as phosphorodiamidate morpholino oligomers (PMOs) using enzyme-linked immunosorbent assay (ELISA)-based techniques are available and have been used for in vivo studies, no such method is available for gapmers to our knowledge. Here, we describe a sensitive ELISA protocol that can be used to quantify the levels of locked nucleic acid (LNA) gapmers in mouse muscle tissue.

Key words Gapmer quantification, Gapmer uptake, LNA, ELISA, Antisense oligonucleotides, Antisense therapy, Pharmacokinetics, Safety, In vivo evaluation, Mouse muscle tissues

1 Introduction

Antisense therapy is one of the most promising strategies currently being developed to treat Duchenne muscular dystrophy (DMD) as well as other genetic disorders [1–4]. This approach uses short, synthetic nucleic acids called antisense oligonucleotides (ASOs) that induce RNase H-mediated transcript knockdown or splicing modulation upon binding to their complementary targets, among other mechanisms. ASOs come in many different chemistries, with the newer generations having enhanced target binding affinities and resistance to degradation by nucleases. The 2′-*O*-methyl phosphorothioate (2′OMe-PS) and phosphorodiamidate morpholino oligomers (PMOs) are two chemistries that have advanced the most in

Kenji Rowel Q. Lim and Quynh Nguyen contributed equally to this work.

Toshifumi Yokota and Rika Maruyama (eds.), *Gapmers: Methods and Protocols*, Methods in Molecular Biology, vol. 2176, https://doi.org/10.1007/978-1-0716-0771-8_17, © Springer Science+Business Media, LLC, part of Springer Nature 2020

therapeutic development for DMD. These modifications, however, do not have the ability to induce RNase H cleavage. Fortunately, this can be overcome by the use of gapmers, which contain a central DNA sequence (that supports RNase H activity) surrounded by modified residues at the 3′ and 5′ ends for enhanced target binding affinity and ASO stability [5].

As previously mentioned, ASO therapy is widely known for its development as a treatment for DMD, and thus serves as a useful example. DMD is an X-linked recessive lethal neuromuscular disorder characterized by progressive muscle degeneration due to loss-of-function mutations in the dystrophin (*DMD*) gene [6]. DMD affects one in 3500–5000 males born worldwide and there is currently no cure for it [7, 8]. Preclinical studies using ASOs for DMD have shown promising results [9]. With the steady advancement of ASOs for DMD therapy, methods to evaluate their safety have to be developed in addition to assessing their efficacy. For instance, the extent of ASO uptake in biological samples has to be quantified in order to have a better idea of the pharmacokinetic and safety profile of the drug. This information is also important to consider in the context of improving the efficacy of an ASO.

Different methods have been developed to quantify ASOs, such as high-performance liquid chromatography (HPLC), liquid chromatography–mass spectrometry (LC/MS), or the enzyme-linked immunosorbent assay (ELISA), with ELISA being considered as the most sensitive approach [10–13]. The original ELISA requires the presence of a hydroxyl group at the terminal end of the nucleotide for ligation of the detection probe; therefore it only works for 2′OMe ASOs and not PMOs [14]. In 2015, Burki and colleagues developed a modified ELISA protocol for detecting PMOs and peptide-conjugated PMOs (PPMOs) in biological samples [10]. In this protocol, PMOs hybridize to complementary probes, which are anchored to avidin-coated 96-well plates via biotin (Fig. 1); the other end of the probe is conjugated to digoxigenin for detection purposes later on. Unbound probes are cleaved by micrococcal nuclease (MNase), which preferentially targets single-stranded DNA. Hybridized probes are detected using an anti-digoxigenin antibody conjugated to an alkaline phosphatase enzyme; a clear substrate is then added, which alkaline phosphatase converts into a fluorescent product whose intensity can be measured using a fluorescence plate reader. A standard curve is required to quantify the amount of ASO in each sample.

The assay has been validated both in vitro and in vivo, and has been proven to be a sensitive and reliable method for PMO quantification in biological tissues. Previous DMD studies have used ELISA to quantify PMOs and PPMOs in biological samples. In a study by Echigoya et al. [15], the concentrations of a PPMO cocktail in serum, liver, skeletal and cardiac muscles of ASO-treated dystrophic dogs were measured using ELISA.

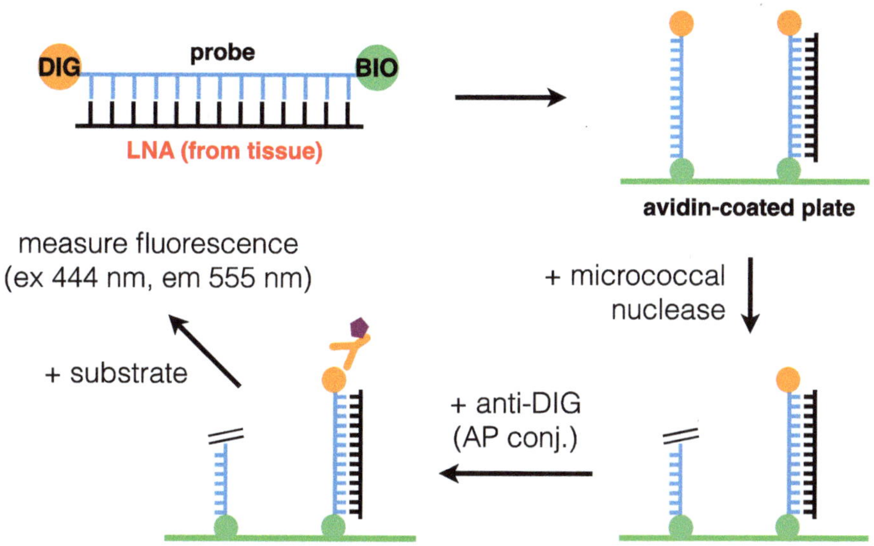

Fig. 1 Overview of ELISA for quantifying LNA gapmers. Samples are incubated with a probe having digoxigenin (DIG) and biotin (BIO) modifications at the 5′ and 3′ ends, respectively; probes are also terminally modified with phosphorothioate backbones. The mixture is allowed to hybridize to an avidin-coated plate, after which micrococcal nuclease treatment is done to cleave unbound probes. Wells are incubated with anti-digoxigenin (anti-DIG) antibodies conjugated to alkaline phosphatase (AP conj.), followed by addition of the AP substrate. The released fluorescence is then measured using the provided excitation (ex) and emission (em) wavelengths, recorded, and analysed for gapmer quantification

Similarly, Lim et al. [16] used ELISA to quantify the levels of PMOs in the sera and muscles of treated dystrophic dogs to evaluate the safety of ASO treatment and the efficiency of PMO uptake in target tissues. No standard protocol for ELISA, however, is available for quantifying gapmers in animal samples. Given the increasing use of gapmers in therapy, with a good number currently in clinical trials, methods to quantify gapmer amounts in biological samples will be needed for assessing gapmer safety and pharmacokinetics. This chapter will outline an ELISA-based protocol we routinely use to quantify gapmers from mouse muscle samples, in particular those of the locked nucleic acid (LNA) chemistry.

2 Materials

2.1 Preparation of Protein Samples

1. Mouse muscle samples, frozen in liquid nitrogen-cooled isopentane and mounted on corks with tragacanth gum.
2. Lysis buffer: RIPA buffer (Thermo Scientific) with cOmplete, Mini, protease inhibitor cocktail, EDTA-free (Roche) (*see* **Note 1**).
3. BCA Protein Assay kit (Thermo Scientific).
4. Equipment: cryostat, vortex, refrigerated centrifuge, spectrophotometer.

2.2 ELISA-Based Quantification of Gapmers

1. LNA gapmers to be detected, used for generating the standard curve.
2. ELISA probes (IDT): designed complementary to the LNA gapmer to be detected, modified at the 5′ end with digoxigenin and at the 3′ end with biotin, and with the first and last 3 nucleotides phosphorothioated (*see* **Note 2**).
3. NeutrAvidin™-coated 96-well plates, black (Thermo Scientific, #15217).
4. Hybridization buffer: 1× TE buffer (Sigma) with 1 M NaCl and 0.1% v/v Triton X-100.
5. Wash buffer: 50 mM Tris–HCl, 150 mM NaCl, 0.1% v/v Tween 20, pH 7.6 in Milli-Q water.
6. Micrococcal nuclease, 2000 U/μl (New England Biolabs, #M0247S).
7. Micrococcal nuclease reaction buffer: 50 mM Tris–HCl, 100 mM NaCl, 5 mM $CaCl_2$, 0.1 mg/ml bovine serum albumin, pH 7.9 in Milli-Q water.
8. Anti-digoxigenin antibody conjugated to alkaline phosphatase (Roche, #11093274910).
9. Blocking buffer: SuperBlock (TBS) blocking buffer (Thermo Scientific), with 0.0025% v/v Tween 20.
10. AttoPhos AP fluorescent substrate system (Promega).
11. Sterile distilled water.
12. Equipment: multichannel pipette, 96-well plate cover, waste bucket, 37 °C incubator, vortex, SpectraMax M3 plate reader (or equivalent).

3 Methods

3.1 Preparation of Protein Samples

1. Using the cryostat, prepare 40–60 20-μm-thick sections of the mouse muscle samples and collect them into sterile 1.5 ml tubes (*see* **Notes 3** and **4**).
2. Add 80 μl of ice-cold lysis buffer into each tube and vortex until homogenized.
3. Incubate the mixture overnight, at least 16 h, at 55 °C.
4. Centrifuge samples at maximum speed for 15 min, at 4 °C.
5. Collect the supernatant and quantify protein content using the BCA assay kit; adjust protein concentration as necessary using the lysis buffer as diluent (*see* **Note 5**).
6. Store extracts at −80 °C until use.

3.2 ELISA-Based Quantification of Gapmers

1. Dilute protein samples to the desired concentration for ELISA using sterile distilled water, preparing a total volume of 220 μl (*see* **Note 6**). Vortex samples to mix.
2. Prepare a 1.5 ml 1 nM stock for the LNA gapmer used as standard. This must have the same concentration of the lysis buffer as used for the diluted protein samples (*see* **Note** 7). Vortex the solution well to mix. The stock can be kept frozen at −20 °C and reused.
3. Prepare the solvent for making the standards by diluting the lysis buffer with sterile distilled water, again at the same concentration as in the diluted protein samples (*see* **Note 8**).
4. Prepare a dilution series of the stock LNA gapmer at the following concentrations: 500, 250, 125, 62.5, 31.3, 15.6, 7.8, and 0 pM. To prepare the 500 pM standard, take 250 μl of the 1 nM stock and mix with 250 μl of the solvent from **step 3** in a 1.5 ml tube. Vortex well to mix. From this, take 250 μl, transfer to a fresh 1.5 ml tube, and mix with another 250 μl of solvent to make the 250 pM standard, and so on. Once all standards are prepared, transfer 220 μl of each into fresh 1.5 ml tubes.
5. Prepare 0.5 nM of the ELISA probe using hybridization buffer as the diluent:
   (a) Make a 100 nM stock first: 1.5 μl of 100 μM probe + 1498.5 μl hybridization buffer.
   (b) Then, prepare the 0.5 nM working solution: 100 μl of 100 nM probe + 19.9 ml hybridization buffer.
6. Add 220 μl of the 0.5 nM probe solution to each of the tubes containing the LNA gapmer standards or samples. Vortex to mix, then incubate at 37 °C for 30 min.
7. Wash the avidin-coated plate by pipetting 200 μl of wash buffer into each well, and then flicking the solution out into a waste bucket. Repeat three times.
8. When the incubation in **step 6** is done, pipette 200 μl of each standard- or sample-probe mixture to the wells of the plate in duplicate (*see* **Note 9**).
9. Incubate at 37 °C for 30 min, and then wash the plate as in **step** 7.
10. Prepare 0.1 U/μl of micrococcal nuclease in the reaction buffer by adding 1.5 μl of the enzyme in 30 ml of the reaction buffer. Add 200 μl of this solution to each well.
11. Incubate at 37 °C for 1 h, and then wash the plate as in **step** 7.
12. Prepare a 1:5000 diluted solution of the anti-digoxigenin antibody in blocking buffer (*see* **Note 10**). Add 200 μl of this solution to each well.

13. Incubate at 37 °C for 30 min, and then wash the plate as in **step 7**.
14. Add 150 μl of the AttoPhos substrate into each well. Incubate at 37 °C for 30 min (*see* **Note 11**).
15. Using the SpectraMax M3 plate reader or an equivalent capable machine, read the fluorescence of the plate at the following parameters: 444 nm excitation, 555 nm excitation, with a cutoff at 530 nm.
16. For quantification, construct a standard curve and then use the equation from this to determine the LNA gapmer concentrations of the samples.

4 Notes

1. We typically add 1 tablet of the protease inhibitor to 7 ml of RIPA buffer.
2. For 25- and 30-mer PMOs, we used probes with the first and last 7 nucleotides phosphorothioated. The change in number was to accommodate the smaller length of LNA gapmers. We also order probes from the manufacturer as a 100 μM solution in TE buffer at pH 8.0.
3. The number of sections will vary depending on the size of the muscle; we based our count on the number we collect from the tibialis anterior muscle of 8–12-week-old mice.
4. We focus on muscle samples in this protocol, but the method can be adapted for other tissues such as the liver.
5. We usually dilute our protein samples to a stock concentration of 2 mg/ml.
6. This concentration will have to be experimentally determined. We recommend preparing a dilution series of a representative sample from an animal treated with the LNA gapmer and using it for ELISA. The results from this first experiment should be informative of what dilution to use for the rest of the samples.
7. For example, if the protein samples were diluted 1:10 (22 μl protein + 198 μl water), then the stock concentration must contain lysis buffer also diluted 1:10.
8. Usually 3 ml of solvent is sufficient.
9. If working with volumes less than 100 μl, use a more appropriate avidin-coated plate (Thermo Scientific, #15117).
10. For a full 96-well plate, 4 μl of antibody in 20 ml blocking buffer is sufficient.

11. From this point on, the plate has to be kept in the dark until it will be read. It is also recommended to read the plate as soon as possible once the incubation period is over.

References

1. Echigoya Y, Lim KRQ, Trieu N, Bao B, Miskew B, Vila MC, Novak JS, Hara Y, Lee J, Touznik A, Mamchaoui K, Aoki Y, Si T, Nagaraju K, Mouly V, Maruyama R, Duddy W, Yokota T (2017) Quantitative antisense screening and optimization for exon 51 skipping in Duchenne muscular dystrophy. Mol Ther 25(11):2561–2572. https://doi.org/10.1016/j.ymthe.2017.07.014
2. Lee J, Yokota T (2013) Antisense therapy in neurology. J Pers Med 3:144–176. https://doi.org/10.3390/jpm3030144
3. Nguyen Q, Yokota T (2019) Antisense oligonucleotides for the treatment of cardiomyopathy in Duchenne muscular dystrophy. Am J Transl Res 11(3):1202–1218
4. Nguyen Q, Yokota T (2017) Immortalized muscle cell model to test the exon skipping efficacy for Duchenne muscular dystrophy. J Pers Med 7:E13. https://doi.org/10.3390/jpm7040013
5. Juliano RL, Carver K (2015) Cellular uptake and intracellular trafficking of oligonucleotides. Adv Drug Deliv Rev 87:35–45. https://doi.org/10.1016/j.addr.2015.04.005
6. Nakamura A, Takeda SI (2011) Mammalian models of Duchenne muscular dystrophy: pathological characteristics and therapeutic applications. J Biomed Biotechnol 2011:184393. https://doi.org/10.1155/2011/184393
7. Emery AEH (1991) Population frequencies of inherited neuromuscular diseases-a world survey. Neuromuscul Disord 1:19–29. https://doi.org/10.1016/0960-8966(91)90039-U
8. Mendell JR, Shilling C, Leslie ND, Flanigan KM, Al-Dahhak R, Gastier-Foster J, Kneile K, Dunn DM, Duval B, Aoyagi A, Hamil C, Mahmoud M, Roush K, Bird L, Rankin C, Lilly H, Street N, Chandrasekar R, Weiss RB (2012) Evidence-based path to newborn screening for duchenne muscular dystrophy. Ann Neurol 71:304–313. https://doi.org/10.1002/ana.23528
9. Lim KRQ, Yokota T (2018) Invention and early history of exon skipping and splice modulation. Methods Mol Biol 1828:3–30. https://doi.org/10.1007/978-1-4939-8651-4_1
10. Burki U, Keane J, Blain A, O'Donovan L, Gait MJ, Laval SH, Straub V (2015) Development and application of an ultrasensitive hybridization-based ELISA method for the determination of peptide-conjugated Phosphorodiamidate Morpholino oligonucleotides. Nucleic Acid Ther 25(5):275–284. https://doi.org/10.1089/nat.2014.0528
11. Bigelow JC, Chrin LR, Mathews LA, McCormick JJ (1990) High-performance liquid chromatographic analysis of phosphorothioate analogues of oligodeoxynucleotides in biological fluids. J Chromatogr 533:133–140
12. Leeds JM, Graham MJ, Truong L, Cummins LL (1996) Quantitation of phosphorothioate oligonucleotides in human plasma. Anal Biochem 235(1):36–43. https://doi.org/10.1006/abio.1996.0088
13. Griffey RH, Greig MJ, Gaus HJ, Liu K, Monteith D, Winniman M, Cummins LL (1997) Characterization of oligonucleotide metabolism in vivo via liquid chromatography/electrospray tandem mass spectrometry with a quadrupole ion trap mass spectrometer. J Mass Spectrom 32(3):305–313. https://doi.org/10.1002/(SICI)1096-9888(199703)32:3<305::AID-JMS482>3.0.CO;2-R
14. Yu RZ, Baker B, Chappell A, Geary RS, Cheung E, Levin AA (2002) Development of an ultrasensitive noncompetitive hybridization-ligation enzyme-linked immunosorbent assay for the determination of phosphorothioate oligodeoxynucleotide in plasma. Anal Biochem 304(1):19–25. https://doi.org/10.1006/abio.2002.5576
15. Echigoya Y, Nakamura A, Nagata T, Urasawa N, Lim KRQ, Trieu N, Panesar D, Kuraoka M, Moulton HM, Saito T, Aoki Y, Iversen P, Sazani P, Kole R, Maruyama R, Partridge T, Si T, Yokota T (2017) Effects of systemic multiexon skipping with peptide-conjugated morpholinos in the heart of a dog model of Duchenne muscular dystrophy. Proc Natl Acad Sci U S A 114:4213–4218. https://doi.org/10.1073/pnas.1613203114
16. Lim KRQ, Echigoya Y, Nagata T, Kuraoka M, Kobayashi M, Aoki Y, Partridge T, Maruyama R, Takeda S, Yokota T (2019) Efficacy of multi-exon skipping treatment in Duchenne muscular dystrophy dog model neonates. Mol Ther 27(1):76–86. https://doi.org/10.1016/j.ymthe.2018.10.011

INDEX

Toshifumi Yokota and Rika Maruyama (eds.), *Gapmers: Methods and Protocols*, Methods in Molecular Biology, vol. 2176, https://doi.org/10.1007/978-1-0716-0771-8, © Springer Science+Business Media, LLC, part of Springer Nature 2020

O

P

Q

R

W

X

Z

MIX
Papier aus verantwortungsvollen Quellen
Paper from responsible sources
FSC® C105338

If you have any concerns about our products,
you can contact us on
ProductSafety@springernature.com

In case Publisher is established outside the EU,
the EU authorized representative is:
Springer Nature Customer Service Center GmbH
Europaplatz 3, 69115 Heidelberg, Germany

Printed by Libri Plureos GmbH
in Hamburg, Germany